Springer Handbook of Auditory Research

Volume 77

The ASA Press

ASA Press, which represents a collaboration between the Acoustical Society of America and Springer Nature, is dedicated to encouraging the publication of important new books as well as the distribution of classic titles in acoustics. These titles, published under a dual ASA Press/Springer imprint, are intended to reflect the full range of research in acoustics. ASA Press titles can include all types of books that Springer publishes, and may appear in any appropriate Springer book series.

Darlene R. Ketten • Allison B. Coffin
Richard R. Fay • Arthur N. Popper
Editors

A History of Discoveries on Hearing

Springer

ASA PRESS

Editors
Darlene R. Ketten
Biology Department
Woods Hole Oceanographic Institution
Woods Hole, MA, USA

Allison B. Coffin
Department of Integrative Physiology and Neuroscience
Washington State University
Vancouver, WA, USA

Richard R. Fay
Department of Psychology
Loyola University Chicago
Chicago, IL, USA

Arthur N. Popper
Department of Biology
University of Maryland
College Park, MD, USA

ISSN 0947-2657 ISSN 2197-1897 (electronic)
Springer Handbook of Auditory Research
ISBN 978-3-031-41319-3 ISBN 978-3-031-41320-9 (eBook)
https://doi.org/10.1007/978-3-031-41320-9

This Springer imprint is published by the registered company Springer Nature Switzerland AG
The registered company address is: Gewerbestrasse 11, 6330 Cham, Switzerland

Paper in this product is recyclable.

The Acoustical Society of America

On 27 December 1928, a group of scientists and engineers met at Bell Telephone Laboratories in New York City to discuss organizing a Society dedicated to the field of acoustics. Plans developed rapidly, and the Acoustical Society of America (ASA) held its first meeting on 10–11 May 1929 with a charter membership of about 450. Today, ASA has a worldwide membership of about 7000.

The scope of this new Society incorporated a broad range of technical areas that continues to be reflected in ASA's present-day endeavors. Today, ASA serves the interests of its members and the acoustics community in all branches of acoustics, both theoretical and applied. To achieve this goal, ASA has established Technical Committees charged with keeping abreast of the developments and needs of membership in specialized fields, as well as identifying new ones as they develop.

The Technical Committees include acoustical oceanography, animal bioacoustics, architectural acoustics, biomedical acoustics, computational acoustics, engineering acoustics, musical acoustics, noise, physical acoustics, psychological and physiological acoustics, signal processing in acoustics, speech communication, structural acoustics and vibration, and underwater acoustics. This diversity is one of the Society's unique and strongest assets since it so strongly fosters and encourages cross-disciplinary learning, collaboration, and interactions.

ASA publications and meetings incorporate the diversity of these Technical Committees. In particular, publications play a major role in the society. *The Journal of the Acoustical Society of America* (JASA) includes contributed papers and patent reviews. *JASA Express Letters* (JASA-EL) and *Proceedings of Meetings on Acoustics* (POMA) are online, open-access publications, offering rapid publication. *Acoustics Today*, published quarterly, is a popular open-access magazine. Other key features of ASA's publishing program include books, reprints of classic acoustics texts, and videos. ASA's biannual meetings offer opportunities for attendees to share information, with strong support throughout the career continuum, from students to retirees. Meetings incorporate many opportunities for professional and social interactions, and attendees find the personal contacts a rewarding experience. These experiences result in building a robust network of fellow scientists and engineers, many of whom become lifelong friends and colleagues.

From the Society's inception, members recognized the importance of developing acoustical standards with a focus on terminology, measurement procedures, and criteria for determining the effects of noise and vibration. The ASA Standards Program serves as the Secretariat for four American National Standards Institute Committees and provides administrative support for several international standards committees.

Throughout its history to present day, ASA's strength resides in attracting the interest and commitment of scholars devoted to promoting the knowledge and practical applications of acoustics. The unselfish activity of these individuals in the development of the Society is largely responsible for ASA's growth and present stature.

From left to right: Arthur and Helen Popper, Catherine and Richard Fay

Dedication for "A History of Discoveries on Hearing"

This book is dedicated to Arthur N. Popper and Richard R. Fay, two giants in their own right for their research accomplishments, but equally well known for creating what is arguably the most remarkable resource for the field of auditory research, the Springer Handbook of Auditory Research, better known as the SHAR series. SHAR is considered "the definitive resource in the field" and since 1992 has provided authoritative syntheses of fundamental and emerging topics in auditory science. These statements are true, but they do not convey the genius and diligence of the two wizards who envisioned and brought to life 77 books over the last 30 years. This team includes their spouses, Helen Popper and Cathy Fay, who were not just supportive of Art and Dick, but active partners in the series.

Before, retiring from their academic positions, Art and Dick were role models and mentors for countless students and researchers, counseling them with wisdom and compassion. They also organized memorable conferences. The Evolutionary Biology of Hearing (Springer, 1992) was a product of one such conference. This book was a precursor of the SHAR books and was published the same year as the first SHAR volumes. Many of the authors in this 77th volume attended that meeting, and it is rather poignant that both that volume and this focused on comparative aspects of hearing. Like the SHAR series, Art and Dick embody unbeatable legacies as scientists, teachers, colleagues, and friends. We are saddened by the passing of Dick Fay last year but expect Art to take pride in future SHAR volumes and in seeing their legacy continue.

Darlene R. Ketten, Woods Hole, MA, USA
Allison B. Coffin, Vancouver, WA, USA

Series Preface

Following this page is the preface we published in volume 1 of the Springer Handbook of Auditory Research back in 1992. As anyone reading the original preface will note, we have far exceeded our original expectation of eight volumes. Indeed, with this, the final SHAR book edited by Fay and Popper, we will have published 77 volumes.

We are very proud that there seems to be consensus, at least among our friends and colleagues, that SHAR has become an important and influential part of the scientific auditory literature. While we have worked hard to develop and maintain the quality and value of SHAR, the real value of the books is thanks to the numerous authors who have given their time to write outstanding chapters and to our many co-editors who have provided the intellectual leadership to the individual volumes. We have worked with a remarkable and wonderful group of people, many of whom have become great personal friends for both of us. We also continue to work with a spectacular group of editors at Springer, many of whom have moved on in the publishing world to become senior executives. We are particularly grateful to Dr. William Curtis, now retired from Springer, for his "relentless" encouragement and support for SHAR both as our editor and later as he moved upward in the leadership of Springer.

In fact, the truth is that the series would never have been possible without the support of our families, and we want to continue to dedicate all SHAR books to our wives, Catherine Fay and Helen Popper, and to our children, Michelle Popper Levit, Melissa Popper Levinsohn, Christian Fay, and Amanda Fay Sierra. Our families have been immensely patient as we developed and worked on this series. We thank them and state, without doubt, that this series could not have happened without them. We also dedicate SHAR to our next generation of (potential) auditory researchers – our grandchildren – Ethan and Sophie Levinsohn; Emma Levit; Nathaniel, Evan, and Stella Fay; and Sebastian Sierra.

As we end our work on SHAR, we want to note that the series will not end. We have "turned the series over" to two great friends and colleagues, Allison B. Coffin and Joseph A. Sisneros. They are already on the way to produce new SHAR volumes. We look forward to seeing how they move SHAR forward as they continue to produce a high-quality series for years to come.

Finally, we want to thank Springer for enabling us to undertake this series and thank all of our friends and colleagues who have contributed so much to making SHAR the success that it has become.

Preface 1992

The Springer Handbook of Auditory Research presents a series of comprehensive and synthetic reviews of the fundamental topics in modern auditory research. The volumes are aimed at all individuals with interests in hearing research including advanced graduate students, post-doctoral researchers, and clinical investigators. The volumes are intended to introduce new investigators to important aspects of hearing science and to help established investigators to better understand the fundamental theories and data in fields of hearing that they may not normally follow closely.

Each volume presents a particular topic comprehensively, and each serves as a synthetic overview and guide to the literature. As such, the chapters present neither exhaustive data reviews nor original research that has not yet appeared in peer reviewed journals. The volumes focus on topics that have developed a solid data and conceptual foundation rather than on those for which a literature is only beginning to develop. New research areas will be covered on a timely basis in the series as they begin to mature.

Each volume in the series consists of a few substantial chapters on a particular topic. In some cases, the topics will be ones of traditional interest for which there is a substantial body of data and theory, such as auditory neuroanatomy (Vol. 1) and neurophysiology (Vol. 2). Other volumes in the series deal with topics that have begun to mature more recently, such as development, plasticity, and computational models of neural processing. In many cases, the series editors are joined by a co-editor having special expertise in the topic of the volume.

Richard R. Fay (Deceased), Chicago, IL, USA
Arthur N. Popper, College Park, MD, USA

SHAR logo by Mark B. Weinberg, Potomac, Maryland, used with permission.

Volume Preface

The past 76 SHAR volumes have focused on recent findings and ideas that shape what we know about the auditory system, with a few interesting digressions to other, but related, sensory systems. Never, however, has SHAR focused on the history of auditory neuroscience. While examining the whole history of the field would take volumes, and no doubt be fascinating, we start with one aspect of that issue in this volume – the history of hearing studies in multiple animal groups.

The extent of history we cover varies by taxa. For many of these groups or taxa, it is possible to trace observations of their hearing to Aristotle. In other cases, such as studies of hearing in bats, scientists did not start asking questions about their hearing and sounds until centuries later. Indeed, extensive research on the subject of bat hearing did not start until the 1930s when new technology revealed that bats hunt at night using ultrasonic signals. It was another 20 years before there was the revelation that dolphins similarly rely on biosonar, the knack of sensing their environment through generating ultrasonic sounds and analyzing the echoes. Thus, while the span of bioacoustic history varies by taxa, the focus of the chapters in this book is the age of discovery for each group and to convey how the early history of each unfolded.

One of the motivations for this book comes from our observation that many investigators today do not delve into the historical studies of species that interest them. By contrast, all the authors in this volume appreciate that once people start exploring older (often pre-internet) literature, they discover the ingenuity and brilliance of much of the work done and how often the questions asked today were first tackled decades ago, with some major questions answered and often more probing questions proposed. Investigators today may find that earlier investigators not only recognized today's interesting questions but made very important, insightful contributions in areas that are still relevant and valuable.

Chapter 1, by Arthur N. Popper, Darlene R. Ketten, and Allison B. Coffin, considers the purposes of this unique SHAR volume and provides an overview of the entire content.

Chapter 2 by Ronald R. Hoy focuses on hearing in diverse insect species. Due to the breadth of this taxa, Hoy did not cover what is known about every insect group but rather provides an overview that discusses the history of studies on some of the more interesting and challenging to study insects, including crickets, cicadas, and giant hissing cockroaches.

This is followed in Chap. 3 by Olav Sand, Arthur N. Popper, and Anthony D. Hawkins with a discussion of fishes, the largest vertebrate group. The authors point out that several ancient scholars assumed fishes could hear, but it was not until early in the twentieth century that fish hearing was clearly scientifically demonstrated. The chapter then discusses major findings in fish hearing from the beginning of the twentieth century until the 1970s.

Chapter 4 by Peter M. Narins, H. Carl Gerhardt, and Jakob Christensen-Dalsgaard discusses hearing by amphibians. They point out that from prehistoric times humans have certainly been aware of the existence of sounds of frogs and toads. Indeed, the significance of sounds for mating was noted by Aristotle (400 BCE). While broadly covering amphibian hearing and sound production history, the authors also review how pioneering work by three key individuals inspired a current legion of modern bioacoustics researchers.

Geoffrey A. Manley continues the path along the phyletic tree to consider lizards in Chap. 5. Although many early papers focused on lizard ear anatomy, the absence of accurate tools severely limited progress in how they function. After 1950, new techniques permitted better detailed anatomical descriptions, revealing the remarkable variations among the auditory papillae of different lizard families and recognition that these variations were useful for systematics. From the late 1960s, much new anatomical and physiological work ensued from several labs, setting the stage of continued work today.

In Chap. 6, Robert J. Dooling and Georg M. Klump examine the history of hearing studies in birds. As in many other taxa, the earliest reference to hearing in birds can be traced back to Aristotle who mused about the parallels between birdsong and human language, a theme that continued through Darwin into present times. Technological advances in the late 1800s set the stage for studies of bird hearing, but it reached a second peak in the 1950s when blending of operant conditioning, psychophysics, and signal-detection methods resulted in an explosion of rigorous studies on all aspects of bird sounds and hearing.

Chapter 7 by Douglas Wartzok and Darlene R. Ketten focuses on marine mammals. While sound production and evidence for hearing in water by marine mammals were yet again noted and investigated by Aristotle, only in the twentieth century did we become aware of the remarkable hearing abilities of many of these animals and the extraordinary importance of sound in their lives. Much of our knowledge of marine mammal sound use was achieved in the last 70 years. Aquatic echolocation in particular had to wait, as with bats, for technology that allowed us to do not only experiments with captive animals but also record and observe animals in the wild.

As discussed by James A. Simmons and Andrea Megela Simmons in Chap. 8, there is a fascinating history of bat hearing research in which investigators "skirted around" the idea that bats have outstanding hearing abilities. Indeed, this was despite evidence that bats hear and use ultrasonic frequencies which humans can only detect by specialized "bat detector" devices. Theories about how bats sense their environment considerably predate the actual experimental demonstration of echolocation in the 1930s. These studies opened up the field of biosonar. The early work on detecting bat echolocation signals led to further research on bat processing of the echoes (echolocation), which inspired development of models and theories exploring the intersection of biosonar acoustics and auditory research.

In mammals, there is a long history of study of central auditory processing. In Chap. 9, Yi Zhou and H. Steven Colburn provide a review of the substantial historical context for human studies of the central auditory system, starting with early ideas of the Greeks and Romans and proceeding through the shaping of ideas up to the 1960s. Their chapter draws from a variety of approaches including neuroanatomy, neurophysiology, and mathematical modeling to capture the empirical and theoretical conceptualizations around signal processing in the human central nervous system.

Finally, we had hoped to have a chapter from our most prolific, multi-species researcher, Aristotle, who contributed to nearly every chapter in the book. However, unfortunately, it seems this is his busy field season!

Darlene R. Ketten, Woods Hole, MA, USA
Allison B. Coffin, Vancouver, WA, USA
Richard R. Fay (Deceased), Chicago, IL, USA
Arthur N. Popper, College Park, MD, USA

Contents

Contributors

Jakob Christensen-Dalsgaard Institute of Biology, University of Southern Denmark, Odense, Denmark

Allison B. Coffin Department of Integrative Physiology and Neuroscience, Washington State University Vancouver, Vancouver, WA, USA

H. Steven Colburn Department of Biomedical Engineering, Boston University, Boston, MA, USA

Robert J. Dooling Department of Psychology and Program in Neuroscience and Cognitive Science, University of Maryland, College Park, MD, USA

H. Carl Gerhardt Division of Biological Sciences, University of Missouri, Columbia, Columbia, MO, USA

Anthony D. Hawkins Loughine Ltd, Kincraig, Blairs, Aberdeen, UK

Ronald R. Hoy Department of Neurobiology and Behavior, Cornell University, Ithaca, NY, USA

Darlene R. Ketten Biology Department, Woods Hole Oceanographic Institution, Woods Hole, MA, USA

Georg M. Klump Carl von Ossietzky University Oldenburg, Division Animal Physiology and Behavior, Department for Neuroscience, School of Medicine and Health Sciences and Cluster of Excellence Hearing4all, Oldenburg, Germany

Geoffrey A. Manley Cochlear and Auditory Brainstem Physiology, Department of Neuroscience, School of Medicine and Health Sciences, Cluster of Excellence "Hearing4all", Carl von Ossietzky University Oldenburg, Oldenburg, Germany

Peter M. Narins Departments of Integrative Biology and Physiology, and Ecology and Evolutionary Biology, University of California Los Angeles, Los Angeles, CA, USA

Arthur N. Popper Department of Biology, University of Maryland, College Park, MD, USA

Olav Sand Department of Biosciences, University of Oslo, Oslo, Norway

Andrea Megela Simmons Department of Cognitive, Linguistic, & Psychological Sciences, Carney Institute for Brain Science, Brown University, Providence, RI, USA

James A. Simmons Department of Neuroscience, Carney Institute for Brain Science, Brown University, Providence, RI, USA

Douglas Wartzok Department of Biological Sciences, Institute of Environment, Florida International University, Miami, FL, USA

Yi Zhou College of Health Solutions, Arizona State University, Tempe, AZ, USA

Chapter 1
A History of Discoveries on Hearing: An Overview

Arthur N. Popper, Darlene R. Ketten, and Allison B. Coffin

1.1 Introduction to the Volume

This volume in the *Springer Handbook of Auditory Series* (SHAR) is the last produced under the editorship of Arthur N. Popper and Richard R. Fay,[1] the founding editors of the series.[2] It is fitting that this is the last volume under their guidance since comparative hearing and the history of hearing research were major interests of both Dick and Art. They long felt that it is of immense importance to know, understand, and value pioneering studies of prior researchers upon whose work current researchers depend. Thus, the goal of this book is to provide a series of chapters that examine the most probing formative studies that led to our current understanding of hearing across taxa.

While past SHAR volumes have occasionally highlighted historically significant hearing research, historical mentions in previous volumes are largely incidental to

[1] This chapter is dedicated to the memory of Richard R. Fay who passed away before this book was completed.

[2] The series will, however, continue under the senior editorship of Allison B. Coffin and Joseph A. Sisneros.

A. N. Popper
Department of Biology, University of Maryland, College Park, MD, USA
e-mail: apopper@umd.edu

D. R. Ketten (✉)
Biology Department, Woods Hole Oceanographic Institution, Woods Hole, MA, USA
e-mail: dketten@whoi.edu

A. B. Coffin
Department of Integrative Physiology and Neuroscience, Washington State University Vancouver, Vancouver, WA, USA
e-mail: Allison.coffin@wsu.edu

D. R. Ketten et al. (eds.), *A History of Discoveries on Hearing*, Springer Handbook of Auditory Research 77,
https://doi.org/10.1007/978-3-031-41320-9_1

the emphasis on recent findings. Thus, our goal with this volume is that each chapter includes a description of seminal work, from the earliest recorded observations to significant mid-twentieth-century scientific results on hearing in taxa that are the focus of each chapter.

Early work on hearing goes back as far as classic Greek and Roman studies by Aristotle and others. There are also seminal studies from sixteenth- to early twentieth-century scientists such as Lazzaro Spallanzani, Gustav Retzius, Santiago Ramón y Cajal, and Hermann von Helmholtz. Much of this earlier work is not well known to modern investigators. Even work in the first 75 years of the twentieth century is similarly unknown or, in some cases, dismissed because it is "old."

Our approach with this volume is that each chapter includes discussions of important early papers, theories, primary experiments, results, and conclusions. The chapters also discuss additional important papers that were outgrowths of the founding research which may confirm, advance, or even reverse the original work.

However, we learned, as our colleagues drafted chapters, that the approach taken for one taxon could be quite different from that for another. In some cases, the "history" of observations and studies was as old, literally, as Babylonia. Several chapters trace the beginning to Aristotle's prolific observations of both land and aquatic animals. Indeed, Aristotle might arguably be considered the first functional morphologist, tying together anatomy, behavioral observations, and theories on functions. For a few chapters, the truly significant research does not begin until the mid-twentieth century, underscoring how technological developments, such as infra and ultrasonic detectors, may be to advancing our knowledge of—to us—a vast, unheard world of sounds for other animals.

One significant aspect of this volume is that every chapter makes clear how earlier research approaches and results are truly fundamental to our understanding of hearing. Equally important, many of the works convey the beauty and excitement of observation, documentation, and synthesis in personal terms that are rarely discussed in today's literature. Thus, a goal of this volume is to introduce, review, and put into perspective older, but exemplary, and sometimes bold and extraordinary, studies by investigators that form the basis of our knowledge and on-going research questions asked today.

In conclusion, the significance of this volume is that it not only shares information about important, innovative research but also serves as an introduction to some of the history of auditory neuroscience investigations that continue to be of immense value, even in the twenty-first century.

1.2 Volume Overview

1.2.1 Insects

Most of this volume focuses on vertebrates. However, Chap. 2 on invertebrates by Ronald R. Hoy provides an uncommon view into the perception and use of sound among some of the most numerous creatures on Earth. The chapter discusses the interest of ancient civilizations in insects as well as how modern technology has aided new discoveries. Although insects and vertebrates have the same fundamental needs to detect and analyze sound, hearing in vertebrates and invertebrates evolved along very different paths. Insects are among the smallest terrestrial animals known to detect and process sounds. Scale matters for hearing. Some larger insects detect acoustic pressure, but this is less common for miniscule insects, most of which detect particle motion. The consequences for structure and function of hearing organs are profound. Equally important in comparing insect and vertebrate hearing is the polyphyletic origins of the insect ear versus the likely monophyletic origin of the vertebrate ear. Vertebrate ears are head-locked, with each ear behind each eye and above the jaw. Insect ears, however, evolved in ways that resulted in a wide range of placements. They are found nearly anywhere on an insect's body, including on appendages. These differences provide rich fodder for comparative studies of hearing at all levels of biological organization, from molecular to evolutionary and phylogenetic.

1.2.2 Fishes

Chapter 3 by Olav Sand, Arthur N. Popper, and Anthony D. Hawkins focuses on bony fishes. The chapter discusses the major findings in the history of fish hearing studies and highlights the contributions of several of the major contributors from the beginning of the twentieth century until the 1970s. By the early twentieth century, it became clear that fishes could indeed hear and that the critical auditory sensory end organs are the otolithic organs of the inner ear. It was observed that the species possessing Weberian ossicles or other structures linking the swim bladder to the inner ear have higher frequency and more sensitive hearing than species lacking such features. It was also shown that the sound-sensitive otolith organs detect particle motion, and that gas-filled chambers may enhance auditory sensitivity by transforming sound pressure to particle motion. Experiments in the 1970s demonstrated that fishes can discriminate between sounds from different directions, but it is not yet clear whether fishes are able to localize a sound source without moving through the sound field. This chapter discusses the major findings in the history of fish hearing studies through the 1970s, including questions that remain unanswered and deserve attention in the present day.

1.2.3 Amphibians

Chapter 4 by Peter M. Narins, H. Carl Gerhardt, and Jakob Christensen-Dalsgaard points out that from prehistoric times, humans have certainly been aware of sounds produced by frogs and toads. In fact, the significance of sounds for mating was noted by Aristotle. The middle ear of frogs was first described in 1676, but the first detailed anatomical studies of their inner ear appeared only in the second half of the nineteenth century as a result of advances in microscopy and histology techniques. Technological advances in the early twentieth century, including development of electronics, recording devices, and computers, facilitated the development of modern amphibian bioacoustics by pioneers in the field. For example, W. Frank Blair showed how differences in the physical properties of calls could promote speciation. Murray Littlejohn conducted experiments proving that call differences alone allowed female frogs to choose conspecific mates. Robert Capranica was the first to electronically synthesize frog calls and to electronically modify them for use as stimuli in acoustic playback experiments. Capranica also employed behavioral studies to inform the analysis of neural processing and was instrumental in establishing neuroethology as a powerful research strategy. Thus, Chap. 4 provides a detailed review of how the work of these three individuals inspired a legion of modern bioacoustics researchers who have made major contributions to the field.

1.2.4 Lizards

Chapter 5 by Geoffrey A. Manley continues along the phyletic tree to consider lizards. Although many early anatomical papers touched on lizard ears, the absence of appropriate tools severely limited progress in understanding their functional aspects. After 1950, new techniques permitted more detailed anatomical descriptions and revealed the remarkable variations among the auditory papillae of different lizard families as well as a recognition of their usefulness for systematics. From the late 1960s, in addition to new anatomical and physiological work, several laboratories undertook recordings from the auditory nerve. The various studies led to a new understanding of the relationships between structure and function and of the evolutionary and functional significance of large differences across lizard families. The structural characteristics of lizard papillae enabled modeling of how hair cells couple to produce both increased sensitivity and frequency selectivity in the auditory nerve. As with earlier chapters, these studies on lizards inform many present-day experiments.

1.2.5 Birds

In Chap. 6, Robert J. Dooling and Georg M. Klump examine the history of hearing studies in birds. The earliest reference to bird hearing can be traced back to Aristotle, who mused about parallels between birdsong and human language, a theme that has continued through Darwin into present times. Technological advancements in the late 1800s and early 1900s set the stage for the first experimental hearing studies on birds in the 1930s–1950s and paralleled technically those performed on humans. In the 1950s, blending operant conditioning, psychophysics, and signal-detection methods resulted in an explosion of studies on many aspects of hearing in birds. Tests on a wide variety of avian species provided a solid contrast with knowledge about hearing in humans including comparisons of absolute and differential thresholds for frequency, intensity, and temporal changes. Soon after, more elaborate studies, often motivated by the anatomical foundations of hearing in birds, advanced our understanding of how hearing functions in the real world, such as in sound source localization and with the cocktail party effect. Despite a relatively narrow perceptual frequency range and closely spaced ears, birds evolved an extremely sophisticated auditory system that matches well with the earliest speculations arising from observations on the widespread use and complexity of their learned vocalizations.

1.2.6 Marine Mammals

In Chap. 7, Douglas Wartzok and Darlene R. Ketten discuss marine mammal hearing. Like bats, marine mammals operate in relatively dark environments. Both evolved well before humans and developed ears capable of echolocaton. Aristotle recognized these animals are mammals and wrote of their extensive acoustic abilities. Dolphin and porpoise ears were recognized by eighteenth- and nineteenth-century investigators as being specialized for high frequencies with associated fatty tissues eventually shown to be low impedance pathways that conduct water-borne sound to the inner ear. However, much about whales, dolphins, and porpoises remained a mystery until the mid-twentieth century when devices to record underwater sound became available. Open water field recordings revealed marine mammals use sound to detect environmental features, navigate, communicate, and find prey and predators. In the 1950s, behavioral work with captive dolphins, seals, and sea lions showed they produce and hear sounds ranging from infrasonic to ultrasonic frequencies. From the 1960s onward, live animal experiments focused on behavioral and electrophysiological studies, particularly echolocation abilities. The first bottlenose dolphin audiogram was obtained in 1967; in 1970, the first porpoise audiogram. Researchers in the early 1970s also employed electrophysiological techniques to measure dolphin evoked potentials. Scientists in Europe and the US also investigated in air and in water hearing of several species of pinnipeds during this time. Russian scientists were pursuing similar studies on marine mammals, but not until the end of the Cold War was research from both sides shared freely. Further research has broadened the field to more species and initiated studies on hearing in free-ranging animals, but many of the original puzzles are not yet solved and are being investigated by a new cohort of researchers.

1.2.7 Bats

Bats are another of those groups for which the bulk of our knowledge begins in the mid-twentieth century. In Chap. 8, James A. Simmons and Andrea Megela Simmons note that early investigators "skirted around" the idea that bats have outstanding hearing, even though no human could hear the sounds. Indeed, despite theorizing and a few notable direct studies in the eighteenth and nineteenth centuries to determine how bats sensed their environments, it was not until the mid-twentieth century that the term "echolocation" was created. Echolocation studies in the 1950s in turn inspired development of models and theories exploring the intersection of acoustics (the bat's biosonar sounds), auditory research (reception of echoes), and ultrasonic signal processing. Some early models and theories of echolocation drew upon concepts from contemporaneous radar and radio communication technology, and in turn, these technologies have since benefitted by biomimetic designs derived from bat research. Some studies emphasized the connection between the use of frequency-modulated sounds by bats and the development of "chirp" radar, which employed a receiver matched to the broadcast. The idea of a matched-filter receiver, which displays the cross-correlation of echoes to broadcasts, is still prominent in some modern models of echolocation. Other early models drew upon comparisons between echo processing and human perception of pitch. Modern models of echolocation strive to identify common parameters that govern the operation of radar systems, human psychoacoustics, and the behavioral performance of echolocating bats.

1.2.8 Mammalian Central Processing

There is a long history of study of central auditory processing in mammals, particularly in humans. In Chap. 9, Yi Zhou and H. Steven Colburn provide an overview of the history of our understanding of the central auditory system, starting with early ideas of the Greeks and Romans and proceeding through the 1960s. The history of studies of the central auditory system involves a combination of theoretical conceptualizations, anatomical studies, and physiological and psychophysical measurements. The chapter covers the breadth of approaches, focusing particularly on neuroanatomy, neurophysiology, and mathematical modeling. The goal is to describe the empirical and theoretical concepts that focus on the central neural processing of sounds. Although psychophysical studies are important and provide the basis for our knowledge of auditory capabilities, these studies are not explicitly reviewed in this chapter. Instead, modeling of sensory and central processing are discussed in detail to explain how processing in the succeeding stages of the auditory system relates to overall hearing abilities.

1.3 Final Thoughts

Clearly, this volume only touches the surface of the history of hearing research. The editors, all of whom have a deep "affection" for comparative issues, chose to explore history from a comparative perspective, allowing readers to draw parallels across taxa and experimental approaches. While we could have taken other approaches, such as the history of studies of the inner ear or the history of research on the auditory central nervous system or perhaps the history of psychoacoustics, we chose a comparative approach since there is elegance and great value in understanding how researchers developed ways to understand how hearing in various taxa differ and how each hears and uses sound. Therefore, we would argue that investigators may benefit from a better understanding of comparative issues and the use of diverse taxa to ask a whole range of questions regarding hearing, perhaps even finding new "ears" and hearing abilities in species yet to be investigated.

Perhaps future SHAR volumes will explore the history of other aspects of hearing, but, for Arthur Popper, after 77 volumes, this is his last volume as senior editor for SHAR. He trusts that Dick Fay, his great colleague, dear friend, and invaluable collaborator of over 50 years (and co-editor of SHAR) would agree that it is time to bequeath SHAR to younger colleagues!

Chapter 2
Insect Hearing: Selected Historical Vignettes

Ronald R. Hoy

2.1 Introduction

Insects are common, numerous, and specific. Most are also miniscule in size as compared to vertebrates. Yet some insects can make audible, often startlingly loud, sounds when signaling to one another.

Insects have been around for hundreds of millions of years. When the first humans arose in Africa, a mere one or two million years ago, the calls of insects pierced their emerging consciousness serenaded by cicadas by day and crickets and katydids by night. Sonic insects have shared ecological habitats with humans since humans evolved and their presence has been noted in the cultural artifacts and practices of ancient civilizations that persist today, as will be seen in this chapter.

The sounds that insects make are signals that serve the same adaptive function as do language and music in humans. They are communication signals for survival that mediate reproductive behavior. The sense of hearing serves as a key surveillance channel for monitoring the external world, especially for the countless nocturnally active insects and it serves the same adaptive purpose as it does for birds and frogs. However, the evolutionary origins of hearing organs in insects took multiple and diverse phylogenetic pathways, as opposed to simpler route in the vertebrates, which has been described as closer to one "big bang" event (e.g., Clack et al. 2016). This multiple origin led to a diversity of hearing organs that have novel anatomic, biomechanical, and neurophysiological "design features" that enable insects to hear a wide variety of sonic signals of salience.

This volume in the Springer Handbook of Auditory Research (SHAR) series is devoted to "A History of Discoveries on Hearing," and this chapter on insects will tackle sonic/acoustic activity from two viewpoints. First, is a reflection on how

R. R. Hoy (✉)
Department of Neurobiology and Behavior, Cornell University, Ithaca, NY, USA
e-mail: rrh3@cornell.edu

D. R. Ketten et al. (eds.), *A History of Discoveries on Hearing*, Springer Handbook of Auditory Research 77,
https://doi.org/10.1007/978-3-031-41320-9_2

human observers/listeners have interpreted the phenomenon of insect hearing anthropomorphically. Second, only when these anthropocentric suppositions were squared with the ground-truth of what and how the insects, themselves, hear did we realize the remarkable diversity of hearing organs in form and function that evolved in insects. Our understanding of how insects hear is still very much a work in progress. While certain fundamental issues were recognized by Greek philosophers, critical studies on insect hearing awaited scientific inquiry until the nineteenth century. Indeed, most issues were not clarified until the twentieth century, and many new discoveries continue to be made, aided by twenty-first-century technical tools.

This historical survey is purposely highly selective—it will be presented as a series of case histories focused on the most recognizable of sonic insects that will be familiar to a naturalist or curious layperson. The editorial guideline for the timeframe unifying these chapters is from antiquity to around 1970–1980, when the SHAR volumes picked up the story of comparative hearing, over the course of its 77 volumes. I have tried to respect this historical interval, but break from it when necessary to maintain the integrity of narrative flow, especially when referring to review articles on salient and timely issues.

At the same time, considering the extraordinary diversity of insects, it is impossible to cover all species and all the ways in which these species have "reinvented" hearing in any single chapter. Thus, rather than attempt to tackle the extreme breadth, this chapter will focus on a select species and groups that give the greatest representation of the diversity of insect hearing and the extraordinary range of "ears" they use for sound detection.

2.1.1 Three Cautionary Notes in Comparing Hearing in Terrestrial Vertebrates with Insects

The reader should keep in mind three issues when comparing insects to terrestrial vertebrates in terms of bioacoustics and auditory behavior. (1) Insects are really small—in addressing the sense of hearing, size matters; (2) biologists tend to make implicit anthropocentrically based expectation that auditory sensitivity is largely a matter of detecting differences in acoustic pressure of an impinging sound wave; and (3) the *evolutionary origins* of insect hearing organs differ markedly, yielding great diversity in structure as well as function.

2.1.1.1 Body Size Constrains Structure and Function of Hearing Organs in Insects

First and most obvious, the body size of an insect can be 4–6 orders of magnitude smaller, both in surface area and volume, than the typical vertebrate. Naturally, insect ears are even smaller and on the micro or even nanoscale. This difference imposes important constraints on the physics of both sound production and sensory detection (Bennet-Clark 1971, 1991). The constraints of small size on the production of acoustic signals in both intensity level and spectral frequency or timbre of the sound emitted. Size also imposes limits for sensing sound, especially sensing air pressure with a tympanum, as was noted by Bennet-Clark in a series of important studies (1991, 1999). Despite these constraints, many sonic insects can hear as well as make sounds that can be distractingly loud to the human ear. Insects deploy a remarkable number of morphological adaptations for amplifying sound level (Bennet-Clark 1999; Yack 2004).

2.1.1.2 Insect Hearing Organs Can Detect Pressure Differences in Acoustic Airflow as Well as Particle Velocity Depending on Species and Body Size

It is not surprising that the eighteenth-century natural history approach, as well as the twentieth-century scientific study of insect bioacoustics, was biased toward species whose sonic emissions or calls are clearly audible to human observers. Insects, like humans, are terrestrial creatures meaning that their sense of hearing is a matter of detecting acoustic signals that propagate through air. Not until the mid-twentieth century did instruments become available to reveal the infra-and ultrasonic sensitivities of the insect ear and Doppler laser vibrometry to measure mechanical responses to airflow.

The acoustics of insect hearing has historically been studied in larger species like crickets, grasshoppers, moths, and cicadas whose hearing organs feature externally visible tympanic membranes. This reinforces the expectation that such ears are pressure or pressure-difference detectors (Roeder 1967; Larsen and Michelsen 1978), by obvious analogy to vertebrate eardrums. It is not surprising that hearing in insects has been interpreted as the detection of changes in air pressure, as in vertebrate hearing, which is valid for larger, tympanate insects. However, as will be discussed below, many minutely small insects can also hear—but by detecting velocity in propagating airflow, not pressure (Menda et al. 2019).

2.1.1.3 Evolutionary Origins

Finally, there is the vitally important matter of evolutionary origins. The ear of all vertebrates appears to have evolved from a common origin (Webster et al. 1992; Clack et al. 2016) and whether or not the term "monophyletic" is invoked to describe the origin of the vertebrate ear it is true that essentially all vertebrate ears are found on opposite sides of the head (Manley 1973; Allin 1975). However, in the late nineteenth and twentieth centuries, when naturalists and entomologists had better tools to investigate insect hearing, it was discovered that insect hearing organs could be located just about anywhere on the body including on appendages, thorax, abdomen, as well as on the head, clearly orthogonal to the monophyletic pattern of vertebrates (Yack 2004). What accounts for these differences?

In terrestrial vertebrates, the organization of outer-to-middle-to-inner ear is obligatory because of the need for acoustic impedance matching in transferring acoustic energy of sound waves to the outer ear of animal and thence into ponderable vibrations in the fluid that bathes the cochlea of the mammalian inner ear. Impedance matching is solved by the middle ear bone(s) that are interposed between an air-facing eardrum and the fluid-filled cochlea. However, in sonic insects, like cicadas, crickets, and grasshoppers, that possess an external tympanal membrane (outer ear), this thin cuticular membrane is directly apposed internally to an air-filled tracheal chamber which means that the external and internal face of the tympanal membrane is air-to-air, obviating the necessity for a middle ear (Fullard and Yack 1993).

The insect's auditory sensory organ is a type of mechanoreceptive chordotonal organ, specialized for hearing. It is where auditory transduction occurs and is analogous to the vertebrate "inner ear." This kind of ear is directly attached to the air-filled tracheal tube, if not directly to the tympanal membrane itself (Yack 2004). This air-to-air coupling of external to inner ear obviates the need for a middle ear in insects that possess tympanal membranes for hearing.

Similarly, where the external acoustic biosensor is a simple cuticular hair that protrudes into the air and is internally innervated by bipolar sensory neuron(s), the transfer of acoustic energy (air particle movements) is direct from air-to-hair-to-neuron and no intervening structure is needed. This simplicity of functional mechanics removes a severe constraint in the task of hearing, even for tympanate insects, by comparison with terrestrial vertebrates.

Lacking the biomechanical constraints of hearing in the vertebrate ear, the evolution of hearing organs in insects has resulted in a multitude of locations where insect ears can be found, to nearly anywhere on the body. The insect ear has evolved multiple times in the Class Insecta and independent evolution has occurred even within Orders and Suborders, especially in the Orthoptera (crickets, katydids, grasshoppers) and the Lepidoptera (moths and butterflies) (Song et al. 2020). These twentieth and twenty-first century papers underscore that the structure-function understanding of insect hearing is still very much a work in progress and should be kept in mind when reading the rest of this chapter. My account will focus on selected examples of sonic insects, which have historical interest.

2.2 The Bioacoustics, Evolution, and Neuroethology of Calling Song in Crickets

Crickets and their songs have been the focus of attention not only by biologists but have figured in popular culture for centuries. Charles Dicken's (1846) "Cricket on the Hearth™" and Walt Disney's Jiminy Cricket in "Pinocchio™" are but two of many testaments from literature and film.

Crickets have co-existed/cohabited with humans from prehistory onward because humans built their villages and towns in what were once open fields and forests, natural habitats for crickets, so that crickets are always within "earshot" since the beginning of humankind. These mostly nocturnal insects and their loud, incessant calling are conspicuously present during the warm seasons in temperate climates and even longer in the tropics. They are large enough to capture by hand and easily kept in captivity where some societies and cultures treasure their calls and cage them for their songs, just like canaries and finches are household companions.

Thus, it is not surprising that when naturalists and biologists selected animals for auditory studies, crickets were among the first chosen—in season, they are always "at hand." As it turned out, studying crickets was not only convenient but a strategically fortuitous choice of insect for comparative studies in bioacoustics and hearing science.

Regrettably, space limits the discussion in this chapter to the field crickets (Family Gryllidae, Genus *Gryllus*) and tree crickets (Family Oecanthidae, Genus *Oecanthus*) to the exclusion of grasshoppers (Family Acrididae) and katydids (Family Tettigoniidae), thus neglecting at least two-thirds of the ubiquitous, species-rich, singing Orthoptera. Historically, however, the sheer breadth of studies of cricket bioacoustics including evolution, taxonomy, behavior genetics, and neuroethology of calling behavior have been conducted over multiple levels of analysis, ranging from molecular and cellular to phylogenetics to neural circuits, and ultimately, to behavior and sensory ecology, warranting their focus in this brief review Entire books have been devoted to cricket bioacoustics, behavior, and neuroethology (Huber et al. 1989; Gerhardt and Huber 2002) and were reviewed in the SHAR series (Hoy et al. 1998; Pollack et al. 2016). Hence, this section will cover the topic very selectively because, arguably, acoustic behavior in crickets is the ideal model system for the study of hearing in insects.

2.2.1 The Scientific Study of Acoustic Behavior in Crickets: The First Model System in Insect Bioacoustics

The nineteenth century was the golden age of comparative anatomy enabled by advances in microscopy and development of histological techniques and staining reagents, including vital dyes and silver staining methods that permitted studies of hearing organs at cellular level resolution. The first careful histological study of the cricket tympanal hearing organ, the "crista acustica," was performed by Schwabe (1906; also replicated by Michel 1974). The spindly auditory sensory cells, scolopidia, are anatomically arranged as a linear proximal to distal array along the long axis of the tibial segment of the leg itself. Much later, when it became possible to make neurophysiological recordings from single cells was it shown that the scolopidial cells of the crista acustica of crickets and katydids were tonotopically organized, similar to the inner hair cells of the mammalian cochlea (Oldfield 1982).

Schwabe's work was known to the late nineteenth-century Slovenian entomologist, Johannes Regen, who was studying the behavior of the field crickets that sang in the capacious, verdant fields surrounding his laboratory.

2.2.2 *Regen and Playback Experiments*

While it may have seemed obvious to naturalists that male crickets "sing" as part of their mating behavior (Darwin 1871, no experimental approaches to confirm this presumption. But it was to come a few decades later. In 1876, Alexander Graham Bell had invented the telephone and within a few decades, telephone installations became common in Europe. Regen's (1913) experiments demonstrated that he was an ingenious early adapter of the telephone for scientific exploration. His experiment involved capturing sexually mature, singing males and female field crickets, *Gryllus campestris*, from nearby fields, and bringing them into his laboratory.

Regan maintained his male and female crickets in separate rooms of his lab building. When Regan heard a male commencing to sing its calling song, he telephoned his assistant who was patiently waiting in another room, distantly located and acoustically isolated, tending his collection of female crickets. Regen then placed the speaker mouthpiece of his phone next to the calling male cricket. When the assistant took the call, he placed his phone's earpiece receiver a short distance away from a female cricket contained within a large, sand-filled arena in which it could walk freely.

The assistant observed the female crickets invariably responded by walking directly to the phone's earpiece and crawling around and over it, presumably in search of the calling male. This clever experiment demonstrated that sound, and sound alone, was sufficient to bring both sexes together around the calling male. In a single stroke, other sensory cues such as odors, tactile touch, and vision as possible cues for attraction were eliminated as necessary. Just as Alexander Graham Bell rang his assistant Thomas Watson with his famous message, "Mr. Watson, come here, I want to see you," so too did Regen's male cricket send a compelling message to a remotely located female cricket to come to the phone, perhaps ironically with a similar urgency! Both Bell and Regen had completed a remote auditory communication loop.

Furthermore, when Regan amputated the tibial joints of the female cricket's forelegs it no longer responded to calls, even though the insect's ability to walk was unimpaired. Tibial amputation acoustically deafferents the insect by removing its hearing organ (crista Acustica) which is contained within its foreleg tibial segment. Regen's experiments were pioneering in his use of an electronic loudspeaker (the telephone's) to broadcast acoustic signals to assess their behavioral relevance and potency to change the behavior of the animal hearing the signal.

Regan's experiments were the forerunner of the playback paradigm to measure phonotaxis—the orientation or tracking movements made by the receiver animal in response to an acoustic playback. Whereas Regan had to use live, calling male crickets to provide the acoustic stimulus, experimentalists, from the 1950s onward, could substitute tape recordings of real animals or electroacoustically generated analogs for playback. By the 1970s, acoustic signals could easily be generated from electronic waveform generators controlled by computers. Phonotaxis trials remain a robust and powerful methodology in bioacoustic research.

Indeed, Regen's experiments in Slovenia inspired, after the two World Wars, historically important "schools" of cricket and orthopteran neurobiology and behavior in Europe, especially Germany because Franz Huber single-handedly founded the electrophysiological study of the song generating networks of field crickets and whose work influenced the careers of hundreds of subsequent cricketeers, including myself. Huber's pioneering work (Huber 1962; Huber et al. 1989; Gerhardt and Huber 2002) established the cricket song system as a model system for investigating the inherently innate temporal pattern or rhythm of species-specific mating calls of males and how conspecific females hear and perceive mating calls. The sheer diversity of species-specific temporal pattern (pulse rhythm) in cricket mating calls combined with the tonal simplicity of song pulses, themselves, provoked the possibility that not only was the motor production of the call rhythm under innate, genetic control, but so too was the perception (recognition) of temporal pattern also innate, presumably under genetic control. His lab and others to follow contributed to the important question of the central auditory processing of songs by tackling how the female cricket's audio-motor neural system discriminates her own species-specific temporal patterns from those of other species, which leads her to localize and "steer" her to the male (Gerhardt and Huber 2002; Hedwig et al. 2018; Schoeneich and Hedwig 2019).

2.2.3 More Playback Experiments

These neuroethological studies that eventually led to the key understanding that a neural network within the male cricket's central nervous system (CNS) generated the species-specific temporal pattern of sound pulses (Alexander 1962; Huber 1962). The Central Pattern Generator "revolution" of the 1960s (Wilson 1961; Bentley 1969) that led to Bentley, Hoy's genetic analysis. In fact, I was drawn to study the auditory behavior of crickets having been inspired by two of the founders of insect neuroethology, the aforementioned Franz Huber and Kenneth Roeder, whose pioneering work on auditory behavior in moths will be described in detail in the following pages.

My laboratory at Cornell University (Ithaca, NY) deployed an auditory playback paradigm to measure the phonotactic steering response of tethered crickets, suspended in mid-air and induced to fly in a laminar wind flow (Moiseff et al. 1978). In this experiment, crickets (both males and females) could sustain wing-flapping (fictive flight) for minutes at a time and respond to directional sounds played from loudspeakers placed to the right or left of the suspended insect. In brief, we confirmed that fictively flying female crickets responded to playback of both tape-recorded calling songs from conspecific males as well as live song, by making asymmetrical movements with their appendages and wings toward the speaker that would steer them toward the song.

This confirms the behavior of walking crickets to playback experiments that are to Regan's time as well as more systematic studies (Walker 1957, 1962). However, when the acoustic stimulus was switched from cricket-like carrier frequency (ca. 5 kHz) to ultrasonic frequencies (25–100 kHz) the fictively flying crickets abruptly responded with reflexively fast steering movements of its appendages that would steer the cricket away from the ultrasound source (Moiseff et al. 1978). This action is an auditory startle (ASR), in this case escape, response that is characteristic of many animals with a sense of hearing, from invertebrates to humans (Eaton 1984). In crickets and moths (see Sect. 2.3), acoustic startle occurs only during flight behavior and reflects evasive behavior in response to the ultrasonic echolocation calls of insectivorous bats (Roeder 1967; Moiseff et al. 1978). Subsequent work by talented students at my Cornell lab showed that many genera and species of nocturnally active, flying insects have evolved ultrasound-sensitive hearing organs for detecting bats that relentlessly hunt insects on the wing, by echolocation (Hoy 1992).

2.2.4 The Cricket's Tympanal Hearing Organ

The cricket's hearing organ and auditory system are typical for insects commonly known to possess a sense of hearing (Hoy 1998; Yack 2004). Such hearing organs are characterized by three traits: (1) a pair of external tympanal membranes, which in crickets and katydids are located in the tibial segment of their prothoracic forelegs and visible by eye or under a magnifier; (2) the internal wall of the tympanal membrane abuts upon an air-filled chamber or tube, part of the insect's tracheal respiratory system; and (3) sensory innervation by a chordotonal mechanosensory organ that can attach directly to the tympanum or indirectly through attachment to the adjacent air-filled trachea. The sensory unit of chordotonal organs is called a scolopidium, which is multicellular complex and may contain one or more bipolar sensory neurons, along with cap cells, support cells, and glial cells.

The respiratory tracheae play a central role in determining the directional properties of hearing in crickets, as shown by the pioneering work of Axel Michelsen and his bioacoustics laboratory in Odense, Denmark (Michelsen 1998). In brief, a four-input tubular system of tracheae, of which the tracheal tubes that adjoin the auditory receptor organ, the *crista acustica,* is part, imparts directional sensitivity to hearing by generating air pressure differences within the tubes. This was shown by Larsen and Michelsen (1978), in their Doppler laser vibrometry studies (Michelsen 1998).

2.2.5 Evolutionary Convergence of Tympanal Hearing Organs: Crickets and Their Parasitic Flies

Male crickets sing to attract females for mating, but their loud calling songs are also heard by parasitoid flies of the genus *Ormia* (Cade 1975). Interestingly *Ormia* flies evolved a tympanal hearing organ to hear crickets (Robert et al. 1992) nicely demonstrating the principle of convergent evolution. However, to localize the cricket based on directional acoustic cues, such as time and intensity, *Ormia* evolved an entirely novel mechanism for sound localization that surpasses the performance of its host in every way (Miles et al. 1995; Robert et al. 1996). The directional properties of hearing are fascinating but beyond the scope of this chapter. However, the bioacoustic principles "invented" in *Ormia* for sound localization, mechanical coupling, has turned out to inspire acoustic engineers to mimic its design principles for producing small, nanofabricated, directional microphones (Miles et al. 2009). Insect ears are fertile ground for physicists and engineers who deploy the strategy of biomimicry to "invent" useful biosensors in the field of biomedical engineering

2.2.6 The Genetic Control of Cricket Song

I was fortunate to stand on the shoulders of the founders of cricket bioacoustics. My postdoctoral mentor, David Bentley, at the University of California Berkeley, studied cricket song for his PhD. under the guidance of Richard Alexander, at the University of Michigan, after which Bentley did a postdoc on the central rhythm generating neural network in Franz Huber's lab in Germany. Alexander had earlier shown (1962) that the calling song of male crickets is species-specific with respect to the temporal pattern of sound pulses in the call. The call rhythm was as predictable and stereotyped as any morphological trait (Alexander 1962) and was therefore genetically controlled. Bentley and I tackled the degree to which song rhythms were controlled genetically by hybridizing species of crickets that had distinctly different song rhythms and recording and measuring the call rhythms of the F-1 and backcross hybrid males (Bentley and Hoy 1972).

Later, I became interested in how female crickets discriminate and recognize the songs of conspecific males which I pursued in my own laboratory. We deployed audio playback of natural recordings and electroacoustically synthesized models of calling songs to study the genetic control of cricket song rhythms, generally. We were able to induce hybridization between field cricket species with distinctly different call rhythms (Hoy and Paul 1973; Hoy et al. 1977).

As indicated earlier, field crickets, like many insects, live only a few months and so must mate and reproduce without ever having contact with the parental generation (Alexander 1957, 1962). Presumably, this means that the act of species-specific communication—singing and perceiving its species-specific calling song—must be largely specified by genetic instructions. This precludes the role of learning song characteristics since there is no overlap with its progenitors, unlike longer lived birds and mammals, in which learning can occur because of overlapping generations. Thus, the cricket song system is an ideal model for investigating species-specific communication signaling that is primarily determined by nature (genetics) uncontaminated by nurture (learning from conspecifics).

2.2.7 The Acoustico-Motor Linkage in Species-Specific Phonotactic Behavior in Crickets

Regan's phonotaxis experiments showed that female crickets are attracted to the sound of the calling song of a male cricket but the degree to which her phonotaxis is dependent on the species of calling cricket would not be clarified until the 1950s and the studies of R.D. Alexander (1962) in field crickets and T.J. Walker (1957, 1962) in tree crickets. Their studies demonstrated that it is the temporal pattern of sound pulses in calling song that differentiates the calls of one species from another. It is the pulse rhythm of a calling song that the conspecific female detects and discriminates in making her decision to approach one call type over another call when both are simultaneously broadcast in a phonotaxis experiment. Of course, carrier frequency of the sound pulses also influences behavior but is less consequential because in both field and tree crickets, the spectral bandwidth of congeneric male calls shows considerable overlap and is shared among males of different species (Walker 1957, 1962; Alexander 1962), which became clear only after the commercial availability of the sound spectrograph in the late 1940s and 1950s.

The implications of this work led to a natural question of how the central auditory system processes the temporal pattern of species-specific calls and the degree to which call rhythms are under genetic control or whether some kind of learning is involved, as was known from the dialects of songbirds (Marler and Tamura 1964). In brief, how do female crickets become "tuned-in" and attracted to the temporal pattern of the calls of a conspecific male, ignoring the call of an equally audible heterospecific male whose call possesses a different temporal pattern of sound pulses? Learning was minimized if not simply dismissed because there is no overlap between parental generations and their offspring.

Especially in temperate climates, the parental generation completes its entire reproductive cycle in one season/year (lifespan is about 4–5 months, including 2 months of sexual, reproductive maturity as adult insects) and eggs are laid in the ground/vegetation to overwinter after the adult parental generation dies. It is not until the spring or summer of the next year that the eggs hatch and the immature instar larvae attain reproductive adulthood without ever hearing the calls of the male parent species—unlike birds, in which there is parent–offspring overlap and opportunities for immature males to "learn" from parental males. Undeniably, the acoustic behavior—both song production and its perception and "recognition" are under firm genetic control. Thus, the calling behavior of crickets—production of species-specific call rhythms as well as the auditory perceptual processing of call rhythm—is a model system in which to study the neurobiological mechanisms that underlie the species-specific behavioral linkage that exists between conspecific male and female.

Such a linkage underlies auditory communication not just in crickets. It may exist for frogs, birds, and mammals, at least to some degree, as well as all acoustically communicating invertebrates. Thus, the cricket song system, a model for genetically determined communication coupling, cleanly separate from ontogenetic influences and learning. The detailed, elegant work of Bertil Hedwig et al. (2018; Schoeneich and Hedwig 2019) indicates that within the CNS of female crickets an identifiable network of interneurons serves as an auditory filter that is tuned to the temporal characteristics of its species-specific calling song. Other hypotheses such as template matching, a form of corollary discharge, remain speculative, absent experimental evidence (Hoy 1974).

2.3 Hearing Insects II: Moths and Neuroethology

Among the 1950s contemporaries of pioneering cricketeer, Franz Huber, were the noted insect physiologist, Kenneth Roeder at Tufts University (Boston, MA) and his colleague, acarologist (one who studies parasitic mites) Asher Treat, at the City College of New York. Huber and Roeder became spokesmen and co-founders of the emerging field of invertebrate neuroethology—a merger of ethology and neurobiology, the neural analysis of an insect's behavior, as performed in ecologically valid, natural habitats. Roeder's 1967 book, "Nerve Cells and Insect Behavior," is a classic, a canonical neuroethological text.

Noctuid moths are large (cricket size or more) nocturnally active insects that are commonly seen on summer nights, flying around porch lights or higher above, flying around street lamps. These large moths first attracted Treat's attention because they possess a pair of large, transparent membranes, one on each side of the insect's metathoracic body wall and normally hidden when the wings are folded in resting moths. Treat specialized in the tiny parasitic mites that infest moths. In particular, he was fascinated by the mites that infested noctuids because these mites formed family colonies that lived within a large, air-filled, thoracic chamber beneath the moth's wings. A large, transparent membrane forms the external wall of the chamber.

Even earlier, a German entomologist studying noctuid moths, had observed these membranes (Eggers 1919). He conjectured that the membranes served an auditory function because they looked like a vertebrate/mammalian eardrum. Aware of this, Treat presumed that he was studying ear mites, which inhabited/infested the air-filled chamber, internal and adjacent to the eardrum, and that this is where they lived their entire life cycle.

Even more intriguing, Treat was struck by the fact that these mites only and always infested one ear chamber, never two. His curiosity about the auditory functionality of noctuid eardrums led him to perform behavioral experiments that showed noctuid moths in fact responded to sound mid-flight with brief, reflex-like turns at short reaction times (Treat 1955).

Such observations begged for rigorous physiological experiments which led Treat to collaborate with Roeder, a noted expert in insect sensory neurophysiology. Together, they demonstrated that the noctuid moth's hearing organ was sensitive only to sounds at ultrasonic frequencies. This led to their joint report in *Science* (Roeder and Treat 1957) that the moth's thoracic membranes indeed functioned as eardrums that were connected to an auditory organ consisting of just two scolopale sensory cells and an auditory nerve that projected to the CNS. Roeder recorded the nerve's auditory responses and discovered it responded to sound only at ultrasonic frequencies from 20 kHz to at least 60 kHz. From this, Roeder and Treat surmised that the Noctuid moth's ear is tuned to hear the biosonar calls of predaceous, insectivorous bats, which were known to include moths, including noctuids, in their diets. This minimalistic, two sensory-cell, ultrasound-sensitive hearing organ likely evolved as a special adaptation to predatory pressure from insectivorous bats, that hunt by emitting biosonar, echolocation signals (Roeder 1967).

In a classic and innovative series of studies, Roeder went on to study moth hearing as it occurs *au naturel,* out of doors and under the night sky, in his backyard and adjacent fields. He deployed powerful searchlights to cast intense beams of light high into the open sky which attract luxophilic flying moths. The phototactically attracted moths became illuminated and visible when they flow into the beams of the searchlight, at which time, Roeder directed beams of ultrasound at the illuminated moths. He stimulated the moths with playback of pulses of ultrasound within the bandwidth of Northeastern bats about 25–75 kHz, mainly *Myotis lucifigus* and *Eptiscus fuscus* (Roeder 1967).

In his experiments, Roeder precisely controlled where and when he wanted to stimulate the moths with ultrasound by mounting powerful loudspeakers on the end of long poles, thrust high into the air, to broadcast beams of ultrasound. Moths are attracted to light and so fly into these beams of light illuminating them against the background of the black, night sky, for photography. When Roeder stimulated high-flying moths with ultrasound, the moths abruptly steered away from the location of the speaker—this is an acoustic startle or evasive response (Roeder 1967).

As a devoted physiologist, Roeder augmented his field behavioral observations with neurophysiological recordings from minimally dissected moths from a recording setup on a table in his yard. He had built a little physiology lab in a shed in the yard—and that is why Kenneth Roeder is considered a pioneering neuroethologist! He was able to detect the presence of a bat chasing moths flying overhead by listening to the neural discharge of spikes from the auditory nerve of his lightly dissected, physiologically prepared moths whenever foraging bats homed in on the moths that were attracted to the searchlight beams—in effect, Roeder had "invented" an ultrasound-sensitive biological microphone!

Treat was especially delighted with his collaboration with Roeder because it helped solve his conundrum: why did moth ear mites inhabit only one ear chamber, but never both (Treat 1955)? Treat and Roeder found that ear mite infestation resulted in physiological deafness because the mites eat and destroy the moth's "inner ear," including its auditory nerve. But if the mites only devoured one ear the semi-deafened moth could still detect and perform evasive movements to escape from echo-ranging bats, permitting survival of both moth and mites. If mites infested both ears, the moth would be totally deaf—in effect, turning it into an insect version of the Titanic, for both mites and moths, but a meal for the bat (Roeder 1967).

2.3.1 *All Night-Flying Insects Have a Bat Problem*

Roeder's foundational study opened the eyes of a new generation of neuroethologists to investigate auditory behavior in other insects. In the 1970s and early 1980s, graduate students and postdocs in my lab followed Roeder's lead by investigating ultrasound-sensitive hearing in other night-flying insects. This led to multiple discoveries of ultrasound-sensitive ears in unexpected places and in surprising species, such as crickets (Moiseff et al. 1978), praying mantises (Yager and Hoy 1986), beetles (Forrest et al. 1995;Yager and Spangler 1997), katydids (Faure and Hoy 2000), and in many more moth species (Fullard and Yack 1993), and even in an acoustic-parasitic fly (Rosen et al. 2009). That may seem surprising until one knows that this fly's host species are field crickets that they locate by hearing their mating calls (Cade 1975) and home in on the wing, at night, when male crickets sing and bats hunt bugs. Hence, to locate its host, a singing male cricket, the parasitic fly must share a female cricket's predation risk from bats (Rosen et al. 2009). These phenomena have been reviewed in previous SHAR volumes (Hoy 1992).

2.4 Cockroaches Are Great Escape Artists, But Can They Hear?

Cockroaches are among the largest insects (one to several inches long) and infamously invade human dwellings, cohabiting with humans ever since humans sought protection from the elements in closed buildings. It might surprise readers that they are mentioned in a chapter on insect sonic behavior and hearing because they are not known to produce sound. In fact, early in the study of insect hearing, cockroaches were suspected to have a sense of hearing (Pumphrey and Rawdon-Smith 1936a, b).

Ironically, cockroaches lack recognizable tympanal membranes, unlike their sonically active insect relatives, crickets, cicadas, and katydids. The most familiar and common species of cockroaches such as the American cockroach (*Periplaneta americana),* German cockroach *(Blatella germanica), and the oriental* cockroach (*Blatta orientalis)* are very well known to humans as pests because they inhabit human dwellings. Cockroaches are notorious and ruefully hard to catch and kill, to the point that frustrated housekeepers have attributed preternatural senses to escapees and, in particular, a suspicion that they can hear approaching threats.

In fact, cockroaches are extremely sensitive to air currents that perturb sensory hairs on their abdominal cercal organs (Pumphrey and Rawdon-Smith 1936a). Instead of tympanal ears cockroaches possess a pair of "hairy" appendages, cerci, on the tips of their abdomen, to which hearing was attributed. Biologists have also long admired the ability of cockroaches to detect local disturbances and avoid capture, even in dim light where vision is minimized, but their cercal sensory organs were regarded as short-range, "wind" detectors, not proper ears, with eardrums like orthopteroid insects such as crickets, katydids, and grasshoppers.

In the lab, where carefully controlled, gentle puffs of wind can be generated by a diaphragm-driven device, such as the cone of a low frequency speaker (woofer), a single puff or a brief puff-train is sufficient to startle resting roaches and send them scuttling rapidly away from the source of airflow (Camhi et al. 1978). The distance separating insect and the stimulus source is usually very short, on the order of up to 10 cm, but far beyond the range of actual touch or tactile contact. The roach's mechanosensory organs, its cerci studded with fine sensory hairs, sense the bulk movement of air particles emitted from a nearby source, such as from acoustic audio speakers, placed close but not touching, and broadcasting low frequency tones, up to about 250 Hz (Pumphrey 1940).

As would be revealed in the 1970s, experimental work showed that cercal filiform sensilla of cockroaches (socketed cuticular hairs) are extremely sensitive to airflow. Further experimentation showed that the filiform hairs clearly sense the mechanical perturbations of air as shown by Juergen Tautz (1977, 1979), who studied the socketed filiform sensory hairs of moth caterpillars. These investigations made it evident that insects with mechanosensory organs bearing long, fine filamentous sensory hairs can detect airborne signals that are due to the movements of air particles and not from substrate vibrations or direct tactile touch.

The cercal organs of a roach bristle with dozens of long, thin, microscopic hairs that project into the air. These hairs are innervated by bipolar sensory receptor cells which serve as acoustic sensors. Thus, sensory hairs provide atympanate insects with another mechanism for sensing acoustic airflow; they detect the oscillatory movements of air particles that constitute the fluid flow of the propagating sound wave (Markl and Tautz 1975). These studies indicated that the role of filiform sensory hairs is to detect low frequency sounds over a few cm distance as well as responses to air puffs generated at slightly greater distances, and of course to respond to direct touch. Thus, cuticular hairs became associated with very short distance airflow detectors. In summary, although the commonest cockroach species have long been suspected to hear sound, based on biomechanics (Shaw 1994) and neurophysiology (Camhi et al. 1978) definitive behavioral evidence is still lacking.

2.4.1 Supersized Sonic Roaches: The "Hissing Cockroaches" of Madagascar

In the 1960s, the entomologist Louis Roth, at the Natick Laboratory Army Research, Development, and Engineering Center, (in search of the perfect insecticide, no doubt) in Massachusetts, famously maintained a cosmopolitan menagerie of cockroaches which he willingly shared with curious biologists. One such species, *Gromphadorhina portentosa* (the Madagascar hissing cockroach), has long captured the attention of entomologists (Roth and Willis 1960) as well as the public at large. This roach plays an outsized role in contemporary cultural awareness simply because it is nearly as big as a house mouse (2–4 inches long) and loudly announces its "displeasure" by audibly loud, startling hissing sounds when it is disturbed or picked up. This belies their usual docile nature—after a few disturbance hisses when touched, they do not run away when picked up and held in one's palm where they remain in docile repose; a quality that makes them a children's favorite at public science centers and museums.

However, *Gromphadorhina*, unlike all other roaches, also exhibits remarkable sonic and auditory behavior in its social interactions. During courtship and mating, male roaches make temporally patterned hissing sounds in the presence of females while competing males emit loud aggressive hisses at each other. Their acoustic behavior is like that seen in other auditive insects, like crickets and katydids. Yet, scrupulous visual and microscopic anatomical examination indicate that *G. portentosa* is atympanate, and moreover, its short, stubby abdominal cercal organs are grossly reduced in size and its sensory hairs are extremely sparse in number and short in length, by comparison with other cockroach species. However, this roach unequivocally emits audible acoustic signals in its social behavior (mating and territorial(, and it hears contextually salient and distinctive social hisses, based on evidence from recordings made from behaving roaches and electroacoustic analogs of actual songs (Fraser and Nelson 1984; Clark and Moore 1994).

G. portentosa produces a loud, noisy, hissing sound by forcibly expelling air through its extensive respiratory tracheal system and released out of a specialized pair of horn-shaped spiracles on the second abdominal segment that are morphologically adapted (tuned) to produce a noisy hissing sound over a wide spectral range that "peaks" at 5 kHz (Barth 1968; Nelson 1979).

Remarkably, *G. portentosa's* sound producing mechanism resembles that of a vertebrate's, given its respiratory system origins. This hiss is produced by forcing air through the roach's respiratory spiracular system located in its second abdominal segment that terminates to the outside in a reed-like valve (Roth and Willis 1960; Barth 1968; Nelson 1979). Muscles controlling the opening and closing of this valve allow this insect to modulate its sonifications, mainly through amplitude modulation (Nelson 1979; Nelson and Fraser 1980). While several large insects like cicadas or praying mantises emit audible disturbance sounds when picked up or perturbed, *G. portentosa* has also adapted its sonic system to subserve mating and territorial behavior in its social communication (Nelson 1979; Clark and Moore 1994).

Social hissing signals are amplitude modulated in courtship/mating and aggressive (male-male) contexts. While no obvious tympanal membrane has been found yet nor reported, histological and functional modifications of its chordotonal subgenual organs in several of its legs suggest a possible auditory function (Fraser and Nelson 1984). Neural recordings made from the leg nerve indicate sensitivity to sound in a frequency band ranging from 1 kHz to about 8 khz, with a broad peak from 4 to 6 kHz.

These data are consistent with Shaw's "missing link" hypothesis (1994). Shaw investigated hearing in the common house roach, *P. americana*, by focusing on mechanoreceptive subgenual organs in the legs of cockroaches, instead of the sensory hair-studded cercal organs. He proposed that these subgenual (beneath the "knee") organs are acoustic "missing links" that connect them, evolutionarily, to the tympanal scolopophorous hearing organ of crickets and katydids, which these sonic orthopteran insects also possess. In addition, they have evolved a separate scolopophorous organ that is associated with the foreleg tympanal membrane. Moreover, the orthopteran subgenual (SGO) and tympanal organs (TO) are adjacent to one another, lying beneath the "knee," in the forelegs of crickets (Michel 1974).

Shaw pointed out the remarkable sensitivity of SGOs to very low levels of acoustic airflow (1994) in making his argument for auditory function. It seems quite plausible that their SGO may explain hearing in the hissing cockroach, where its subgenual organs are modified from the typical SGO, in other roach species) by possessing many more sensory scolopales. The hissing cockroach is a testament to the remarkable and biodiverse themes that are presented in studying the acoustic behavior of insects. Among all cockroach species, it is unique in its respiratory system-derived sonic emitting mechanisms for social communication (Nelson 1979; Fraser and Nelson 1984). Importantly, hissing cockroaches are endemic to and found only in Madagascar. They represent an example of isolated island endemism giving rise to animals uniquely different from related continental species. Apparently, all members of the endemic genus *Gromphadorhina* emit hissing sounds, which is a unique sound among the world's cockroaches, so far as known (Roth and Willis 1960).

2.5 The Familiar, Aggravating, Hum of Mosquitos

Unquestionably, the relationship between humans and mosquitos has a very long history, ranging from annoyance to fear because of the habit of some species to "bite" humans. Mosquito bites are at least irritating and a nuisance but at worst cause serious illness and death due to the viruses and parasites that are carried in the salivary glands of infected mosquitoes and transmitted to human hosts. Viruses such as Yellow Fever, Zika, Malaria, and Dengue are but a few infamous examples of mosquito-borne diseases.

Since time immemorial humans have been aware of a mosquito's immediate presence by (barely) hearing its low-pitched, humming flight sound when flying around their heads especially annoying at bedtime, in the still of night. A mosquito's extremely soft flight tones are inaudible at distances of an arm's length, even in silence, but become perceptible at very close range as when it flies near its victim's head and ear, in preparation to land and bite. The fundamental frequency of the flight tone ranges from 300 to 800 Hz.

Mosquitos have always drawn the curiosity of entomologically inclined naturalists. An occupational interest possibly motivated the Baltimore physician and naturalist, Christopher Johnston (1855), to investigate mosquitos. He is credited with discovering the fly's donut-shaped hearing organ, the eponymously named Johnston's Organ (JO), at the base of each antenna. Johnston performed skillful anatomical studies of the mosquito JO and ascribed auditory functionality to it without any empirical proof, writing "…that the male should be endowed with superior acuteness of the sense of hearing appears from the fact that he must seek the female for sexual union either in the dim twilight or the dark night where nothing but her sharp humming noise can serve him as a guide" (Johnston 1855). As a medical doctor, Johnston is likely to have had first-hand experience with major epidemics of Yellow Fever and malaria, which were already known to be associated with having been bitten by mosquitos.

2.5.1 Are the Low-Pitched Tones of a Flying Mosquito Mating Signals?

Johnston's famous conjecture about mosquito hearing stimulated the imagination of other mid-nineteenth-century scientists and naturalists. Just a few decades after Johnston's paper was published it caught the attention of the scientist-inventor and naturalist, Hiram Maxim. Maxim was responsible for setting up the first night-time electrically illuminated exhibition in upstate NY, from the roof of adjacent buildings overlooking the exhibition site.

The next morning following a night of illuminated revelry and after the dynamo generator motors had been switched off, Maxim (1901) noticed that wherever the motors were located, the floor was carpeted with huge numbers of carcasses of mosquitos, all males. He astutely inferred from this that the loud, persistent, low-pitched hum from the dynamos lighting the exhibit grounds attracted the mosquitos by their sound. Maxim surmised that the sound of his dynamo motors emitted pitched tones resembling the flight tone of female mosquitos, on the wing, to which the males would be attracted for mating. Maxim, and earlier physicist A.M. Mayer (1874), demonstrated that male mosquitos could be attracted to tuning forks, for example, concert A or middle C (440 and 512 Hz respectively) both of which fall within 300–800 bandwidth of wingbeat frequency for female mosquitos such as *Aedes aegypti* and *Anopheles* species, both disease carriers.

It was not until the mid-twentieth century, that a definitive laboratory study was made of the mosquito's phonotactic behavior (in this case, *Aedes aegypti)* by the entomologist, Louis Roth (1948), and of the cockroaches mentioned above. Roth, like his predecessors, used tuning forks for simple experiments, but he also deployed newly available electronic audio oscillators and high-fidelity loudspeakers to demonstrate that sound in the frequency range of a female mosquito's wingbeats was sufficient to attract caged males to electronic and mechanical sound sources, in the absence of actual live females. Crucially, Roth found that amputating the antennae eliminated the mosquito's response to sound.

A few preliminary, crude, recordings were just being made in the 1950s from the JO and auditory nerves, setting the stage for a physiological approach to investigate behavior (Tischner 1953; Tischner and Schief 1955). However, the actual behavioral role of the mosquito's wingbeat as an acoustic signal in mating as well as truly rigorous biomechanical measurements of the sensitivity of the mosquito's JO hearing organ would not come until 50 years later (Goepfert and Robert 2001; Gibson and Russell 2006; Cator et al. 2009).

2.6 Loudest for Last: Cicada Songs

Since prehistoric times, humans and their hominid progenitors have likely associated the warmest months of the year with buzzing, crackling, whirring sounds that are emitted from male cicadas, calling from tall bushes or high above, in trees, during the hottest hours of summer days. Their songs are easily recognized as coming from individual males, but often nearby male cicadas in adjacent bushes or high in trees sing together, forming a deafening chorus. They are among the largest of the sonic insects and the loudest to inhabit human soundscapes during the day. No historical account of insect bioacoustics is complete without including cicadas.

Given the prehistoric fossil evidence as well as recent genomic research, it is safe to assume that all the sonic insects that ensonify our twenty-first-century environment are related, at least to the Family level, to ancient lineages of sonic insects that sang in the rise and sang out the fall of past ancient human populations and their civilizations. Clearly, cicadas have penetrated the consciousness of ancient cultures prominently enough to have left bountiful cultural footprints, in both cultural practice and as materially solid, symbolic objects in the form of jewelry and other statuary (Aldred 1971; Lim 2001), and they have been the muses for poets and philosophers for centuries.

Cicadas are large insects, fitting comfortably in the palm of one's hand and they are conspicuously the loudest. Like other sonic insects, it is the male that emits persistent, hours long calling songs. They can make their presence known by the sound produced by individuals, singing high in trees, but they undeniably dent even the most distracted human consciousness when males form local aggregations or congregations in adjacent bushes and trees, where they produce a chorus of ear-shatteringly intense sonic calls. Chorus size can range from dozens to hundreds and thousands, and in the case of 13- and 17-year-old cicadas, to millions of calling males. It seems likely that humans have always associated the sound of cicadas with summertime.

2.6.1 Cicadas as Human Fetish Objects

Cicadas were always among the most prominent animals that inhabited the auditory, if not visual, scene of humankind's earliest civilizations, especially in temperate and tropical climates of the Mediterranean and the Near and Far East, during the hot, humid summer, Hence, they drew the attention of citizenry at all levels of society—they became fetishized. Whereas virtually all other sonic insects live their life cycles within only one seasonal year, some cicadas have a prolonged larval period in which they live and grow through multiple larval molts that span multiple years, from a few to seventeen, before emerging above ground and molting into their sexually mature, sonically active adult form. The recurrent cycles of death and larval emergence from the ground and metamorphosis into adults was a behavior well known in ancient civilizations and to their sages. The cicada's life cycle consisting of clamorously loud and very brief adult stage quickly followed soon by death and apparent "rebirth" after some years impressed all within earshot. Cicadas became fetishized as evidence of resurrection and came to symbolize immortality—a power attributed only to gods and other deities—in ancient civilizations as distantly separated in space and time as the Far East, Middle East (Egypt), and Western Europe. Cicada-shaped jewelry, sculpture, and art testify to their being spiritual or religious symbols. Such artifacts have been found in tombs and other burial sites (Aldred 1971; Lim 2001).

While cicadas may have lost their mystical religious symbolism today, they are still valued and regarded as tokens of good fortune and long life in some cultures. In fact, the ever-curious Greek philosopher and sage, Aristotle (1910), notably commented on cicadas. Aristotle was a noted naturalist and was well aware of the cicada's periodic life cycle and cited them as a symbol of immortality. His interest in them was not just naturalistic and spiritual, for he prized the tastiness of cicada grubs, dug from the ground and roasted. The ancient Greeks were not alone in valuing cicadas as a food, and in the tropical climes of Africa and Asia, they are even now featured as summer snacks and serve seasonal staples in food markets. Cicada grubs and adults are large and easy to dig up or catch. Hungry humans find adults by hearing them, since the adult male cicada's calling song is to predators a "dinner bell."

2.6.2 The Scientific Study of Cicada Bioacoustic and Hearing

Even in Renaissance and post-Renaissance academic circles and naturalist societies, the sonic insect that has historically attracted the most attention is the cicada, likely because its singing activities are limited to daylight hours, on sunny days in the hottest weeks of the summer. They are large, easily captured, and identified. However, most of the auspiciously sonic insects are crepuscular or strictly nocturnal creatures and hence mostly heard but not seen, making them far less easy to observe and study before artificial lighting was invented. In Myers' (1929) "Insect Singers," he cites obscure, early literature that the sound-making (tymbal) organs of cicadas were described by in the eighteenth century by Reaumur. Published descriptions of the cicada hearing organ, featuring a conspicuously large and transparent tympanal membrane, came much later with the detailed histological and anatomical studies of Vogel (1923). Clearly, even by early in the twentieth century, there was a notable literature on cicada auditory behavior.

Cicadas of both sexes possess a pair of typical chordotonal hearing organs, one beneath each wing. It consists of a clearly visible, externalized, eardrum that opens internally into a large air-filled chamber, and is innervated by a scolopophorous auditory sensory organ consisting of about 1000 sensory scolopidia (Young and Hill 1977). The tympanic membranes ("eardrums") are located in the abdomen and reside immediately adjacent to the smaller tymbal organ. The air-filled tympanal chamber is part of the air-filled abdominal sound radiator. The eardrum is attached to its scolopophorous auditory organ by a thin, stiff, apodeme, of cuticular origin that is interposed between the tympanic membrane and the auditory organ. Obviously, there must be a decoupling of auditory function from the intense acoustic emissions, given the anatomical proximity of tymbal membrane (sound emitter) and tympanal membrane (sound receiver) and it is thought that there is an efferent decoupling mechanism analogous to the stapedial reflex in the human cochlea to protect the sensitivity of the hearing organ, and indeed the auditory and tymbal tensor nerves contain both afferent and efferent axons (Wohlers et al. 1979). However, the mechanistic biomechanical and neural circuitry details are unknown and remain to be worked out.

2.7 The Age of Neurophysiology and Auditory Physiology

While functional morphology and behavioral studies were necessary to demonstrate the behaviorally adaptive functionality of hearing and ears, auditory function is incomplete without knowledge of how the biomechanics of auditory stimulation results in mechano-acoustic transduction into the neural activity, from the auditory nerve to the brain. There could be no auditory neuroscience until the invention of electrophysiology. That did not come about until the twentieth century and is identified with pioneers such as C.S. Sherrington and E. Adrian in Britain, A. Forbes, H. Gasser, and J. Erlanger in America, and H. Berger and O. Foerste in Germany.

Edgar Adrian (Lord Adrian) and his supervisor, Keith Lucas, at Cambridge University (UK), were among the first physiologists to make recordings from sensory nerves in both vertebrates and invertebrates. Indeed, Adrian and Sherrington became the first neurophysiologists to become Nobel laureates, in 1932. Their pioneering work, and that of their American contemporaries, Herbert Gasser and Joseph Erlanger, at Washington University, St. Louis, who years later would themselves become Nobel laureates, set the stage for auditory neuroscience to emerge. Electrophysiological recording techniques exploded in labs, worldwide, including those groups that made recordings from auditory nerves in both vertebrates and invertebrates, including insects.

Starting in the 1930s, the great pioneer of auditory science, Ernest Glen Wever investigated the physiology of the grasshopper hearing organ with his colleagues and students, C.W. Bray and J.A. Vernon. They published papers on the acoustic sensitivity of the insect auditory nerves and organs of several species of *Orthoptera*, including crickets and katydids (Wever and Bray 1933; Wever 1935; Wever and Vernon 1957). Beginning in 1928, Wever and his colleagues published numerous first recordings from the auditory nerves of animals, a menagerie that included pigeons, frogs, lizards, as well as many mammals including cats and bats.

The fact that Wever also made the earliest auditory nerve recordings in insects is a testament to his insatiable curiosity, dedication to a truly comparative approach, as well as his bench skills. In fact, much later in life, Wever coauthored a paper with James Simmons on Brood X, of the 17-year-old cicada, about the possible adaptive significance of the cicada's explosive, simultaneous emergence and singing in painfully loud choruses consisting of millions of insects, to deter predation by birds (Simmons et al. 1971).

2.7.1 Insect Auditory Neurophysiology—Post-World War II and the 1950s

World War II spun off a revolution in electronic devices, for both industrial/research and consumer markets, to the benefit of auditory research science. For bioacoustics, the emergence of a giant consumer audio industry, led by tape recorders and audiophile quality amplifiers and speakers at affordable prices, enabled ethologists as well as physical scientists to more broadly explore bioacoustics and auditory neuroscience.

It is notable that important, post-war neurophysiological studies of insect hearing emerged from the Tokyo laboratory of Yasuji Katsuki and his very accomplished student, Nobuo Suga, in the late 1950s and into the early 1960s (Katuski and Suga 1959; Suga 1966). They made recordings from the auditory tympanal organs as well as from the central nervous system in cicadas, katydids, crickets, and grasshoppers, at the single unit level (Katuki and Suga 1960). They were among the first to record from the CNS, using KCl-filled micropipette electrodes, and they characterized large axon T cells believed to be associated with sound localization in katydids (Suga 1963).

G. A. Horridge (1960) in the UK and Katsuki's lab (Yanagisawa et al. 1967) studied locust hearing and were among the first investigators to show that insect neurons demonstrated true frequency discrimination, thus refuting the then-common belief that insects with tympanate auditory organs could not discriminate the frequency of an acoustic stimulus independently of level, as claimed by Pumphrey (1940).

Upon completing his doctoral training in Katsuki's lab, Suga embarked for post-doctoral work in the USA with Donald Griffin, at Harvard University (Cambridge, MA), and later, with Theodore Holmes (Ted) Bullock, at UCLA (Los Angeles, CA), where he applied his neurophysiological skills to the auditory CNS of bats. He went on to establish a distinguished, life-long, career studying how the bat brain processes biosonar signals in his laboratory at Washington University, in St. Louis. Suga's remarkable span of accomplishments as a neurophysiological recordist can be traced to his PhD studies with Katsuki, on the insect auditory system (Suga 1960, 1961, 1963).

The work of Kenneth Roeder (1967) on moth hearing and auditory physiology that took place in the 1950s and 1960s, discussed in Sect. 2.3, is arguably proto-neuroethological, as a paradigm for the field. As discussed in Sect. 2.2.2, Franz Huber's pioneering neurophysiological work in Germany in the 1960s established the field cricket as a model for insect bioacoustics. The 1970s saw the flowering of the field of neuroethology that ushered in a period of unparalleled investigation of the neurobiological basis of the insect senses of vision and olfaction, as well as audition. The auditory system of every known sonic insect was studied at the level of its neural systems, from sensory organs to the brain. To review this work is far beyond the scope of this retrospective. It is also the subject of two previous volumes of the SHAR series, (Hoy et al. 1998 and Pollack, Mason et al. 2016), to which the interested reader is referred.

2.8 Sizing Up Insect Hearing, a Reprise

I began this chapter by calling attention to how the extremely small size of insects, by comparison to vertebrate animals, can drive differences in the mechanisms of hearing, particularly in the "outer ear," and noting that insects do not need a middle ear. Hearing in small insects like mosquitoes and other small flies makes membrane-constrained outer ears problematic. Instead, these tiny insects deploy thin, filamentous, nanoscale airflow detectors that are sensitive to the velocity of local movements of surrounding air particles that are set into oscillation by the propagating sound waves. These particle-velocity-sensitive hairs or hair-like antennae are innervated by mechanoreceptive sensory cells that perform sensory transduction. Insects like cicadas, grasshoppers, katydids, beetles, moths, and praying mantises that possess the sense of hearing are large enough to deploy tympanal membranes (eardrums) substantial enough to detect pressure differences in the sound wave.

To summarize, hearing in both vertebrates and insects involves sensing an aerial mechano-acoustic disturbance that radiates from the vibrating noise source. The sound wave propagates through oscillatory movements of air particles that impinge upon the animal's outer ear structure, which is set into movement. In vertebrates, the outer ear may be as simple as thinned patch of epithelium/skin or an internalized tympanal membrane—in either case, it is an eardrum. The vertebrate eardrum senses the vibrating airflow as oscillations in air pressure, integrated over its surface. Large sonic insects, such as cicadas, locusts, and crickets also possess tympanal membranes which detect oscillating changes in air pressure, just like vertebrates.

However, minutely small insects, such as mosquitos and drosophila, are insensitive to changes in air pressure but instead detect the velocity of air particle oscillations in the impinging sound wave, which they sense with long, thin cylindrical mechanosensory hairs. Importantly, such sensory hairs can sense particle velocity even at relatively long distances from the source where sound propagates as a pressure wave, the acoustic far-field, and not just in the immediate vicinity of a vibrating source, the near acoustic field.

The ability of a sensory hair to detect air particle movements in the pressure field is limited only by its mechano-neural sensitivity, and its auditory threshold is determinant. As long as its sensitivity is above thermal noise, particle velocity can be neuro-acoustically transduced into action potentials and a sensory hair would be an adequate acoustic organ.

Mosquitos can hear sound in the range of 400–700 Hz over relatively long distances (meters). This is the case for the mosquito antenna-Johnston's organ complex, as elegantly demonstrated in the biomechanical experiments of Goepfert and Robert (2001) and much later, in neurophysiological recordings from the Johnston's organ auditory nerve (Menda et al. 2019). Thus, the customary use of the terms, acoustic "near-field" and "far-field," where the former implies detection of air particle velocity at close range and the latter, detection of air pressure at long range, needs to be reconsidered.

There is a tendency to think of acoustic transducers as being pressure detectors, whether in the world of commercial audio microphones or in comparative bioacoustics labs investigating the animal ear. This is an understandable but implicit anthropocentric bias. In fact, all commercial microphones are based on the human hearing mechanism and are constructed to detect sound pressure. But this is not so for the hearing organs in minute insects (nor in fishes—see Sand et al., this volume).

Retracing the history of discovery in insect hearing, it was conjectured or known by the 1930s, that cockroaches and mosquitos sensed sound with cuticular sensory hairs on cercal organs or antennae, but functionality was thought to be limited to very close distances. Behavioral reactions could be induced by local puffs of wind or near-field acoustic excitation. Long range acoustic sensitivity was assumed to be the province of pressure-sensing tympanal organs, as exemplified in crickets and cicadas. It was not until the twenty-first century that the sensitivity of hairs or antennae was demonstrated at distances where sound is propagated as changes in air pressure, thus demonstrating that sensory hairs detect the velocity of air particles oscillating in the pressure field (Goepfert and Robert 2001; Menda et al. 2019).

2.9 After Words

There remains much to be discovered about insect hearing because of the diversity of hearing organs and mechanisms, especially those in small insects that operate at nanoscale-to-microscale dimensions. They can be studied with new imaging, biomechanical, and neurophysiological tools that are now available. While this chapter has emphasized differences in hearing mechanisms and organs between vertebrates and insects, very recent work points to functional convergence in the biomechanical workings of hearing organs, especially those possessed by katydids (Tettigonidae). This chapter will conclude with some recent studies that are food for thought when considering convergence and divergence in the structure and function of insect hearing organs.

Insect cuticular sensory hairs serve as remote sensors that detect action at a distance. Goepfert and Robert (2001) demonstrated the exquisite sensitivity of the mosquito JO to particle velocity in acoustic flow. However, sound is not the only force acting at a distance that is an adequate stimulus to stimulate sensory hairs. All that is required is a force to stimulate the hair into movement, however slight. Daniel Robert's lab (Clark et al. 2013) has recently shown that an electrostatic force will cause (charged) cuticular hairs to move and generate action potentials. They showed that bumblebees can detect and even learn the configuration of the electrostatic field of flowers and leaves and utilize this information for efficient foraging. This work opens a new window into the sensory ecology of insects—that of terrestrial electroreception—and a new role for sensory hairs. Thus far, it has been shown that sensory hairs are superb models for detecting weak forces at a distance, such as air particle movements in acoustically driven airflow, as well as weak electrostatic force fields—mechanoreceptive hairs can also electroreceptors—as long as they are moved by an external force they can be excited to respond.

It has been known for several decades that the scolopidia cells in a chordotonal hearing organ (such in crickets and katydids) form linear arrays in the organ and that they are organized tonotopically, analogous to that in the vertebrate cochlea (Oldfield 1982). This is an interesting convergence of structure and function in the cricket and katydid ears with vertebrates.

However, an even more interesting story has emerged in the recent investigations of the hearing organs of tropical, South American katydids (tettigonid), *Copiphora gorgonensis* (Montealalegre-Z et al. 2012). The work of Montealegre-Z, also in the Robert lab, on the tympanal hearing organ of this katydids reveals an even greater degree of convergence of structure and function, with mammalian hearing. Not only did they demonstrate tonotopy, but they also discovered traveling waves within a fluid-filled chamber that contains the sensory scolopale cells. Presumably, this sets up a vertebrate-like impedance matching problem and indeed, they discovered a structure called the tympanal plate which seems to act like the vertebrate middle ear (Montealalegre-Z et al. 2012). These discoveries were enabled by micro-CT scanning techniques and laser vibrometry. A new technique for investigating the micromechanics within the miniscule katydid ear with even higher resolution optical coherence tomography vibrometry increases the spatio-temporal resolution of measurement, which reinforces the convergence in vibrational mechanics between insect and vertebrate ear, at the level of micromechanics. It will be interesting to know if a similar structure-function mechanism occurs in the ears of crickets and other sonic insects that have tympanal hearing organs (Vavacou et al. 2021). Traveling waves are also a feature in the ears of locusts, but the parallelism with vertebrates and katydids is far less pronounced (Windmill et al. 2005).

Another interesting parallel between vertebrate hearing and an insect is the discovery of metabolically dependent acoustic amplification and the discovery of spontaneous acoustic emissions in the hearing organs of mosquitos (Goepfert and Robert 2001) and grasshoppers (Koessl and Boyan 1998). There is evidence that efferent control of sensitivity may occur in insect hearing and that the phenomenon of stochastic resonance appears to play a role in enhancing sensitivity in insect hearing (Windmill and Jackson 2016).

I conclude with these "reports from the front," which I hope testify to the value of investigating hearing at nano-to-microscale in insects as well as the continuing value of comparative studies at the neuroethological level of sensory ecology and behavior. It is obvious that such program of basic research studies also provides rich ground to till for the applied acoustical engineering field, through the application of the strategy of biomimicry for discovery. We live in an age of novel applications of audio in the age of the internet, and the need for new designs for microphones and audio transducers is alive and well in the world of the internet. For tomorrow's students: study insect hearing—It has much to offer.

Acknowledgments I am grateful to the 25 grad students and 26 postdocs I was lucky enough to attract to my lab at Cornell and regret that I could not cite the work of every one of them in this chapter. However, this book focuses on the historical roots of insect hearing up to about 1970 which is when I opened my lab. Special thanks are owed to the editors: Arthur Popper, without whose prodding and encouragement, I would not finished this chapter, and Darlene Ketten whose support encouraged me to keep on trucking. Lastly, to my wife Margy who said I was crazy to take this on alone—she was right.

References

Aldred C (1971) Jewels of the pharaohs. Egyptian Jewelry of the dynastic period. Praeger Publishers, N.Y

Alexander RD (1957) The taxonomy of the field crickets of the eastern United States (*Orthoptera: Gryllidae*). Ann Entomol Soc Am 50:584–602

Alexander RD (1962) Evolutionary change in cricket acoustical communication. Evolution 16:443–467
Allin EF (1975) Evolution of the mammalian middle ear. J Morphol 147(4):403–437
Aristotle (1910) Aristotle, History of Animals (Thompson), Bk. 5, Ch. 30, 556b6–14
Barth RH (1968) The mating behavior of *Gromphadorhina portentosa*. Psyche 75:125–131
Bennet-Clark HC (1971) Acoustics of insect song. Nature 234:255–259
Bennet-Clark HC (1991) Size and scale as constraints on insect sound communication. Philos Trans R Soc Lond B Biol Sci 353:407–419
Bennet-Clark HC (1999) Resonators in insect sound production: how insects produce loud, pure tones. J Exp Biol 202:3347–3357
Bentley DR (1969) Intracellular activity in cricket neurons during the generation of behaviour patterns. J Insect Physiol 15:677–684
Bentley DR, Hoy RR (1972) Genetic control of the neural network generating cricket song patterns. Anim Behav 20:478–492
Cade WH (1975) Acoustically orienting parasitoids: Fly phonotaxis to cricket song. Science 190:1312–1313
Camhi JM, Tom W, Volman S (1978) The escape behavior of the cockroach Periplaneta americana - II. Detection of natural predators by air displacement. J Comp Physiol A 128:203–212. https://doi.org/10.1007/Bf00656853
Cator LJ, Arthur BL, Harrington LC, Hoy RR (2009) Harmonic convergence in the love songs of the dengue virus mosquito. Science 323:1077–1079
Clack JA, Fay RR, Popper AN (eds) (2016) Evolution of the vertebrate ear: a paleontological perspective. Springer, New York
Clark DC, Moore AJ (1994) Social interactions and aggression among male Madagascar hissing cockroaches (*Gromphadorhina portentosa*) in groups (Dictyoptera: Blaberidae). J Insect Behav 7(2):199–215
Clark D, Whitney H, Sutton G, Robert D (2013) Detection and learning of floral electic fields by bumblebees. Science 340:6128–6132
Darwin C (1871) The descent of man and selection in relation to sex. John Murray, London
Dickens C (1846) The cricket on the hearth: a fairy tale of home. Seawolf Press. 2018 reprint, Orinda
Eaton R (1984) Neural mechanisms of startle behavior. Springer Science, Business Media, New York
Eggers E (1919) Das thoracale bitympanale organ einer Gruppe der Lepidoptera Heterocera. Zool Jahrb Anat 41:272–373
Faure PA, Hoy RR (2000) The sounds of silence: cessation of singing and song pausing are ultrasound-inducerd acoustic startle behaviors in the katydid *Neoconocephalus ensiger*. J Comp Physiol 186:129–142
Forrest TG, Farris HE, Hoy RR (1995) Ultrasound acoustic startic startle response in scarab beetles. J Exp Biol 198:2593–2598
Fraser J, Nelson MC (1984) Communication in the courtship of Madagascan hissing cockroach. Anim Behav 32:32194–32203
Fullard JH, Yack JE (1993) The evolutionary biology of insect hearing. Trends Ecol Evol 8:248–252
Gerhardt HC, Huber F (2002) Acoustic communication in insects and anurans. University of Chicago Press, Chicago
Gibson G, Russell IJ (2006) Flying in tune: sexual recognition in mosquitoes. Curr Biol 16:1311–1316
Goepfert MC, Robert D (2001) Nanometer-range acoustic sensitivity in male and female mosquitoes. Proc R Soc Lond B 267:453–457
Hedwig B, Schoeneich S, Kostarakos K (2018) A neural circuit for song pattern recognition in the cricket brain. J Acoust Soc Am 143:1858–1858
Horridge GA (1960) Pitch discrimination in Orthoptera demonstrated by responses of central auditory neurons. Nature 185:623–624
Hoy RR (1974) Genetic control of acoustic behavior in crickets—a review. Am Zool 14:1067–1080

Hoy RR (1992) The evolution of hearing in insects as an adaptation to predation from bats. In: Webster DB, Fay RR, Popper AN (eds) The evolutionary biology of hearing. Springer, New York, pp. 115–130
Hoy RR (1998) Acute as a bug's ear. In: Hoy RR, Popper AN, Fay RR (eds) Comparative hearing: insects. Springer, New York, 1–17
Hoy RR, Paul RB (1973) Genetic control of song specificity in crickets. Science 180:4081–4082
Hoy RR, Hahn J, Paul RC (1977) Hybrid cricket auditory behavior: evidence for genetic coupling in animal communication. Science 195:82–84
Hoy RR, Popper AN, Fay RR (eds) (1998) Comparative hearing: insects. Springer Science, Business Media, New York
Huber F (1962) Central nervous control of sound production in crickets and some speculations on its evolution. Evolution 16:429–442
Huber F, Moore TE, Loher W (1989) Cricket neurobiology and behavior. Cornell University Press, Ithaca. ISBN 0-8014-2272-8
Johnston C (1855) Auditory apparatus of the Culex mosquito. Q J Microsc Sci 3:97–102
Katuki Y, Suga N (1960) Neural mechanism of hearing in insects. J Exp Biol 37:379–290
Katuski Y, Suga N (1959) Electrophysiological studies on hearing in common insects in Japan. Proc Jpn Acad 34:623–638
Koessl M, Boyan GS (1998) Acoustic distortion products from the ear of a grasshopper. J Acoust Soc Am 104:326–335
Larsen ON, Michelsen A (1978) Biophysics of the ensiferan ear: the cricket ear as a four-input system. J Comp Physiol 123:217–227
Lim RK (2001) The symbolic importance of insects in jewelry. Trans Am Entomol Soc 127:167–171
Manley GA (1973) A review of some current concepts of the functional evolution of the ear. Evolution 26:608–621
Markl H, Tautz J (1975) The sensitivity of hair receptors in caterpillars of *Baratha brassicae* to particle movement in the sound field. J Comp Physiol 99:79–87
Marler P, Tamura M (1964) Culturally transmitted patterns of vocal behavior in sparrows. Science 146:1483–1486
Maxim HS (1901) Mosquitoes and musical notes. Letter to the Editor. The Times (London). October 28, p 11
Mayer AM (1874) Researches in acoustics. Reprinted in Kessinger's Legacy Reprints. 2015. Kessinger Publishing Co. (LLC), Whitefish
Menda G, Nitzani EI, Shamble PS, Wells A, Harrington LC, Miles RN, Hoy RR (2019) The long and short of hearing in the mosquito, *Aedes aegypti*. Curr Biol 29:709–714
Michel K (1974) Das tympanalorgan von *Gryllus bimaculatus*. Zeitschrift fur Morphologie und Okologie der tiere 77:285–315
Michelsen A (1998) Biophysics of sound localization in insects. In: Hoy RR, Popper AN, Fay RR (eds) Comparative hearing: insects. Springer, New York
Miles RN, Robert D, Hoy RR (1995) Mechanically coupled ears for directional hearing in the parasitoid fly *Ormia ochracea*. J Acoust Soc Am 98:3059–3070
Miles RN, Su Q, Qui W, Shetye M (2009) A low noise differential microphone inspired by the ears of the parasitoid fly, *Ormia ochracea*. J Acoust Soc Am 125:2013–2026
Moiseff A, Pollack GD, Hoy RR (1978) Steering responses of flying crickets to sound and ultrasound. Mate attraction and predator avoidance. Proc Natl Acad Sci U S A 75:4052–4056
Montealalegre-Z F, Jonesson T, Robson-Brown KA, Postles M, Robert D (2012) Convergent evolution between insect and mammalian audition. Science 338:968–971
Myers JG (1929) Insect Singers. G. Routledge, Sons, Ltd., London
Nelson MC (1979) Sound production in the cockroach, *Gromphadorhina portentosa*: the sound-producing apparatus. J Comp Physiol 132:27–38
Nelson MC, Fraser J (1980) Sound production in the cockroach, *Gromphadorhina portentosa*: evidence for communication by hissing. Behav Ecol Sociobiol 6:305–314

Oldfield BP (1982) Tonotopic organisation of auditory receptors in Tettigoniidae (Orthoptera: Ensifera). J Comp Physiol 147:461–46982
Pollack GS, Mason AC, Popper AN, Fay RR (eds) (2016) Insect hearing. Springer, New York
Pumphrey RJ (1940) Hearing in insects. Biol Rev 15:107–133
Pumphrey RJ, Rawdon-Smith AF (1936a) Hearing in insects: the nature of the response of certain receptors to auditory stimuli. Proc R Soc Lond B Biol Sci 121:18–27
Pumphrey RJ, Rawdon-Smith AF (1936b) Sensitivity of insects to sound. Nature 137:990
Regen J (1913) Über die Anlockung des Weibchens von *Gryllus campestris* L. durch telephonisch übertagene Stridulationslaute des Männchens. Akademie Wissenschaften Mathematik Nat Kl Abt I (Wien) 132:81–88
Robert D, Hoy RR, Amoroso J (1992) The evolutionary convergence of hearing in a parasitoid fly. Science 258:1135–1137
Robert D, Miles RN, Hoy RR (1996) Directional hearing by mechanical coupling in the parasitoid fly Ormia ochracea. J Comp Physiol A 179:29–44
Roeder KD (1967) Nerve cells and insect behavior. Harvard University Press, Cambridge
Roeder KD, Treat AE (1957) Ultrasonic reception in the tympanic organ of noctuid moths. J Exp Biol 134:127–158
Rosen MJ, Levin EC, Hoy RR (2009) The cost of assuming the life history of a host: acoustic startle in the parasitoid fly *Ormia ochracea*. J Exp Biol 212:4056–4064
Roth LM (1948) A study of mosquito behavior: an experimental study of the sexual behavior of Aedes aegypti. Am Midl Nat 40:265–352
Roth LM, Willis ER (1960) The biotic associations of cockroaches. Smithsonian Miscellaneous Collections 141, 470 pp
Schoeneich S, Hedwig B (2019) Feedforward discharges couple the singing central pattern generator and ventilation central pattern generator in the cricket abdominal central nervous system. J Comp Physiol 205:881–895
Schwabe J (1906) Beitrage zur Morphologie und Histologie der tympanalen sinnessapparate der Orthoptera. Zoologica 20:1–54
Shaw SR (1994) Detection of airborne sound by a cockroach 'vibration detector': a possible missing link in insect auditory function. J Exp Biol 193:3–47
Simmons JA, Wever EG, Pylka JM (1971) Periodical cicada: sound production and hearing. Science 171:212–213
Song H, Bethoux O et al (2020) Phylogenetic analysis sheds light on the evolutionary pathways toward acoustic communication in Orthoptera. Nat Commun 11:4939
Suga N (1960) Peripheral mechanism of hearing in the locust. Jpn J Physiol 10:533–546
Suga N (1961) Functional organization of two tympanic neurons in noctuid moths. Jpn J Physiol 11:666–667
Suga N (1963) Central mechanism of hearing and sound localization inn insects. J Insect Physiol 9:867–873
Suga N (1966) Ultrasonic production and its reception in some neotropical Tettigoniidae. J Insect Physiol 12:1039–1050
Tautz J (1977) Reception of medium vibration by thoracal hairs of caterpillars of *Barathra brassicae*. J Comp Physiol 118(13):31–31
Tautz J (1979) Reception of particle oscillation in a medium—an unorthodox sensory capacity. Naturwissenschaften 66:452–461
Tischner H (1953) Uber der Gehoersinn von Stechmuecken. Acust 3:335–343
Tischner H, Schief A (1955) Fluggeraiischund Schallwahrnehmungbei *Aedes aegypti* L. (Culicidae). Zool Anz 18(Suppl):433460
Treat AE (1955) The response to sound in certain Lepidoptera. Ann Entomol Soc Am 48:272–284
Vavacou A, Nowotny M, Scherberich J, van der Heijden M (2021) Tuned vibration modes in a miniature hearing organ: insights from a bushcricket. Proc Natl Acad Sci U S A 118:30–49
Vogel R (1923) Über ein tympanales Sinnesorgan, das mutmaßliche Hörorgan der Singzikaden. Z Anat Entwickl-Gesch 67:190–231

Walker TJ (1957) Specificity in the response of female tree crickets (Orthoptera, Gryllidae, Oecanthinae) to calling songs of the males. Ann Entomol Soc Am 50:626–636
Walker TJ (1962) Factors responsible for intraspecific variation in the calling songs of crickets. Evolution 16:407–428
Webster DB, Popper AN, Fay RR (eds) (1992) The mammalian auditory pathway: neuroanatomy. Springer, New York
Wever EG (1935) A study of hearing in the sulfur-winged grasshopper. J Comp Psychol 20:17–20
Wever EG, Bray CW (1933) A new method for the study of hearing in insects. J Cell Comp Physiol 4:79–93
Wever EG, Vernon JA (1957) The auditory sensitivity of the Atlantic grasshopper. Proc Natl Acad Sci 43:346–348
Wilson DM (1961) The central nervous control of flight in a locust. J Exp Biol 38:471–4
Windmill JFC, Jackson JC (2016) Mechanical specializations of insect ears. Insect hearing. In: Pollack GS, Mason AC, Popper AN, Fay RR (eds) Insect hearing. Springer, New York, pp 125–157
Windmill JFC, Goepfert MC, Robert D (2005) Tympanal traveling waves in migratory locusts. J Exp Biol 208:157–168
Wohlers DW, Williams JLD, Huber F, Moore TE (1979) Central projection of fibers in the auditory tensor nerves of cicadas. Cell Tissue Res 203:35–51
Yack JE (2004) The structure and function of auditory chordotonal organs in insects. Microsc Res Tech 63:315–337
Yager DD, Hoy RR (1986) The cyclopean ear: a new sense for the praying mantis. Science 231:727–729
Yager DD, Spangler HG (1997) Behavioral response to ultrasound by the tiger beetle *Cicindela marutha dow* combines aerodynamic changes and sound production. J Exp Biol 200:649–659
Yanagisawa K, Hashimoto T, Katsuki Y (1967) Frequency discrimination in the central nerve cords of locusts. J Insect Physiol 13:633–643
Young D, Hill KG (1977) Structure and function of the auditory system of the cicada, *Cystosoma saundersii*. J Comp Physiol 117:23–45

Chapter 3
Evolution of the Understanding of Fish Hearing

Olav Sand, Arthur N. Popper, and Anthony D. Hawkins

3.1 Introduction

Fishes have no external auditory structures. Nevertheless, several thousand years before the first formal studies of fish audition in the late nineteenth century, a number of observers, including fishers, became aware that fishes can hear. Information on the very earliest observations on fish bioacoustics can be found in several reviews written by some of the leading twentieth-century researchers on fish hearing, including Parker (1918, 1903), von Frisch (1936, 1938a), Kleerekoper and Chagnon (1954), and Moulton (1963). Translations of several of the earlier papers are found in a volume compiled by Tavolga (1976). This chapter focuses on the origin and evolution of the current understanding of the mechanisms of hearing and the hearing capabilities of fishes based upon experiments starting in the nineteenth century.

A related topic of considerable interest and importance, with a similar long history, is sound communication by fishes, but the topic is beyond the scope of this chapter. However, those interested in it are directed to papers by (among others) Moulton (1963), Tavolga (1971, 1977), and Hawkins and Myrberg Jr (1983).

Two other areas, the anatomy of the auditory central nervous system (CNS) and the physiology of the CNS in fishes, are not covered in any depth in this chapter because much of the work has been performed since the late 1970s. For those interested in the anatomy of the CNS, there are excellent reviews by Northcutt (1980)

O. Sand (✉)
Department of Biosciences, University of Oslo, Oslo, Norway
e-mail: olav.sand@ibv.uio.no

A. N. Popper
Department of Biology, University of Maryland, College Park, MD, USA
e-mail: apopper@umd.edu

A. D. Hawkins
Loughine Ltd, Kincraig, Blairs, Aberdeen, UK

D. R. Ketten et al. (eds.), *A History of Discoveries on Hearing*, Springer Handbook of Auditory Research 77,
https://doi.org/10.1007/978-3-031-41320-9_3

and by McCormick (1992, 1999). A paper by Walton et al. (2017) reviews both the anatomy and physiology of the CNS.

3.2 Anatomical Studies of the Inner Ear

Early in the nineteenth century, the German physiologist Ernst Heinrich Weber (1795–1878) published a seminal study of the auditory system of fishes. In his study, Weber (1820) illustrated not only the inner ears of the carp (*Cyprinus carpio*), a close relative of the goldfish (*Carassius auratus*), but also described a series of bones between the swim bladder and the inner ear (Fig. 3.1). Moreover, Weber rightly suggested an auditory function of this bony chain, later termed the Weberian ossicles[1] (see Sect. 3.6.3). Fishes that possess Weberian ossicles are referred to as *otophysans* (or ostariophysans).

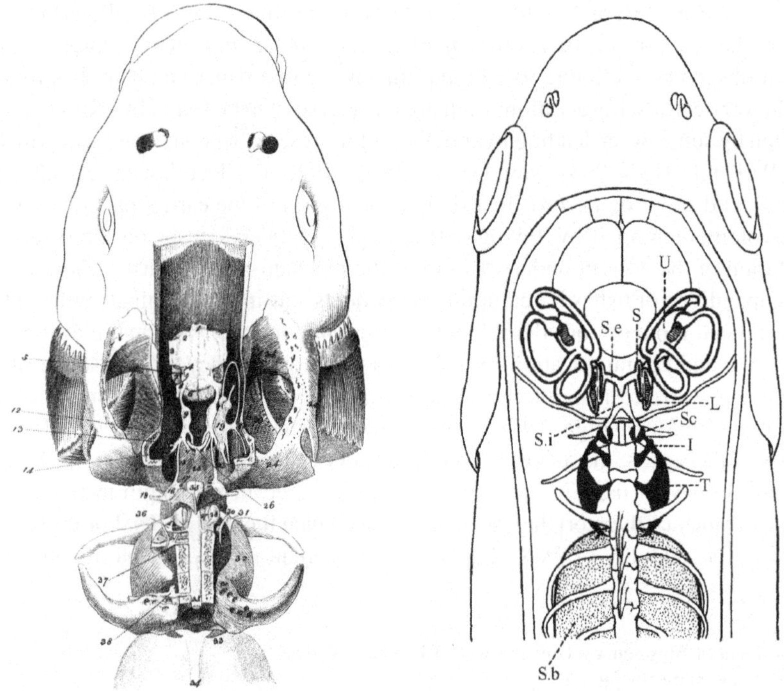

Fig. 3.1 The Weberian ossicles. Left: Original drawing by Weber (1820). Right: Schematic illustration by von Frisch (1936). Labels: I—intercalarium; L—lagena; S—saccule; Sc—scaphium; S.i—sinus impar; S.e—sinus endolymphaticus; S.b—swim bladder; T—tripus; U—utricle

[1]A copy of Weber (1820) can be found at: https://archive.org/details/b22007374/page/42. This includes the complete text (in Latin) and all of the figures.

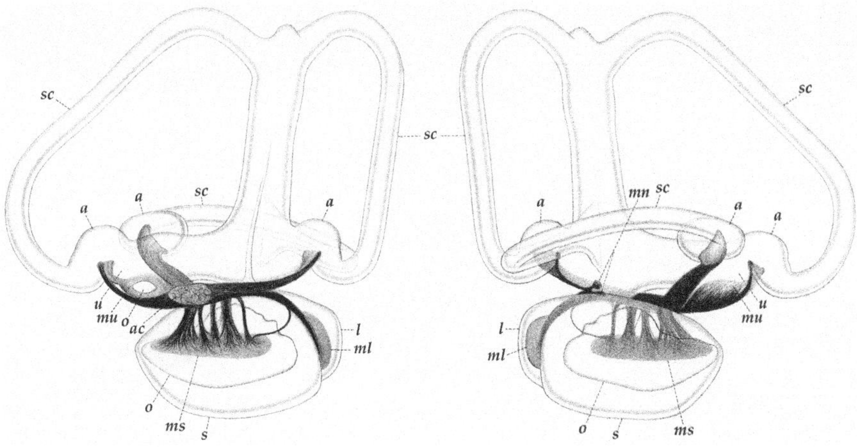

Fig. 3.2 Drawings of the ear of the Atlantic salmon, *Salmo salar*, from plate XIV of Retzius (1881). Left is a medial view (anterior to the left). Right is a lateral view (anterior to the right). Labels: a—ampullae; l—lagena; ml, ms, and mu—epithelia (maculae) of otolith organs; mn—macula neglecta; o—otoliths; r—rami of the eighth cranial nerve to various ear parts; s—saccule; sc—semicircular canals; u—utricle

In the later nineteenth century, the Swedish physician and anatomist Gustaf Retzius (1842–1921) published a set of volumes in which he beautifully illustrated the detailed structure of the ears in many vertebrate species from lampreys to mammals (Retzius 1881). Of these, 48 species were elasmobranchs (cartilaginous fishes) or teleosts (bony fishes) (Fig. 3.2) showing details of the anatomy of fish ears that are of immense value even today. Retzius illustrated the major otolith organs (the saccule, lagena, and utricle), the semicircular canals, and their innervation. He also showed the presence of another organ, the macula neglecta, which consists of two patches of sensory tissue overlain by a gelatinous cupula (mn in Fig. 3.2). The macula neglecta is diminutive in bony fishes and terrestrial vertebrates, but is much larger in many elasmobranch species. In these species, it may be involved in hearing (Tester et al. 1972; Fay et al. 1974; Corwin 1977, 1981).

The morphology of the ear was examined in detail by other investigators in the first part of the twentieth century, with a focus on comparative morphology (de Burlet 1931), although some early workers also asked questions about function of the inner ear in hearing (e.g., Werner 1926). Whereas the earlier workers only had access to studies with the light microscope for anatomical studies, the advent of electron microscopy allowed for ultrastructural examination of the ear in the mid-twentieth century. One of the earliest workers to do this was the Swedish physiologist Åke Flock, who used transmission electron microscopy (TEM) to investigate the structures of the sensory epithelia in the burbot, *Lota lota* (e.g., Flock and Wersäll 1962; Wersäll and Flock 1965; Flock 1970). Flock and his colleague Jan Wersäll demonstrated the structure and the morphological polarization of the ciliary bundle of hair cells, with a single true cilium, the kinocilium, at one end of a bundle

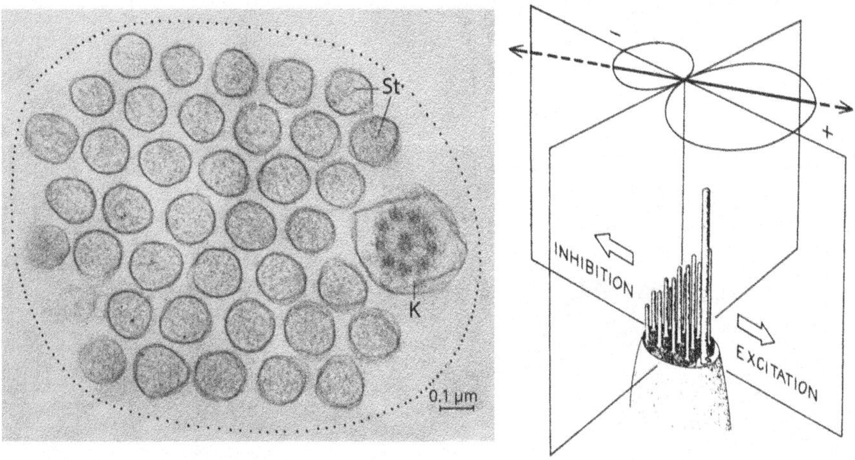

Fig. 3.3 Left: Morphological polarization of the ciliary bundle of hair cells (also see Fig. 3.4). Cross-section TEM of a sensory hair bundle showing a single kinocilium (K) at one end of a bundle of stereocilia (St). Dotted line (inserted using Photoshop) indicates the circumference of the apical surface of the hair cell. (From Flock and Wersäll 1962, with permission). Right: Physiological polarization of hair cells. When the ciliary bundle is bent toward the kinocilium, the hair cell depolarizes and the firing rate from the afferent nerve fiber increases. The hair cells hyperpolarize when bending is in the opposite direction. As indicated by the polar coordinate sketch, the magnitude of the depolarization exceeds hyperpolarization at equal bending in opposite directions. (From Flock 1971, with permission)

of stereocilia (Fig. 3.3 left and Fig. 3.4). He also demonstrated that there was a distinct hair cell orientation pattern in each of the otolith organs, though the saccular data, based on TEM, were shown to be inaccurate in later studies using scanning electron microscopy (SEM) (Popper 1981). Finally, Flock demonstrated that the sensory hair cells of fishes are not only morphologically polarized but also physiologically polarized, meaning that their electrical response is dependent on the bending direction of the ciliary bundles (Fig. 3.3, right) (Flock 1964, 1971). Thus, Flock demonstrated that sensory hair cells are directionally sensitive, a feature of crucial importance regarding the ability to discriminate between sound directions (Sect. 3.8).

Whereas these earlier workers suggested that the hair cell polarization in the ear was important, this was finally demonstrated in physiological studies on the goldfish by the Japanese investigator Taro Furukawa and colleagues (e.g., Furukawa and Ishii 1967; Ishii et al. 1971; Furukawa et al. 1972) and reviewed in an historical context by Furukawa (2002). In the first of a series of papers, Furukawa and Ishii (1967) recorded from single fibers of the eighth nerve to the saccule and demonstrated that different epithelial regions along the saccular epithelium responded to different frequencies. This was the first experimental evidence of a peripheral place mechanism for frequency discrimination in fishes. However, a possible role of such a mechanism in fishes is still largely unexplored.

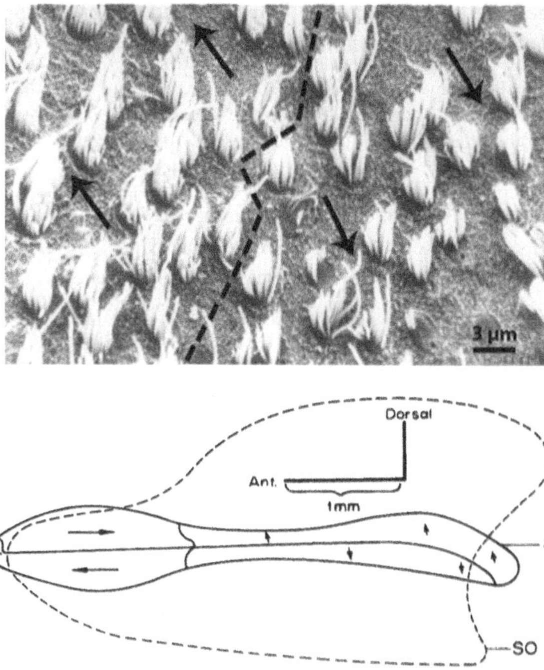

Fig. 3.4 Upper: Hair cell orientation patterns in fishes. Scanning electron micrograph from a lizardfish (*Sardia* sp.). Hair cell orientation is defined as the direction from the center of the bundle of stereocilia and toward the kinocilium. Hair cells to the left of the dashed line are oriented to the upper left, whereas those to the right are oriented in the opposite direction. Lower: The saccular hair cell orientation pattern in the lake whitefish. The saccular (S) macula is partially overlain by otolith (SO), represented by dashed line. Arrows indicate the orientation of the ciliary bundles in each region, with different orientation groups separated by solid lines. (From Popper 1976, with permission)

Furukawa and Ishii (1967) also showed that the hair cells with opposing orientations responded to different phases of the signal, a finding that was later further confirmed in the goldfish (Piddington 1972; Fay 1974b). Furukawa and colleagues also did a series of studies on synaptic function in the ear of goldfish as well on the ultrastructure of sensory hair cells in fishes (e.g., Hama 1969; Hama and Saito 1977).

3.3 Ideas About Fish Hearing Prior to 1900

In his *History of Animals* (Peri ta zôia historiôn), the Greek philosopher Aristotle (384–322 BCE) declared that fishes can hear, although he could not identify any organs of hearing. Some 400 years later, the Roman author and naturalist Pliny the Elder (23–79 CE), in his *Natural History* (Naturalis Historia), also concluded that fishes do not appear to have a hearing organ but can still hear. Later, several Roman authors confirmed the assumption that fishes can hear by presenting anecdotal evidence, as demonstrated by Claudius Aelianus (175–235 CE):

> Those who live by the lake of Marea catch the Sprats there by singing with the utmost shrillness, accompanying their song with the clash of castanets. And the fishes, like women dancing, leap to the tune and fall into the nets spread for their capture. And through their dancing and frolics the Egyptians obtain an abundant catch.[2]

Claudius Aelianus also described dense calcareous objects, later termed otoliths, in several species and linked these structures to hearing. This impressive speculation was not explored further until Casserius (1609) discovered that fishes possess internal ears.

Much later, Izaak Walton (1593–1683), an English writer and avid angler, wrote the most famous of all books on angling, *The Compleat Angler or the Contemplative Man's Recreation* (Walton 1653), which is still in print in a somewhat updated edition (Walton and Cotton 1861). Walton noted:

> And this reason of Sir Francis Bacon has made me crave pardon of one that I laughed at for affirming that he knew Carps come to a certain place, in a pond, to be fed at the ringing of a bell or the beating of a drum. And, however, it shall be a rule for me to make as little noise as I can when I am fishing…

By the eighteenth century, it was well known that the critical organs of hearing in man and other terrestrial vertebrates resided in their inner ears. Consequently, the presence of inner ears in fishes was considered firm evidence for the ability of fishes to hear. For example, Hunter (1782) stated:

> As it is evident that fish possess the organ of hearing, it becomes unnecessary to make or relate any experiment, made with live fish, which only tends to prove this fact.

This view was generally accepted by all the leading physiologists during the nineteenth century, as reviewed by Parker (1903). However, toward the end of the nineteenth century, it became increasingly clear that the inner ear of terrestrial vertebrates has the dual sensory function of hearing and balance (*equilibrium*, i.e., sensitivity to various types of acceleration). Because no traces of the complex structures of the inner ear responsible for hearing in terrestrial vertebrates (e.g., basilar papilla, cochlea) were evident in the fish labyrinth (defined as all parts of the inner ear) (Platt 1983), it soon became a general assumption that the ear was a sense organ only for balance (Lee 1898). This new understanding caused a shift in the common assumption that fishes can hear, and the pendulum quickly swung away to the opposite conclusion. Critical scrutiny of previous reports of hearing in fishes revealed mainly anecdotal evidence, and it was proposed that the observed responses to presumed auditory stimuli may instead have been elicited by visual cues (e.g., Kreidl 1896). This uncertainty catalyzed numerous experiments aimed at clarifying whether fishes can hear.

[2] http://www.attalus.org/translate/animals6.html#30 (see section 30), from AELIAN: ON THE NATURE OF ANIMALS, book 6

3.4 Experimental Studies Demonstrating Hearing in Fishes

The first controlled experimental study on fish hearing was probably performed by the American physiologist G. H. Parker (1903), who developed a method to test hearing in fishes. He built an apparatus he referred to as a "sounding board" attached to the outer wall at one end of a large aquarium. A bass violin string was stretched on the board in such a way that the 40 Hz sound generated by the string's vibrations was transmitted into the aquarium. The study used the mummichog (referred to by Parker as the green killifish), *Fundulus heteroclitus*, a hardy marine species common along the US east coast.

Parker examined innate behavioral responses of mummichogs to the sound from a violin string and found that the fish consistently responded to the sound with rapid movements of their pectoral fins. Whereas this suggested hearing, Parker was concerned that the whole tank was being vibrated by the string, and that this could have resulted in the fish responding to alternative, non-acoustic stimuli. To test this idea, Parker prevented movement of the tank walls by using a tuning fork vibrating at 128 Hz as a source and found that the mummichog again responded to the sound, leading him to conclude that he had demonstrated hearing. He also performed a series of experiments on fish with impaired inner ears and lateral line systems, and the results supported the idea that the fish detected sound via the ears (see Sect. 3.6).

At that time, researchers working on several other species doubted that fishes could hear because no innate responses to sound were observed. However, the German investigator Karl von Frisch (later awarded the Nobel Prize for his work on the dance language of bees) pointed out that observation of innate responses may be an unreliable method for studying sensory abilities in animals. Therefore, von Frisch (1923) questioned the quality of previous studies of hearing in fishes and developed his own method that involved training a fish to behave in an unambiguous manner when it detected a particular sound. Training paired a conditioned stimulus (sound) with an unconditioned stimulus to which the fish responded without training. In the first study employing this method, von Frisch (1923) trained blinded catfish (brown bullhead, *Ameiurus nebulous*) to respond whenever he whistled, by pairing the sound with food. The fish learned to respond quite quickly and were very consistent in their response.

3.5 What Fishes Hear?

Although the studies by Parker (1903) and von Frisch (1923) demonstrated that fishes can hear, the frequency range (bandwidth) of sounds they could hear was still not known. Later, von Frisch and his students addressed this question using a conditioning technique. Indeed, many of the present ideas on fish hearing were first explored by von Frisch, his Dutch student Sven Dijkgraaf and many of Dijkgraaf's students, including Arie Schuijf (Sect. 3.8). Dijkgraaf provided fundamental thinking about both fish hearing and the role of the lateral line, as discussed in Sect. 3.6.

Several of the early studies revealed that responses to sound were evoked most readily in otophysan species (reviewed by von Frisch 1936, 1938a). However, later studies showed that the audible frequency ranges of several species within some non-otophysan taxa are similar to those of some otophysan species (see Sect. 3.5.1.2).

3.5.1 Auditory Frequency Range and Sensitivity

3.5.1.1 Otophysans

The conditioning studies by von Frisch were continued by his student H. Stetter (1929) on two otophysans, the brown bullhead and the European minnow (*Phoxinus phoxinus*). Stetter explored both the frequency range of hearing (using tuning forks) and auditory sensitivity (using a whistle). He demonstrated that the brown bullhead could detect frequencies up to about 13 kHz and as low as 16 Hz, whereas European minnows could hear pure tones up to 5–7 kHz.

In order to examine auditory sensitivity (the lowest sound level detectable by the subject), Stetter used a whistle generating a fundamental tone of 660 Hz and then lowered the sound level until the fish failed to respond. The actual level of the auditory threshold was not established, but when Stetter and von Frisch submerged individual colleagues and students into a large aquarium to compare their ability to detect the same sounds as the European minnow (Fig. 3.5), they concluded that the minnow could hear as well as humans underwater.

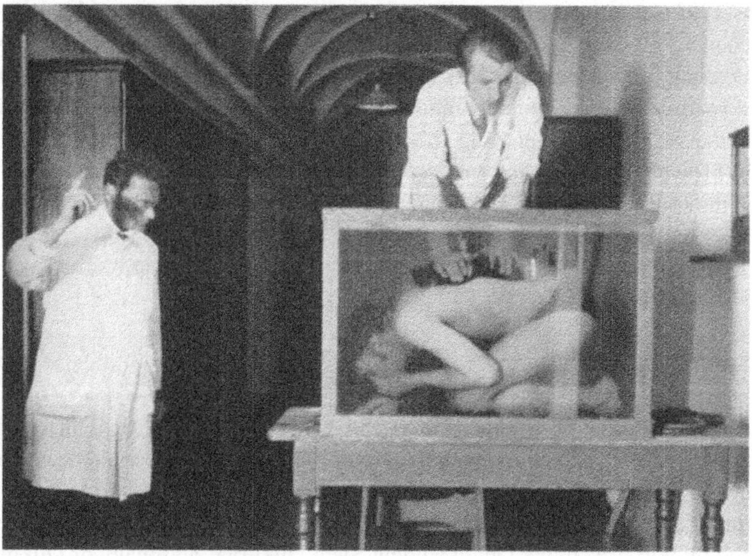

Fig. 3.5 Arrangement for comparing the hearing sensitivity of European minnow to that of a submerged human. Stetter is pushing on the person and von Frisch is overseeing the experiment. (From Stetter 1929, with permission)

Later, von Frisch and Stetter (1932) combined conditioning experiments with ablation of various structures assumed to play a role in sound detection in fish, in order to reveal the sensory components that actually are involved in hearing. This is discussed in Sect. 3.6 and in von Frisch (1936).

3.5.1.2 Non-Otophysan Species with Gas-Filled Chambers Associated with the Ear

A common feature of a number of non-otophysan taxa (which lack Weberian ossicles) including, but not limited to, the Mormyridae (elephant-nosed fishes), Anabantidae (labyrinth fishes), and Clupeiformes (e.g., Clupeidae, the herring family), is close contact between the ears and gas-filled compartments. The anatomy of such couplings and their presumed auditory roles are discussed in Sect. 3.6.3.

The hearing ability of a mormyrid, *Pollimyrus isodori*, was studied by Grete Diesselhorst (1938) who reported high auditory sensitivity with an upper frequency cutoff of about 3.1 kHz. This work was continued by Elisabeth Stipetić, who tested the hearing ability of another mormyrid, *Marcusenius macrolepidotus* (Stipetić 1939), with results almost identical to those for *Pollimyrus isodori*. Stipetić also studied the anatomy of the mormyrid ear in great detail.

Soon after, Hiltrude Schneider (1942) found that several anabantid species had hearing matching that of the mormyrids, with audible frequency ranges up to about 4.7 kHz. Diesselhorst, Stipetić, and Schneider were all students of von Frisch.

The first audiogram for a clupeid, the Atlantic herring (*Clupea harengus*), was reported by the Norwegian physiologist Per S. Enger (1967), based on neurophysiological recordings from the brainstem. The recordings revealed sensitive hearing with an upper frequency limit of about 4 kHz.

A bit earlier, William N. Tavolga and Jerome Wodinsky (1963), working in the Bimini Bahamas marine lab of New York's American Museum of Natural History (Tavolga 2002), examined hearing in several marine species. They found that the hearing ranges of two closely related squirrelfish species, *Holocentrus adscensionus* (longjaw squirrelfish) and *Holocentrus vexillarius* (dusky squirrelfish), differ greatly. Interestingly, both species have anterior projections of the swim bladder in close proximity to the ear, and both make similar sounds that they use in various, and similar, behavioral contexts. However, *H. adsensionus* can detect sound frequencies up to almost 3 kHz whereas *H. vexillarius* only detects sounds to about 1.2 kHz and with poorer sensitivity. The difference in hearing capabilities may be related to differences in the specific distance from the swim bladder to the ear in the two species (Nelson 1955). The relationship between the position of the anterior part of the swim bladder and hearing was further demonstrated in squirrelfish in a study on *Myripristis kuntee*, in which the swim bladder is in intimate contact with the ear. Hearing range and sensitivity were greater in *M. kuntee* than in the aforementioned species, supporting the notion that close proximity between the swim bladder and the ear improves hearing (Coombs and Popper 1979).

3.5.1.3 Non-Otophysan Species Without Gas-Filled Chambers Associated with the Ear

Diesselhorst (1938) also studied the hearing ability of the European eel (*Anguilla anguilla*) and Atlantic mudskipper (*Periophthalmus barbarus*) and found that they could only detect frequencies below 650 Hz. This agrees with later studies revealing an upper audible frequency cutoff usually below 1000 Hz in many of the non-otophysan species studied to date (reviewed by Popper et al. 2003). However, there is substantial variation in hearing ability among non-otophysan species, which is likely related to a possible auditory function of a swim bladder (see Sect. 3.6.3). Moreover, it must be recognized that there are hearing data for only just over 100 of the more than 34,000 species of fish (reviewed in Ladich and Fay 2013).

3.5.1.4 Other Investigations of Fish Hearing

Much of the work on fish hearing in the first half of the twentieth century was done in Europe, except for the studies by Parker and his students. In the second half of the twentieth century, several scientists from the United States entered the fish hearing research scene. Most notably, Tavolga, a biologist, and Wodinsky, a psychologist, approached the question of fish hearing from the perspective of modern psychoacoustics (see Sect. 3.5.1.2).

Basing their designs on the system developed by comparative psychologist M. E. Bitterman and his group (Wodinsky et al. 1962), Tavolga and Wodinsky used a shuttle box (Fig. 3.6) to train fish to perform a conditioned response whenever it heard a sound in order to avoid a weak electric shock (see Tavolga 2002 for an autobiographical history). Tavolga and Wodinsky (1963) combined this method with a "staircase" approach (explained in the legend to Fig. 3.6, lower) to determine hearing sensitivity in nine species of marine fishes. However, they were not able to determine whether the fishes were responding to sound pressure or particle motion (Tavolga and Wodinsky 1965).

Conditioning based on negative reinforcement (electric shock), rather than feeding, proved to be a very efficient method for obtaining sensory thresholds. The most common techniques in later studies have been respiratory conditioning (e.g., Fay 1969) (Fig. 3.7, top) and cardiac conditioning (e.g., Chapman and Hawkins 1973). In such studies, restrained fish are exposed to a sound followed a few seconds later by a mild electric shock. Normal unconditioned awareness or fright responses in fish are reduced heart and respiratory rates. After a few trials, the heart and respiratory rates change if the fish hears the sound, before receiving the shock, thus indicating detection of the sound.

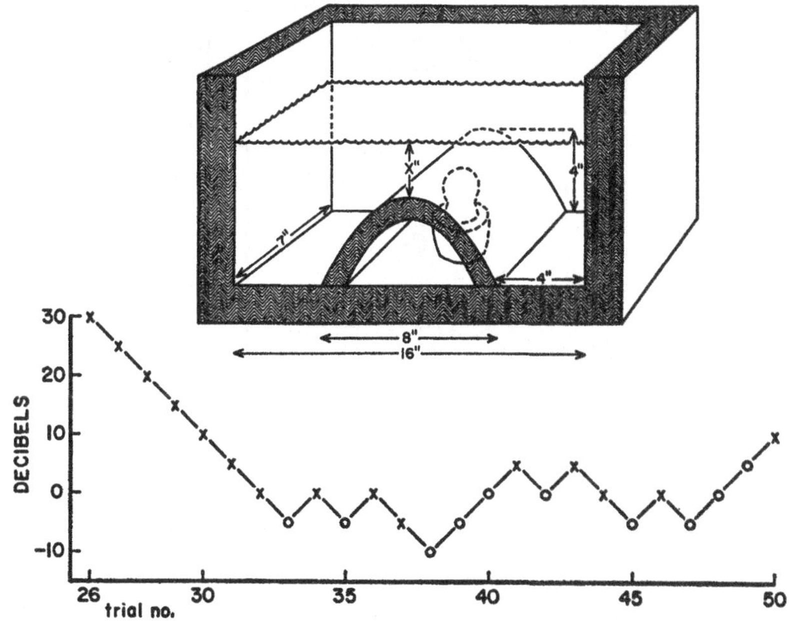

Fig. 3.6 Upper: Shuttle box used by Tavolga and Wodinsky (1963) to test hearing sensitivity in fishes. The box was partially divided into two chambers by a barrier, under which was an underwater speaker. Using electric shock as an unconditioned stimulus, fish were trained to cross the barrier whenever they heard a sound. If the fish heard the sound it would cross the barrier before shock onset, thus avoiding getting a shock. Lower: An example of the staircase method used in the same study to determine auditory thresholds. Data are for the longjaw squirrelfish at 1600 Hz. At the highest sound levels, the fish learned the task. Thereafter, the sound level was lowered for each successive trial resulting in a positive response (indicated by an "x"), until the animal did not respond (o). It then received a shock, and the sound was raised for the subsequent trial. Following the next positive response, the sound level was again lowered, and so on. The threshold was finally estimated as the sound level giving 50% probability for a positive response. In the depicted example, the threshold at 1600 Hz was −1.25 dB re 1 µbar, which, using today's terminology, would be 98.5 dB re 1 µPa. (From Tavolga and Wodinsky 1963, with permission)

3.5.2 Auditory Discrimination

The first published study of frequency discrimination in fishes was done by F. Westerfield (1922) in the United States. She worked with mud minnows (*Umbra limi*) that she trained to discriminate between two sounds with different frequencies produced by a ukulele, one sound meant food and the other a noxious stimulus. Fish were able to discriminate between the sounds, as demonstrated by different behavioral responses to the two sounds. However, the work was rather limited and does not provide full insight into frequency discrimination by fishes.

This was followed by a more extensive study by Stetter (1929), who was the first to systematically explore auditory frequency range, sensitivity, and frequency discrimination in a fish. Stetter trained European minnows to respond to one sound

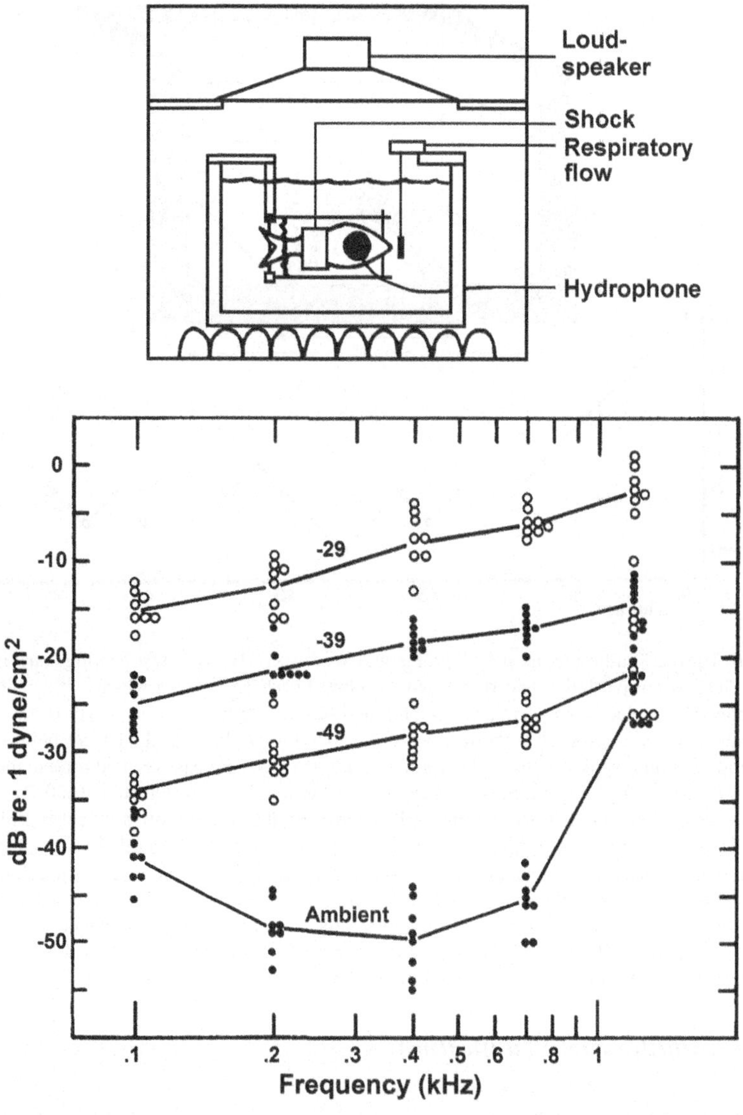

Fig. 3.7 Upper: Arrangement for studying auditory masking in goldfish. Auditory thresholds of submerged, restrained fish were obtained using respiratory conditioning (see Sect. 3.5.1.4). The fish was fitted with shock electrodes and the respiration frequency was recorded by detecting water flow across the mouth. Lower: Sound pressure thresholds from four goldfish determined at four noise conditions (ambient, −29, −39, and −49 dB re1 dyne/cm²/Hz) as a function of the frequency of the test tone. The applied noise was uniform, and the bandwidth was 50–2,000 Hz. Lines connect the means at each noise level. Filled and open symbols separate data obtained at different noise levels. Overlapping points are displaced to the right. Within the tested frequency range, the masking effect was independent of frequency, and 10 dB noise increment elevated the threshold by about 10 dB. (Modified from Fay 1974a, with permission)

with a food reward and then punished, with a tap from a glass rod, responses to another sound that differed significantly in frequency (called the warning sound) from the feeding sound. After a number of trials, the fish learned to respond only to the feeding sound. Stetter then moved the frequency of the warning sound closer to that of the feeding sound until the minnows also started to respond to the warning sound, indicating that the fish failed to detect a significant difference between the two sounds. He found that all minnows could discriminate between frequencies separated by one octave, and that the two best performing animals could discriminate between 290 and 345 Hz.

One problem with Stetter's study, however, was that the sounds to be discriminated were presented 15 min apart, thus requiring memory of the first sound to make the discrimination. In contrast, another student of von Frisch, Th. Wohlfahrt (1936a), examined frequency discrimination in the European minnow by presenting sounds in rapid succession. He showed a much finer discrimination of about six percent difference between tone frequencies.

However, whereas the studies by Stetter (1929) and Wohlfahrt (1936a) demonstrated discrimination between tones, the authors did not control the differences in tone intensity. Thus, it is possible that the animals were discriminating between intensities rather than frequencies.

It was not until the discrimination studies by Jacobs and Tavolga (1967, 1968) that it became clear that fishes can discriminate between both intensity and frequency. Jacobs and Tavolga tested frequency discrimination, using their shuttle box and operant conditioning, applying the concept of the just noticeable difference (JND), or the smallest difference between two signals (e.g., frequency or intensity) that could be discriminated. In these studies, fish were constantly presented with pulsed sounds of the same frequency or same intensity. At some point, the standard pulse was alternated with a pulse of different amplitude or frequency. The fish were trained to respond when they detected any alternation in the signal. The two sounds were then brought closer and closer together until the fish failed to respond, thus indicating that the difference between the sounds had become less than the JND. Results showed a sound intensity JND of about 5 dB from 100 to 1,0000 Hz and a frequency discrimination JND of 4–5 % in the same frequency range (Jacobs and Tavolga 1968). However, studies were restricted to the goldfish, and so applicability to other species, and in particular to non-otophysans, remained an open question.

Tavolga, again cognizant of critical aspects of hearing in terrestrial vertebrates, was the first to examine the focal questions of masking and critical bands in fishes to better understand the function of the hearing mechanisms and capabilities of goldfish, using a modification of the operant conditioning paradigm (Tavolga 1967, 1974). The only other behavioral studies of masking and critical band have been in goldfish (Enger 1973; Fay 1974a; Popper and Clarke 1979), Atlantic cod (Buerkle 1969; Hawkins and Chapman 1975), and Atlantic salmon (Hawkins and Johnstone 1978). Typically, fishes respond to masking by showing an increase in auditory threshold that is related to the level of the masker above threshold, as demonstrated by Fay (1974a) (Fig. 3.7).

It is interesting, from a historical perspective, that Jacobs and Tavolga (1967) were the first since Bigelow (1904) to study hearing in the goldfish. Subsequently, also in the United States, Richard R. Fay measured goldfish hearing using respiratory conditioning (Fig. 3.7) (Fay 1969), followed by a decades long series of studies that considered broad questions about goldfish hearing, from stimulus generalization (Fay 1970) to soundscape analysis (Fay 2009). The extensive data on hearing in goldfish provided by Fay serve as an invaluable basis for comparison with terrestrial species as well as with other fishes (reviewed in Fay and Megela Simmons 1999; Yost et al. 2020).

3.6 The Quest for the Hearing Organ(s)

3.6.1 The Function of the Labyrinth

As noted in Sect. 3.3, the understanding at the end of the nineteenth century that fishes completely lack the labyrinthine structures responsible for hearing in mammals (e.g., a cochlea) undermined the general assumption that fishes could hear. Hence, in parallel with the experiments discussed in Sect. 3.4, that demonstrated that fishes are definitely able to hear, other experiments were performed in order to reveal the sound detecting organs.

Kreidl (1895) was the first to explore experimentally the function of the teleost labyrinth. He supposedly removed the complete labyrinth in goldfish using forceps and observed that equilibrium was greatly disturbed while there was no change in responses to sound. Kreidl concluded that the sense of balance in goldfish resides in the labyrinth, whereas sound is detected by mechanoreceptors in the skin or the lateral line. Kreidl's conclusion was supported by Frederic Lee (1898), who stated, in a paper reviewing the sensory functions of the ear and lateral line, that the sole function of the fish labyrinth is equilibrioception. Similar observations were made by the English geneticist William Bateson (1890) as part of a review of the literature on fish sensory systems and baitfish.

The conclusions by Kreidl, Lee, and Bateson were contradicted by Parker (1903), who disabled the labyrinths in mummichog killifish by cutting the nerves innervating the ears (cranial nerve VIII). Subsequently, the fish had disturbed balance and a complete loss of responses to sound. Parker concluded that the labyrinth serves as a sense organ for both balance and sound and that both the skin and the lateral line are insensitive to sound.

Parker doubted that the very different findings between his study on killifish and the study by Kreidl (1895) on goldfish could be explained by species differences. This inspired Henry Bryant Bigelow to repeat Kreidl's experiments. Bigelow (1904) found that cutting cranial nerve VIII in goldfish rendered the fish insensitive to sound, as Parker (1903) had reported for the killifish. Attempting to find the cause of the discrepancy between his and Kreidl's results, Bigelow carefully repeated

Kreidl's surgical method, by which he removed the semicircular canals and the attached parts of the labyrinth. Following the operation, the goldfish exhibited greatly disturbed balance, but eventually recovered their normal posture, which Bigelow explained by compensatory mechanisms based on sight. The treated fish then displayed normal responses to sound, in agreement with Kreidl's results. However, closer examination showed that this method only removed the semicircular canals and the utricle (together often referred to as the pars superior), whereas the saccule and lagena (the pars inferior), being more deeply embedded in the skull in cyprinids (members of the carp family), were left behind. Hence, Bigelow concluded that the hearing ability in goldfish probably depends on the saccule and the lagena.

Later, Parker and Van Heusen (1917) studied the relative roles of the labyrinth, the lateral line, and the skin for sound detection in the brown bullhead. They impaired the labyrinths by cutting cranial nerve VIII. They also surgically destroyed the lateral line canal organs and desensitized the skin senses with an aqueous solution of magnesium sulfate.

It is interesting to note the technical requirements adopted to achieve well-defined auditory stimuli in these experiments in 1917. At the time, sounds for hearing experiments were often created by bumping objects into the aquarium wall, dropping items into the water, clapping of hands, etc. More refined experiments applied tones from musical instruments, which include harmonics, or pure tones generated by tuning forks. Parker and Van Heusen (1917) were the first to attempt to generate relatively pure tones using electricity. In their experiments, seven frequencies from 43 to 2752 Hz were produced using a series of seven AC-current generators sharing a common shaft driven by a 10 hp (7.5 kW) electric motor. The total construction weighed "not far from half a ton" (450 kg). To avoid vibrations from the apparatus reaching the test site, they placed the generators in a separate basement room about 40 m from the aquarium. The AC-currents drove a submerged telephone with a tight rubber cover, whereas airborne sounds evoking clear responses in normal fish were produced by blowing a whistle.

Parker and Van Heusen (1917), extirpating various tissues that could be involved in sound detection, concluded that airborne sound, entering the water, is only detected by the labyrinth, which is also sensitive to sounds from a submerged source, whereas the skin and the lateral line respond only to the lowest underwater frequencies. It was thus assumed that the lateral line organs are stimulated by low-frequency water vibrations, and that these organs resemble primitive ears. Parker and Van Heusen also reported that the brown bullhead can detect sounds with the labyrinth up to 688 Hz, but not to higher frequencies.

Bigelow's (1904) experiments were the first to show that the various otolith organs may have different functions linked to hearing and the sense of balance. This important conclusion was not, however, tested and verified until Manning (1924) used the same method to remove the semicircular canals and the utricle as did Kreidl (1895) and Bigelow (1904). However, a similar procedure was not feasible for extirpating the lagena and the saccule, which, in goldfish, are closely linked to one another and embedded in the skull just beneath the brain (see Platt 1977 for a comprehensive description of the goldfish auditory system). Rather than cutting the eighth cranial nerves, which would also disconnect the utricle, Manning surgically destroyed the saccular and lagenar otoliths in situ. Using the same equipment for generating underwater sound as Parker and Van Heusen (1917) (see above), Manning observed responses of normal and operated goldfish to frequencies from 43 to 2752 Hz. He concluded that the utricle was responsible for hearing from 43 to 688 Hz, whereas the saccule and lagena, and especially the lagena, were responsible for higher frequency hearing. The results were essentially in agreement with Parker's (1903) and Bigelow's (1904) conclusion that the sense of balance depends on the utricle, whereas the lagena and saccule are the main sound detectors. However, Manning's results also indicated that the utricle, the skin, and the lateral line might all be sensitive to the lowest frequencies tested.

von Frisch and Stetter (1932) later studied the effects of extirpation of the different components of the labyrinth, using the ability of the European minnow to respond to various sound frequencies. Removal of the utricle and semicircular canals only affected equilibrium, whereas hearing remained normal. Moreover, through very delicate and skillfully performed surgery, von Frisch and Stetter (1932) succeeded in removing the saccule and the lagena without damage to other parts of the labyrinth. These fish displayed completely normal postural reflexes, while frequencies above about 150 Hz failed to elicit a response. The fish still responded to lower frequencies with less sensitivity, but at 16 Hz the sensitivity was the same as before the operation. In addition, surgical impairment of the lateral line canal organs along the trunk and on the head did not reduce the sensitivity to the lowest frequencies. Therefore, it was assumed that such low frequencies were detected by sensitive tactile sensors in the skin, and not the lateral line. However, it should be noted that the free neuromasts on the body surface were not considered, and no attempts were made to impair this component of the lateral line system. Furthermore, the utricle was not removed, and so the possibility that the utricle might detect very low frequencies, in addition to controlling balance, was not ruled out.

Subsequently, von Frisch (1938b) applied an even more refined surgical technique and selectively removed either the saccule or the lagena in European minnows (Fig. 3.8). After bilateral removal of the lagena, the hearing sensitivity remained normal for all frequencies tested, as was the case after selective removal of the saccule on both sides. Hence, both saccule and lagena were thought to serve as sensitive sound detectors in the minnow. The study was then extended to juvenile specimens of the Ide (*Leuciscus idus*), another otophysan, confirming the conclusions from the minnow studies.

So far, all the studies on the function of the different components of the inner ear in fishes, from Bigelow's (1904) pioneering study to the refined experiments by von Frisch (1938b), were performed on otophysans. In order to explore if the reported functional organization of the labyrinth might be an exclusive otophysan feature, Dijkgraaf examined the effects of removing either the pars superior or the pars inferior on postural reflexes and hearing in two non-otophysan species, the rock goby (*Gobius paganellus*) (Dijkgraaf 1950) and the black goby (*Gobius niger*) (Dijkgraaf 1952). The experimental approach was to train the gobies to respond to the sound of

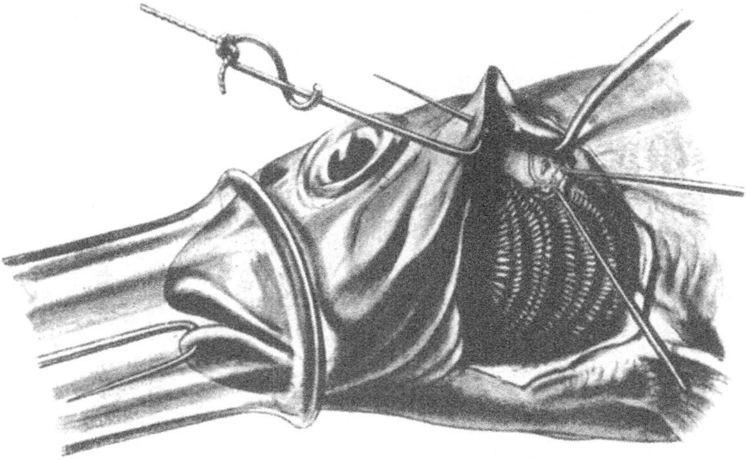

Fig. 3.8 Surgical arrangement for extirpating the lagena and the saccule (pars inferior) in the European minnow. (From von Frisch and Stetter 1932, with permission)

tuning forks using the conditioning method introduced by von Frisch (1923). In both species, bilateral removal of the pars superior abolished all postural reflexes, whereas these were intact after removal of the pars inferior. However, perception of sound was impaired by bilateral elimination of the pars inferior, although the fish showed normal sensitivity to frequencies at and below 100 Hz, which presumably were detected by the skin or lateral line.

The studies on both otophysans and non-otophysans did indicate a general division of labor between the pars superior and pars inferior in teleosts, the former being responsible for the sense of balance and the latter for hearing. Yet, there are probably numerous exceptions to this rule. For example, it now appears that the utricle detects sound up to at least a few hundred Hz in several teleost species, as suggested first by Manning (1924), and may be involved in directional hearing (e.g., Lu et al. 2004; Rogers and Sisneros 2020), and in ultrasound detection in some clupeids (Plachta et al. 2004). Furthermore, in elasmobranchs, all three otolith organs respond to both gravitational stimuli (Lowenstein and Roberts 1949) and low-frequency vibrations up to 120 Hz (Lowenstein and Roberts 1951).

3.6.2 Sound Stimulation of Otolith Organs

Although Parker and his students showed that sound can be detected by otolith organs in fishes, the actual detection mechanism was not understood at the time. It was known that the mass density of soft body tissue is about the same as that of water, whereas the mass density of otoliths and otoconia (found in all non-teleost fishes and terrestrial vertebrates) is much higher. This enables otolith organs to detect gravitation and linear acceleration in all vertebrates. However, it was not obvious that sensitivity of fishes to sound might depend on a related mechanism. In 1950, the Dutch physicist Hessel de Vries and the English zoologist Richard Julius Pumphrey independently pointed out that in a sound field, the fish body will follow the particle motion in the surrounding water. The inertia of the denser otoliths then causes these to lag behind—thus creating relative movements between the otoliths and the sensory maculae.

de Vries (1950) measured the mechanical properties of otolith organs in isolated heads of several fish species. He mounted the heads on a tilting or rotating board with attached X-ray film and measured the otolith movements induced by various acceleration forces (Fig. 3.9). de Vries (1950, 1956) modeled the otolith organs as critically damped, simple harmonic oscillators, and he presented the equations describing the motions of such systems. At frequencies below the natural frequency of the system, the deflection of the otolith relative to the sensory epithelium follows the acceleration of the organ, and the model indicates a working range of otolith organs reaching from zero Hz to the upper frequency limit of hearing. Hence, de Vries concluded that detection of gravity, linear locomotory acceleration, and forced acceleration of a fish in a sound field occurs by the same mechanism and that the transition between these functions is seamless. de Vries' groundbreaking description of the mechanical properties of otolith organs in fishes ought to have been the foundation of later studies of this subject, but was poorly understood at the time.

de Vries (1950) also attempted to calculate the minimum perceptible energy of an otolith organ, but erroneously based the estimation on the assumption that the minimum detectable acceleration is similar to that of the human labyrinth. At the time, it was not known that the labyrinth in fishes is nearly four orders of magnitude more sensitive to linear acceleration than the human labyrinth (Sand and Karlsen 1986). Thus, when very low auditory thresholds were reported during the following two decades, scientists lost confidence in de Vries' model for hearing in the acoustic far field (see below).

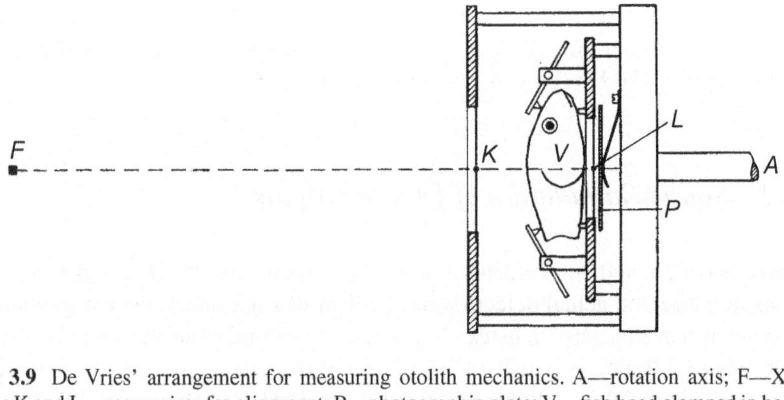

Fig. 3.9 De Vries' arrangement for measuring otolith mechanics. A—rotation axis; F—X-ray tube; K and L—cross wires for alignment; P—photographic plate; V—fish head clamped in holder. (From de Vries 1950, with permission)

The concepts of acoustic near and far fields were introduced to marine bioacoustics by Americans Gerard G. Harris and Willem van Bergeijk (Harris and van Bergeijk 1962; Harris 1964; van Bergeijk 1964). The acoustic near field is the region in which particle motions caused by movements of the sound source and incompressible flow are larger than those caused by the sound pressure-induced compression and rarefaction of the medium. The acoustic far field is the region beyond, in which the sound pressure-induced particle motions dominate. Both investigators (who often collaborated) assumed that far field hearing in fishes was only possible in species possessing a swim bladder produced by that transforms sound pressure into sufficiently high particle motion to stimulate the otoliths (see Sect. 3.6.3). However, Chapman and Sand (1974) later utilized the properties of a free sound field in the ocean to demonstrate that two teleost species without a swim bladder, the flatfishes dab (*Limanda limanda*) and plaice (*Pleuronectes platessa*), detect particle motion and not sound pressure. Stimuli with different ratios between sound pressure and particle motion were obtained by varying the distance to the sound source within the acoustic near field (Fig. 3.10). The lowest auditory thresholds in the flatfishes were below 0.1 nm particle displacement, or close to 10^{-5} m s^{-2} particle acceleration, which allows detection of normal sound intensities in the far field. Later, Hawkins and MacLennan (1976) achieved similar stimulus control using a standing wave acoustic tube fitted with a sound projector at each end and confirmed that the plaice is sensitive only to particle motion.

3.6.3 The Swim Bladder as an Accessory Hearing Organ

As discussed in Sect. 3.5.1.1, early twentieth-century investigators noted that otophysans are generally more sensitive to sound than most non-otophysan species, thus supporting the assumption of an auditory role of the Weberian ossicles as proposed by Weber (1820) (Fig. 3.1). The physical basis for this notion is that a volume of gas pulsates when exposed to oscillating pressure (Alexander 1966a, b). Therefore, when exposed to sound, the surface of a swim bladder may display larger radial motion amplitudes than the water particles in the absence of a gas-filled bladder. The amplified motions may then be transferred to the inner ear via the Weberian ossicles, thus providing an auditory gain and making the intact fish sensitive to sound pressure, though the otolith organs remain sensitive to particle acceleration.

As an extension of the experiments described in Sect 3.6.1, von Frisch and Stetter (1932) were the first to test this hypothesis experimentally. They found that surgical removal of the swim bladder in the European minnow increased the auditory thresholds by up to 20–30 dB. Similar results were obtained for the goldfish by Fay and Popper (1974). An alternative approach to demonstrate an auditory function of the swim bladder would be to explore if the fish is sensitive to sound pressure rather than particle motion. This strategy was followed in Germany by Autrum and Poggendorf (1951) and Poggendorf (1952, in translation in Tavolga 1976) for the brown bullhead.

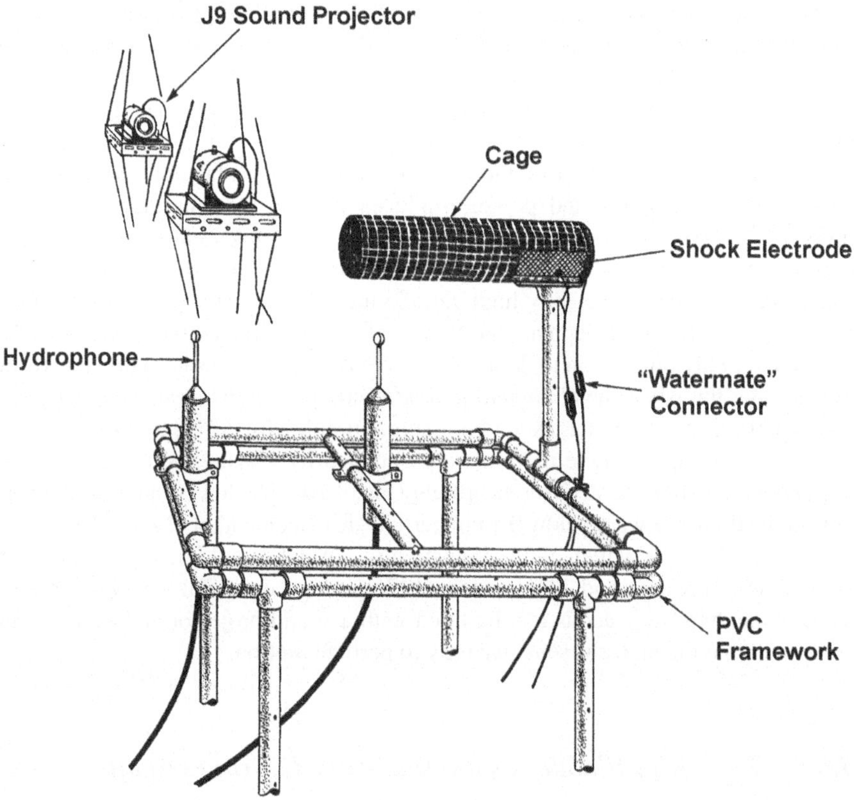

Fig. 3.10 Setup for bioacoustic field experiments under virtually free field condition. The top of the underwater framework with the fish cage was 15 m below the surface, 6 m above the bottom, and 100 m from the shore. (From Chapman and Hawkins 1973, with permission)

Poggendorf used an acoustic tank in which the ratio between sound pressure and particle motion varied between regions. The central part of the tank bottom consisted of an oscillating piston (Fig. 3.11, left), and the fish cage was positioned at various depths above the piston. For the first time in a fish bioacoustic study, sound pressure under water was measured using a hydrophone (a modified piezoelectric microphone), and particle motion was estimated from measurement of the pressure gradient. For a given piston amplitude, sound pressure was lowest close to the surface and increased with depth, whereas the opposite was true regarding particle motion. By employing the conditioning technique introduced by von Frisch (1923), Poggendorf found that the sound pressure thresholds were independent of depth. Hence, he concluded that sound pressure is the relevant stimulus parameter at threshold level, and that the swim bladder has an auditory function in the brown bullhead.

The importance of the Weberian ossicles for the auditory function of the swim bladder was demonstrated by surgical disruption of the chain, thereby increasing

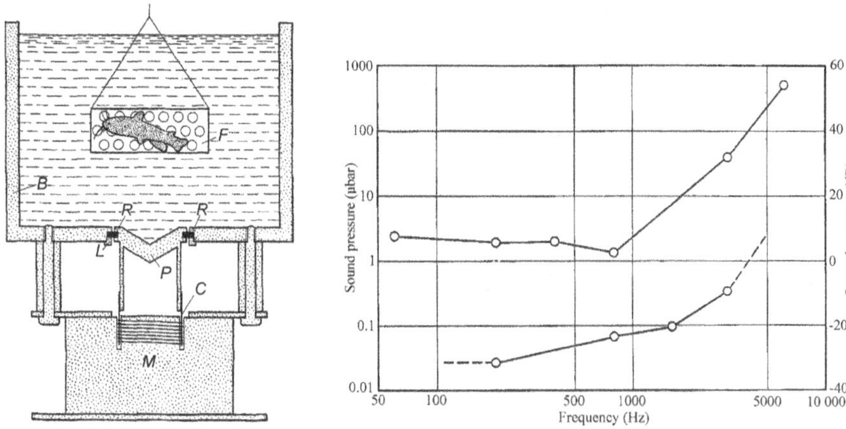

Fig. 3.11 Left: The acoustic tank used by (Poggendorf 1952) to obtain regions with different ratios between sound pressure and particle motion. Labels: B—brass wall; C—coil; F—fish cage; L—lock ring; P—piston; M—magnet; R—rubber washer. Right: Lower audiogram presents mean threshold values from two normal brown bullheads. The upper audiogram shows mean thresholds from the same fish after surgical impairment of the Weberian ossicles. (Both panels modified from Poggendorf 1952, with permission)

auditory thresholds by 30–40 dB (Fig. 3.11, right). However, the fish were still sensitive to sound pressure at threshold levels. Based on this observation, Poggendorf (1952) was the first to suggest that the swim bladder may also be important for the hearing ability of non-otophysan species.

Between 1959 and 1966, British scientist Robert McNeill Alexander published a series of papers on the physical properties and the various physiological functions of the swim bladder in several species of fish. Regarding its auditory role and the mechanical properties of the Weberian ossicles, the most relevant results are reviewed in Alexander (1966a). Alexander also concluded that the very presence of a swim bladder may aid hearing, but without an efficient mode of transmission, the gain will fall off in proportion to the square of the distance to the ear. He suggested that the extra gain at resonance would be modest, due to heavy damping of the swim bladder oscillations. Later, Sand and Hawkins (1973) confirmed this heavy damping by measuring swim bladder resonance in intact, submerged Atlantic cod at different depths.

Alexander also studied the function of the Weberian ossicles in detail. Based on anatomical measurements, he suggested that the mechanical arrangement of ossicles and ligaments in the chain (Fig. 3.1) might provide a considerable amplification of the radial movements of the swim bladder surface. Unfortunately, later researchers have not further explored and tested this assumption.

In Sect. 3.5.1.2, it was noted that some non-otophysan species display audible frequency ranges similar to many otophysans. A common characteristic of these species is the existence of gas-filled chambers in close contact with the ears. For example, in the clupeids, narrow, tubular extensions from the swim bladder end in a gas-filled bulla close to each utricle (Wohlfahrt 1936b; O'Connell 1955). Mormyrids also possess swim bladder-derived, gas-filled chambers in close contact with the ears (von Heusinger 1826; Stipetić 1939; Popper 1981). In anabantids (labyrinth fishes), the gas chamber is an accessory respiratory organ in the form of an air-filled cavity in the head, the labyrinth organ (Das 1928), which is filled with air by the fish gulping air at the surface.

The broad hearing bandwidth in the non-otophysan species referred to above did indicate an auditory role of gas-filled chambers in close contact with the ears and also in the absence of Weberian ossicles, but direct evidence for this hypothesis was lacking. Comparison of the hearing ability with and without gas in the chambers could have provided more direct evidence, but such experiments are difficult in both clupeids and mormyrids due to complex morphologies. However, in the anabantids, such experiments are relatively simple to perform, because the fish fills the labyrinth organ with air by gulping air at the surface. Schneider (1942) took advantage of this feature and compared the hearing ability of four species of labyrinth fish with and without air in the labyrinth organ. Air depletion caused the upper audible frequency cut-off to drop from up to about 4.7 kHz to below 700 Hz (see also Saidel and Popper 1987).

Following these studies, it was still an open question as to whether a swim bladder, with the primary role to provide buoyancy, might improve the hearing ability also in non-otophysan species. Such a function had been suggested by Poggendorf (1952), de Vries (1956), and Alexander (1966a), but was not tested until Enger and Andersen (1967) performed experiments indicating that the Atlantic cod was sensitive to sound pressure in the upper range of its audible frequencies. They recorded microphonic potentials from the inner ear and utilized the properties of the near field of a free sound field in the ocean to alter the ratio of particle motion to sound pressure (see Sect. 3.6.2). Their conclusion was later supported by Chapman (1973) and Chapman and Hawkins (1973) who used cardiac conditioning to determine hearing thresholds in a number of species, including the Atlantic cod, under virtually free field condition in a Scottish loch, using sound projectors positioned at various distances from the fish (Fig. 3.10) (reviewed in Hawkins and Chapman 2020). They showed that four species of Gadidae (the cod family) were sensitive to sound pressure in the upper part of their audible frequency range. In particular, the Atlantic cod responded to sound pressure at threshold level for all frequencies between 50 Hz and about 500 Hz, the highest audible frequency, whereas the thresholds were related to particle motion at lower frequencies.

Evidence of an auditory function of the swim bladder in the Atlantic cod was also provided by Sand and Enger (1973), who recorded microphonic potentials from the same fish with and without gas in the swim bladder. Emptying the swim bladder lowered the upper frequency cutoff and reduced the hearing sensitivity by more than 20 dB in the optimal frequency range (200–300 Hz). However, the sensitivity was not altered at frequencies below 100 Hz. In contrast to this, Chapman and Sand (1974) fitted the flatfish dab, which lacks a swim bladder, with an artificial "swim bladder" in the form of a small rubber balloon placed exterior to the fish just beneath its head. This arrangement shifted the upper frequency cutoff from about 200 to about 350 Hz. The hearing threshold was lowered by about 20 dB at 200 Hz, but by only 3 dB at 30 Hz.

The results from the field experiments discussed above are in agreement with theoretical considerations by Sand and Hawkins (1973), who pointed out that the swim bladder pulsations exceed the free field particle motions only above a certain frequency. Thus, the gain provided by a swim bladder is frequency dependent. Below a certain frequency, which depends on both swim bladder volume and depth, the swim bladder provides no auditory gain.

The evidence supporting an auditory role of the swim bladder in the Atlantic cod is convincing. However, in this species, anterior projections from the swim bladder bring its rostral end rather close to the ear, and it is still a matter of discussion to which degree other non-otophysan species may utilize the swim bladder as an accessory hearing organ. This question can be elucidated by studying the diversity in ear structure and hearing range within single taxa, which has been explored most extensively in the squirrelfishes (Holocentridae). Within this family, the hearing ability correlates clearly with the distance between the swim bladder and the ears. Species with the closest proximity between these structures have a hearing range similar to that of the otophysans, whereas species with a greater distance between the swim bladder and the ear have a hearing range closer to that of the Atlantic cod (Tavolga and Wodinsky 1963; Coombs and Popper 1979). These and other similar results support the argument that adaptations to utilize the swim bladder to enhance hearing have evolved multiple times in fishes (Popper et al. 2003).

3.7 The Lateral Line System

Before the middle of the nineteenth century, the leading view was that the lateral line system in fishes comprised an arrangement of mucous glands. However, based on anatomical evidence, Leydig (1868) proposed that the lateral line system constituted a hitherto unknown sensory system—a "sixth sense." This assumption was demystified by Schulze (1870), who discovered that free neuromasts possess a cupula that is moved by weak water currents. He suggested that both water currents and low-frequency sounds are relevant stimuli for these organs.

The idea that the lateral line system is primarily an accessory hearing organ for detection of low-frequency sound soon became generally accepted (Mayser 1881), and prevailed well into the twentieth century. Major reasons for this view were the similar anatomical features and the common embryological origin of the sensory cells in the lateral line system and the labyrinth, and it has been suggested that the ear arose from an invagination of the cephalic lateral line early in the evolution of fishes (Pumphrey 1950; van Bergeijk 1967).

This acousticolateral hypothesis was originally proposed by Ayers (1892), who argued that the ear and lateral line system develop from the same embryonic placode, and that their innervation arises from the same region of the brain. This notion prevailed for many decades, but later studies, using modern neuroanatomical tracing methods, demonstrate a clear distinction in innervation between the ears and the lateral line system (reviewed in Northcutt 1997; McCormick 1999). The intriguing early history of lateral line research has been thoroughly reviewed by Parker (1903) and Coombs and Bleckmann (2014).

Between the early 1930s and the late 1970s, there was a controversy between two of the leading fish bioacoustics scientists at the time, Sven Dijkgraaf and Willem A. van Bergeijk, regarding the possible function of the lateral line as an accessory hearing organ. Dijkgraaf explained his view in a classical review (Dijkgraaf 1963), whereas van Bergeijk reviewed his a few years later (van Bergeijk 1967).

Although both agreed that the lateral line is stimulated by water motion, they disagreed about the nature of effective stimuli. Based on recordings of microphonic potentials from the lateral line in the mummichog killifish in response to an adjacent, vibrating sphere, Harris and van Bergeijk (1962) argued that the lateral line would be an acoustic detector of water motion within the whole extent of the near field. This view led van Bergeijk (1967) to conclude:

> *...the lateral line organs are organized in a system that is ... capable of near-field hearing...*

At the time, the majority of scientists in this field of research embraced van Bergeijk's view. Within the acoustic near field, the lateral line took care of the low frequencies, while the inner ear responded to the higher frequencies (e.g., Tavolga and Wodinsky 1963). In the acoustic far field, only fishes possessing a swim bladder were assumed able to detect propagating sound waves (van Bergeijk 1964).

Contrary to this view, Dijkgraaf (1963) emphasized that the long wavelengths of the relevant frequencies (15 m at 100 Hz) would cause an almost neutrally buoyant fish and its surrounding water to vibrate with virtually the same phase and amplitude within almost the whole near field of a sound source. Hence, no relative motions would occur between the fish surface and surrounding water, which is the relevant stimulus for the lateral line. Later, the insensitivity of the lateral line system to such sound stimuli was confirmed experimentally (Sand 1981).

Dijkgraaf argued that the lateral line can detect moving and vibrating objects at close range, surface ripples somewhat farther from the source, the turbulent wake of moving objects long after they have passed, and stationary objects at close range. The latter was termed "distant touch" and depends on distortion of the self-generated flow field around a moving fish. The British scientists Eric Denton and John Gray were the first to confirm experimentally that the lateral line may only be stimulated within the innermost part of the acoustic near field, generally within a range of just a few centimeters from the source (Denton and Gray 1982). Coombs and Janssen (1990) later confirmed this finding in the mottled sculpin (*Cottus bairdi*). Denton and Gray (1982) also emphasized that the limited distance range of lateral line sensation protects against masking by the high levels of low-frequency ambient noise in the oceans. The inner ear, on the other hand, is sensitive to such background noise, which is an essential component of the aquatic acoustic environment. The various aspects of lateral line sensation, which is a sensory modality distinctly different from hearing, are reviewed in Sand (1984) and Sand and Bleckmann (2008), and the division of labor between the lateral line system and the auditory system is discussed in Braun and Sand (2014).

3.8 Directional Hearing

Determining the location of a sound source at some distance from an animal is important for locating predators, prey, mates, rivals, etc. Indeed, Pumphrey (1950) suggested that detection of the direction of sound propagation is a fundamental property of hearing and should be included in its very definition. However, whereas terrestrial vertebrates possess this ability, it is not yet fully clear whether fishes are able to determine the direction to a sound source from a distance.

Early experiments on directional responses of fishes to sound were carried out in a small laboratory tank by Parker (1912). During the continuous production of repetitive underwater sounds, several fish species gradually moved away from the sound source, leading Parker to suggest that sound could influence the direction of fish locomotion.

Later, von Frisch and Dijkgraaf (1935) carried out behavioral studies on European minnows at the edge of a lake. The fish were rewarded with food when they came close to a sound projector. However, there was no definite orientation of the fish toward the sound source and it was not possible to establish whether this species was able to locate sound sources beyond a few body lengths. Reinhardt (1935), after conducting experiments in a tank, also concluded that the European minnow and the brown bullhead lacked a directional acoustic sense for sounds or vibrations.

Despite these indeterminate results, the American physiologist Donald Griffin (1950) concluded that fishes may be attracted to or repelled by a sound source. Subsequently, Kleerekoper and Chagnon (1954) observed that creek chub (*Semotilus atromaculatus*) were able to locate a source of vibration, most likely oriented by the intensity gradient, in an experimental tank.

Still, van Bergeijk (1964), in a very influential paper, concluded that fishes were not likely to be able to determine sound directions using the mechanisms employed by terrestrial vertebrates, such as detection of differences in the intensity, time of arrival, and phase at the two ears. He argued that, in fishes, the differences in the binaural cues used for localization in air would be miniscule because their ears are very close together and the speed of sound is about 4.3 times faster in water than in air. van Bergeijk also assumed that hearing by fishes in the acoustic far field involved the detection of sound pressure, utilizing the swim bladder, and that this single sound pressure detector would stimulate both ears simultaneously and equally. Because he assumed that the information conveyed to the two ears would be identical in fishes, he concluded that only the lateral line system was able to locate sound sources, operating only in the acoustic near field. However, as pointed out by Dijkgraaf (1963) and as discussed in Sect. 3.7, the lateral line may only detect acoustic near field motions at a distance of just a few centimeters from the source.

Subsequent experiments supporting the existence of directional hearing in fishes were carried out in the sea by Olsen (1969) and Schuijf et al. (1972). Discrimination between sounds coming from different directions was confirmed by Chapman and Johnstone (1974) and by Schuijf (1975) (Fig. 3.12), who showed that some teleost species could detect changes in the propagation direction of sound in the horizontal plane. Chapman and Johnstone (1974) also showed that the degree of masking of a pure tone by broadband noise from a separate sound projector became reduced as the angle between the two sound projectors was increased. It was later shown that the Atlantic cod can discriminate between spatially separated sound sources in the median vertical plane (Hawkins and Sand 1977), between frontally incident and caudally incident sounds (Schuijf and Buwalda 1975), and even between diametrically opposed sound sources in both the median vertical and transverse vertical planes (Buwalda et al. 1983).

Obviously, these field experiments contradict van Bergejk's hypothesis that fishes lack a directional auditory sense in the acoustic far field. Pumphrey (1950) and de Vries (1950) had suggested earlier that the otolith organs effectively detect particle motion, a vector quantity with both magnitude and direction, and Dijkgraaf (1960) later proposed that the otolith organs are inherently directional in their response to sounds. This suggestion was ultimately supported by anatomical studies (Sect. 3.2) showing that the individual hair cells within the otolith organs are directionally orientated and organized in patterns in which the direction of the axis of

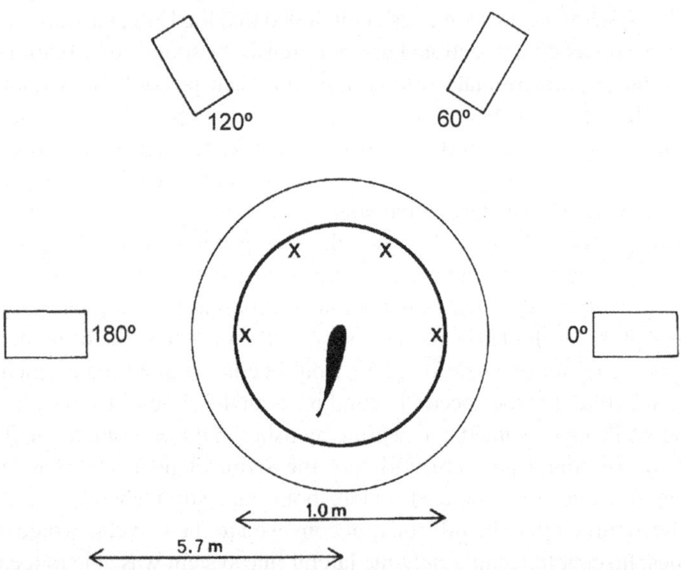

Fig. 3.12 Example of behavioral testing of directional discrimination. The fish moved within a netting cage and was trained to orient toward the food dispenser (x) in line with the active sound transducer. However, this approach cannot unambiguously demonstrate that the fish perceives the actual location of the sound source. (Redrawn from Schuijf 1975, with permission)

optimal sensitivity varies along the macula (e.g., Dale 1976; Enger 1976; Popper 1976). Thus, the fish brain may determine the direction of particle movements of the incident sound by vectorial weighing of the input from different regions of the sensory maculae. This was the underlying hypothesis that motivated both the SEM studies of the ultrastructure of fish ears discussed in Sect.3.2, and the electrophysiological studies of regional differences in directional sensitivity of otolith organs described below. All current models of directional hearing in fish are based on such a mechanism. See Popper et al. (1988) and Rogers et al. (1988) for a discussion of vectorial weighing.

When the hypothesis of directional hearing based on vectorial weighing of inputs from different populations of hair cells emerged, it seemed difficult to provide experimental support, due to the inevitably complex sound fields in small laboratory tanks described by US physicist Antares Parvulescu (1964). However, Enger et al. (1973) eliminated the problem of making directionally well-defined stimuli in small tanks by vibrating the fish in air, which simulates the effect of the kinetic sound component in water. They recorded microphonic potentials as the fish was vibrated in different directions (Fig. 3.13) and confirmed that the otolith organs of the haddock (*Melanogrammus aeglefinus*) were directionally sensitive. Later, the vibration method was refined and used by others (e.g., Sand 1974; Hawkins and Horner 1981; Fay 1984), and recordings from afferent nerve fibers from different parts of the ear showed regional differences in directional sensitivity within the sensory maculae.

Piddington (1972) reported that the goldfish can discriminate between compressions and rarefactions in a sound field. Soon after, Schuijf (1976) pointed out that

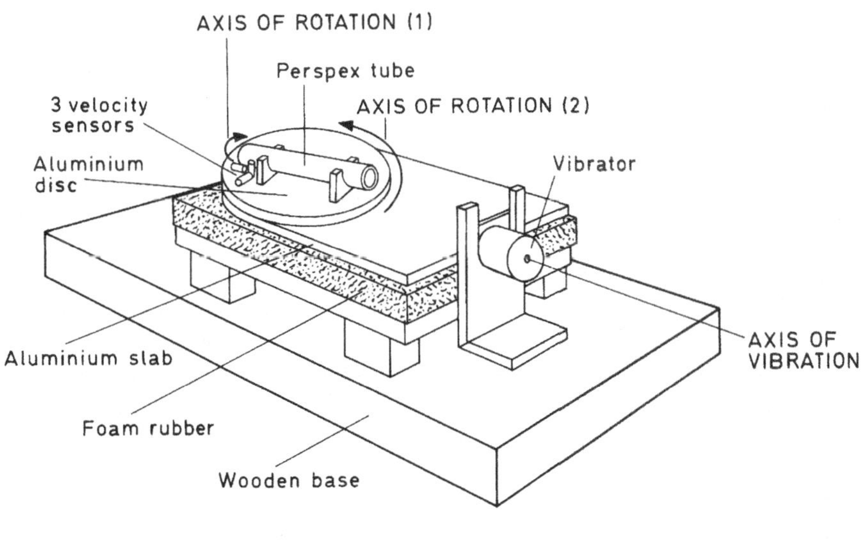

Fig. 3.13 The first vibration table used to simulate the kinetic sound component by vibrating the fish in air. The direction of vibration was altered by rotating the tube with a fish in the horizontal plane. More refined designs were later used in numerous studies of directional hearing in fish. (From Enger et al. 1973, with permission)

information about propagation direction of sound may be obtained by comparing the phase of particle motion with the phase of sound pressure sensed via the swim bladder.

Later, Buwalda et al. (1983) confirmed that phase relationship could be used for resolving sound direction. In this study, the phase relationship was controlled by using pairs of opposing sound projectors to generate standing waves in the ocean. In an additional study, Schuijf and Hawkins (1983) showed that Atlantic cod could even discriminate between sound sources at different distances, as the phase difference between particle motion and sound pressure varies with distance.

The emerging picture based on the phase model of directional hearing is that species with a swim bladder may have an acoustical sense of space. However, experiments have yet to demonstrate that fishes are able to perceive the actual location of a sound source from a distance without moving through the sound field, unlike mammals. Instead, the behavioral studies have only revealed that some species of fish are able to *discriminate between* sounds from different directions. Kalmijn (1997) suggested that a moving fish may still locate a source, although not instantaneously, by keeping a constant angle between its body axis and the incident particle motion detected by the inner ear. This strategy will guide the fish to all types of sound sources, i.e., monopoles, dipoles, and combinations including higher order sources.

The field of directional hearing in fish is more comprehensively reviewed by Sand and Bleckmann (2008), Rogers and Zeddies (2008), and Hawkins and Popper (2018). Despite the numerous experimental and theoretical studies on directional hearing in fish for more than 100 years, the mechanism by which fishes localize sound sources is still one of the great enigmas of fish hearing.

3.9 Conclusions

The intent of this chapter is to highlight the history of work on fish hearing, with primary emphasis on a substantial number of papers from the nineteenth century and the first 75 or so years of the twentieth century. The selected papers are not only formative (as well as informative), but they often are also germane to current thinking on fish bioacoustics. Many of the issues raised, the answers provided, and the thinking about the results directly address current research areas, including hearing thresholds, directional hearing, masking, effects of anthropogenic sound, and so on.

Furthermore, it is clear that the earlier investigators recognized technical problems that we are still dealing with today, including how to do hearing measures, tank acoustics, and various other methodological issues. Most of these topics are discussed in detail by Popper et al. (2019). Numerous current questions that need to be addressed to better understand fish bioacoustics rest on understanding many of the earlier studies discussed in this historical review.

Acknowledgments We dedicate this chapter to mentors and close friends and to present and future researchers in the fascinating field of fish bioacoustics. We also dedicate this chapter to (in chronological order of their initial contributions): G. H. Parker, K. von Frisch, S. Dijkgraaf, P. S. Enger, W.N. Tavolga, A. Flock, W. A. van Bergeijk, and R. R. Fay. We are grateful to Dr. Darlene Ketten for her excellent editing and suggestions that substantially improved this chapter.

Compliance with Ethics Requirements Olav Sand declares that he has no competing interest. Anthony Hawkins declares that he has no competing interest. Arthur Popper declares that he has no competing interest.

References

Alexander RM (1966a) Physical aspects of swimbladder function. Biol Rev 41:141–176

Alexander RM (1966b) Structure and function in the catfish. J Zool 148:88–152

Autrum H, Poggendorf D (1951) Messung der absoluten Hörschwelle bei Fischen (*Amiurus nebulosus*). Naturwissenschaften 38:434–435

Ayers H (1892) Vertebrate cephalogenesis. II. A contribution to the morphology of the vertebrate ear, with a reconsideration of its functions. J Morphol 6:1–360. https://doi.org/10.1002/jmor.1050060101

Bateson W (1890) The sense-organs and perceptions of fishes; with remarks on the supply of bait. J Marine Biol Assoc UK 1:225–256. https://doi.org/10.1017/S0025315400072118

Bigelow HB (1904) The sense of hearing in the goldfish *Carassius auratus* L. Am Nat 38:275–284

Braun CB, Sand O (2014) Functional overlap and nonoverlap between lateral line and auditory systems. In: Coombs C, Bleckmann H, Fay RR, Popper AN (eds) The lateral line system. Springer, New York, pp 281–312

Buerkle U (1969) Auditory masking and the critical band in Atlantic cod (*Gadus morhua*). J Fish Res Board Can 26:1113–1119

Buwalda R, Schuijf A, Hawkins A (1983) Discrimination by the cod of sounds from opposing directions. J Comp Physiol 150:175–184

Casserius I (1609) Pentaestheseion, Hoc Est De Quinque Sensibus Liber, Venice

Chapman C (1973) Field studies of hearing in teleost fish. Helgoländer wissenschaftliche Meeresuntersuchungen 24:371–390

Chapman CJ, Hawkins A (1973) A field study of hearing in the cod, *Gadus morhua* L. J Comp Physiol 85:147–167

Chapman C, Johnstone A (1974) Some auditory discrimination experiments on marine fish. J Exp Biol 61:521–528

Chapman C, Sand O (1974) Field studies of hearing in two species of flatfish *Pleuronectes platessa* (L.) and *Limanda limanda* (L.) (Family Pleuronectidae). Comp Biochem Physiol A Physiol 47:371–385

Coombs S, Bleckmann H (2014) The gems of the past: a brief history of lateral line research in the context of the hearing sciences. In: Coombs C, Bleckmann H, Fay RR, Popper AN (eds) The lateral line system. Springer, New York, pp 1–16

Coombs S, Janssen J (1990) Behavioral and neurophysiological assessment of lateral line sensitivity in the mottled sculpin, *Cottus bairdi*. J Comp Physiol A 167:557–567

Coombs S, Popper AN (1979) Hearing differences among Hawaiian squirrelfish (family Holocentridae) related to differences in the peripheral auditory system. J Comp Physiol A 132:203–207. https://doi.org/10.1007/bf00614491

Corwin JT (1977) Morphology of the macula neglecta in sharks of the genus *Carcharhinus*. J Morphol 152:341–362. https://doi.org/10.1002/jmor.1051520306

Corwin JT (1981) Peripheral auditory physiology in the lemon shark: evidence of parallel otolithic and non-otolithic sound detection. J Comp Physiol 142:379–390

Dale T (1976) The labyrinthine mechanoreceptor organs of the cod *Gadus morhua* L. (Teleostei: Gadidae). Norwegian J Zool 24:85–128
Das BK (1928) The bionomics of certain air-breathing fishes of India, together with an account of the development of their air-breathing organs. Philos Trans R Soc London Ser B 216:183–219
de Burlet HM (1931) Über die Gliederung des häutigen Labyrinthes. Z Anat Entwicklungsgesch 94:54–67. https://doi.org/10.1007/bf02117997
de Vries H (1950) The mechanics of the labyrinth otoliths. Acta Otolaryngol 38:262–273. https://doi.org/10.3109/00016485009118384
de Vries H (1956) Physical aspects of the sense organs. Prog Biophys Biophys Chem 6:207–264
Denton EJ, Gray JA (1982) The rigidity of fish and patterns of lateral line stimulation. Nature 297:679–681
Diesselhorst G (1938) Hörversuche an Fischen ohne weberschen Apparat. Z Vgl Physiol 25:748–783
Dijkgraaf S (1950) Untersuchungen uber die Funktionen des Ohrlabyrinths bei Meeresfischen (Investigatrions into the functions of the labyrinth of marine fishes). Physiologia Comparata et Oecologia 2:81–106
Dijkgraaf S (1952) Über die Schallwahrnehmung bei Meeresfischen. Z Vgl Physiol 34:104–122
Dijkgraaf S (1960) Hearing in Bony Fishes. Proc R Soc London Ser B 152: 51–54
Dijkgraaf S (1963) The functioning and significance of the lateral-line organs. Biol Rev 38:51–105
Enger PS (1967) Hearing in herring. Comp Biochem Physiol 22:527–538
Enger PS (1973) Masking of auditory responses in the medulla oblongata of goldfish. J Exp Biol 59:415–424
Enger P (1976) On the orientation of hair cells in the labyrinth of perch (*Perca fluviatilis*). In: Schuijf A, Hawkins AD (eds) Sound reception in fish. Elsevier, Amsterdam, pp 49–62
Enger PS, Andersen R (1967) An electrophysiological field study of hearing in fish. Comp Biochem Physiol 22:517–525

Enger P, Hawkins A, Sand O, Chapman C (1973) Directional sensitivity of saccular microphonic potentials in the haddock. J Exp Biol 59:425–433
Fay RR (1969) Behavioral audiogram for the goldfish. J Aud Res 9:112–121
Fay RR (1970) Auditory frequency generalization in the goldfish (*Carassius auratus*). J Exp Anim Behav 14:353–360
Fay RR (1974a) Masking of tones by noise for the goldfish (*Carassius auratus*). J Comp Physiol Psychol 87:708–716
Fay RR (1974b) Sound reception and processing in the carp: saccular potentials. Compar Biochem Physiol Part A 49:29–42
Fay RR (1984) The goldfish ear codes the axis of acoustic particle motion in three dimensions. Science 225:951–954
Fay RR (2009) Soundscapes and the sense of hearing of fishes. Integ Zool 4:26–32
Fay RR, Megela Simmons A (1999) The sense of hearing in fishes and amphibians. In: Fay RR, Popper AN (eds) Comparative hearing: fish and amphibians. Springer-Verlag, New York, pp 269–318
Fay RR, Popper AN (1974) Acoustic stimulation of the ear of the goldfish *(Carassius auratus)*. J Exp Biol 61:243–260
Fay RR, Kendall JI, Popper AN, Tester AL (1974) Vibration detection by the macula neglecta of sharks. Comp Biochem Physiol A Mol Integr Physiol 47:1235–1240
Flock A (1964) Structure of the macula utriculi with special reference to directional interplay of sensory responses as revealed by morphological polarization. J Cell Biol 22:413–431
Flock Å (1970) Transduction in single hair cells in the lateral line organ. Neurosci Res Prog Bullet 8:492–495
Flock Å (1971) Sensory transduction in hair cells. In: Loewenstein WR (ed) Principles of receptor physiology. Springer Berlin Heidelberg, Berlin/Heidelberg, pp 396–441. https://doi.org/10.1007/978-3-642-65063-5_14

Flock Å, Wersäll J (1962) A study of the orientation of the sensory hairs of the receptor cells in the lateral line organ of fish, with special reference to the function of the receptors. J Cell Biol 15:19–27

Furukawa T (2002) A recollection of early electrophysiological studies of goldfish hearing. Bioacoustics 12:109–111. https://doi.org/10.1080/09524622.2002.9753665

Furukawa T, Ishii Y (1967) Neurophysiological studies on hearing in goldfish. J Neurophysiol 30:1377–1403

Furukawa T, Ishii Y, Matsuura S (1972) An analysis of microphonic potentials of the sacculus of goldfish. Japan J Physiol 22:603–616

Griffin DR (1950) Underwater sounds and the orientation of marine animals, a preliminary survey. DTIC Document

Hama K (1969) A study on the fine structure of the saccular macula of the goldfish. Z Zellforsch Mikrosk Anat 94:155–171

Hama K, Saito K (1977) Fine structure of the afferent synapse of the hair cells in the saccular macula of the goldfish, with special reference to the anastomosing tubules. J Neurocytol 6:361–373

Harris GG (1964) Considerations on the physics of sound production by fishes. In: Tavolga WA (ed) Marine bio-acoustics. Pergamon, New York, pp 233–247

Harris GG, van Bergeijk WA (1962) Evidence that the lateral-line organ responds to near-field displacements of sound sources in water. J Acoust Soc Am 34:1831–1841

Hawkins AD, Chapman CJ (1975) Masked auditory thresholds in the cod, *Gadus morhua* L. J Comp Physiol 103:209–226

Hawkins AD, Chapman CJ (2020) Studying the behaviour of fishes in the sea at Loch Torridon, Scotland. ICES J Marine Sci. https://doi.org/10.1093/icesjms/fsaa118

Hawkins A, Horner K (1981) Directional characteristics of primary auditory neurons from the cod ear. In: Tavolga WN, Popper AN, Fay RR (eds) Hearing and sound communication in fishes. Springer, New York, pp 311–328

Hawkins AD, Johnstone ADF (1978) The hearing of the Atlantic salmon, *Salmo salar*. J Fish Biol 13:655–673

Hawkins AD, MacLennan DN (1976) An acoustic tank for hearing studies on fish. In: Schuijf A, Hawkins AD (eds) Sound reception in fish. Elsevier, Amsterdam, pp 149–170

Hawkins AD, Myrberg AA Jr (1983) Hearing and sound communication underwater. In: Lewis B (ed) Bioacoustics, a comparative approach. Academic Press, New York, pp 347–405

Hawkins AD, Popper AN (2018) Directional hearing and sound source localization by fishes. J Acoust Soc Am 144:3329–3350. https://doi.org/10.1121/1.5082306

Hawkins A, Sand O (1977) Directional hearing in the median vertical plane by the cod. J Comp Physiol 122:1–8

Hunter J (1782) XXIII. Account of the organ of hearing in fish. Philos Trans R Soc 72:379–383. https://royalsocietypublishing.org/doi/pdf/10.1098/rstl.1782.0025

Ishii Y, Matsuura S, Furukawa T (1971) An input-output relation at the synapse between hair cells and eighth nerve fibers in goldfish. Japan J Physiol 21:91–98

Jacobs DW, Tavolga WN (1967) Acoustic intensity limens in the goldfish. Anim Behav 15:324–335

Jacobs DW, Tavolga WN (1968) Acoustic frequency discrimination in the goldfish. Anim Behav 16:67–71

Kalmijn AJ (1997) Electric and near-field acoustic detection, a comparative study. Acta Physiol Scand Suppl 638:25–38

Kleerekoper H, Chagnon EC (1954) Hearing in fish, with special reference to *Semotilus atromaculatus atromaculatus* (Mitchill). J Fish Res Board Can 11:130–152. https://doi.org/10.1139/f54-012

Kreidl A (1895) Ueber die Perception der Schallwellen bei den Fischen. Archiv für die gesamte Physiologie des Menschen und der Tiere 61:450–464. https://doi.org/10.1007/bf01662070

Kreidl A (1896) Ein weiterer Versuch über das angebliche Hören eines Glockenzeichens durch die Fische. Archiv für die gesamte Physiologie des Menschen und der Tiere 63:581–586. https://doi.org/10.1007/bf01789552

Ladich F, Fay RR (2013) Auditory evoked potential audiometry in fish. Rev Fish Biol Fish 23:317–364
Lee FS (1898) The functions of the ear and the lateral line in fishes. Am J Physiol--Legacy Content 1:128–144
Leydig F (1868) Ueber Organe eines sechsten Sinnes: Zugleich als Beitrag zur Kenntniss des feineren Baues der Haut bei Amphibien u. Reptilien. vol 34. Druck v. E. Blochmann & Sohn,
Lowenstein O, Roberts T (1949) The equilibrium function of the otolith organs of the thornback ray (*Raja clavata*). J Physiol 110:392–415
Lowenstein O, Roberts TD (1951) The localization and analysis of the responses to vibration from the isolated elasmobranch labyrinth; a contribution to the problem of the evolution of hearing in vertebrates. J Physiol 114:471–489
Lu Z, Xu Z, Buchser WJ (2004) Coding of acoustic particle motion by utricular fibers in the sleeper goby, *Dormitator latifrons*. J Comp Physiol A 190:923–938. https://doi.org/10.1007/s00359-004-0550-3
Manning F (1924) Hearing in the goldfish in relation to the structure of its ear. J Exp Zool 41:5–20
Mayser P (1881) Vergleichende anatomische Studien über das Gehirn der Knochenfische. Zeitschrift für wiss. Zoologie. Bd. XXXVI
McCormick CA (1992) Evolution of central auditory pathways in anamniotes. In: Webster DB, Fay RR, Popper AN (eds) The evolutionary biology of hearing. Springer, New York, pp 323–350
McCormick CA (1999) Anatomy of the central auditory pathways of fish and amphibians. In: Fay RR, Popper AN (eds) Comparative hearing: fish and amphibians. Springer, New York, pp 155–217. https://doi.org/10.1007/978-1-4612-0533-3_5
Moulton JM (1963) Acoustic behaviour of fishes. In: Busnel RG (ed) Acoustic behaviour of animals. Elsevier, Amsterdam, pp 655–693
Nelson EM (1955) The morphology of the swim bladder and auditory bulla in the Holocentridae. Fieldiana. Zoology 37:121–137
Northcutt RG (1980) Central auditory pathways in anamniotic vertebrates. In: Popper AN, Fay RR (eds) Comparative studies of hearing in vertebrates. Springer, New York, pp 79–118
Northcutt RG (1997) Evolution of gnathostome lateral line ontogenies. Brain Behav Evol 50:25–37
O'Connell CP (1955) The gas bladder and its relation to the inner ear in *Sardinops caerulea* and *Engraulis mordax*. Fish Bull 56:505–533
Olsen K (1969) Directional responses in herring to sound and noise stimuli. Paper presented at the International Council for the Exploration of the Sea.
Parker GH (1903) The sense of hearing in fishes. Am Nat 37:185–204
Parker GH (1912) Sound as a directing influence in the movements of fishes. Bulletin of the US Bureau of Fisheries. US Government Printing Office
Parker GH (1918) A critical survey of the sense of hearing in fishes. Proc Am Philos Soc 57:69–98
Parker GH, Van Heusen AP (1917) The reception of mechanical stimuli by the skin, lateral-line organs and ears in fishes, especially in *Amiurus*. Am J Physiol 44:463–489
Parvulescu A (1964) Problems of propagation and processing. In: Tavolga WN (ed) Marine bioacoustics. Pergamon Press, Oxford, pp 87–100
Piddington R (1972) Auditory discrimination between compressions and rarefactions by goldfish. J Exp Biol 56:403–419
Plachta DT, Song J, Halvorsen MB, Popper AN (2004) Neuronal encoding of ultrasonic sound by a fish. J Neurophysiol 91:2590–2597. https://doi.org/10.1152/jn.01200.2003
Platt C (1977) Hair cell distribution and orientation in goldfish otolith organs. J Comp Neurol 172:283–287. https://doi.org/10.1002/cne.901720207
Platt C (1983) The peripheral vestibular system of fishes. In: Northcutt RG, Davis RI (eds) Fish neurobiology, vol 1. University of Michigan Press, Ann Arbor, pp 89–123
Poggendorf D (1952) Die absoluten Hörschwellen des Zwergwelses (*Amiurus nebulosus*) und Beiträge zur Physik des Weberschen Apparates der Ostariophysen. Z Vgl Physiol 34:222–257. https://doi.org/10.1007/bf00298202

Popper AN (1976) Ultrastructure of the auditory regions in the inner ear of the lake whitefish. Science 192:1020–1023

Popper AN (1981) Comparative scanning electron microscopic investigations of the sensory epithelia in the teleost sacculus and lagena. J Comp Neurol 200:357–374. https://doi.org/10.1002/cne.902000306

Popper AN, Clarke NL (1979) Non-simultaneous auditory masking in the goldfish, *Carassius auratus*. J Exp Biol 83:145–158

Popper AN, Rogers PH, Saidel WM, Cox M (1988) Role of the fish ear in sound processing. In: Atema J, Fay RR, Popper AN, Tavolga WA (eds) Sensory biology of aquatic animals. Springer, New York, pp 687–710

Popper AN, Fay RR, Platt C, Sand O (2003) Sound detection mechanisms and capabilities of teleost fishes. In: Collin SP, Marshall NJ (eds) Sensory processing in aquatic environments. Springer-Verlag, New York, pp 3–38

Popper AN, Hawkins AD, Sand O, Sisneros JA (2019) Examining the hearing abilities of fishes. J Acoust Soc Am 146:948–955

Pumphrey R (1950) Hearing. Symp Soc Exp Biol 4:3–18

Reinhardt F (1935) Über Richtungswahrnehmung bei Fischen, besonders bei der Elritze (Phoxinus laevis L.) und beim Zwergwels (*Amiurus nebulosus* Raf.). J Comp Physiol A Neuroethol Sens Neural Behav Physiol 22:570–603

Retzius G (1881) Das Gehörorgan der Fische und Amphibien. In: Das Gehörorgan der Wirbelthiere, vol 1. Wallin, Stockholm

Rogers LS, Sisneros JA (2020) Auditory evoked potentials of utricular hair cells in the plainfin midshipman, *Porichthys notatus*. J Exp Biol:jeb.226464. https://doi.org/10.1242/jeb.226464

Rogers PH, Zeddies DG (2008) Multiple mechanisms for directional hearing in fish. In: Webb JF, Fay RR, Popper AN (eds) Fish bioacoustics. Springer Science+Business Media, LLC, New York, pp 233–252

Rogers PH, Popper AN, Hastings MC, Saidel WM (1988) Processing of acoustic signals in the auditory system of bony fish. J Acoust Soc Am 83:338–349

Saidel WM, Popper AN (1987) Sound reception in two anabantid fishes. Comp Biochem Physiol 88:37–44

Sand O (1974) Directional sensitivity of microphonic potentials form the perch ear. J Exp Biol 60:881–899

Sand O (1981) The lateral-line and sound reception. In: Tavolga WN, Popper AN, Fay RR (eds) Hearing and sound communication in fishes. Springer, New York, pp 459–480

Sand O (1984) Lateral line systems. In: Bolis L, Keynes RD, Maddrell SHP (eds) Comparative physiology of sensory systems. Cambridge Univ. Press, Cambridge, pp 3–32

Sand O, Bleckmann H (2008) Orientation to auditory and lateral line stimuli. In: Webb JF, Fay RR, Popper AN (eds) Fish bioacoustics. Springer Science+Business Media, LLC, New York, pp 183–222

Sand O, Enger PS (1973) Evidence for an auditory function of the swimbladder in the cod. J Exp Biol 59:405–414

Sand O, Hawkins AD (1973) Acoustic properties of the cod swim bladder. J Exp Biol 58:797–820

Sand O, Karlsen HE (1986) Detection of infrasound by the Atlantic cod. J Exp Biol 125:197–204

Schneider H (1942) Die Bedeutung der Atemhöhle der Labyrinthfische für ihr Hörvermögen. J Comp Physiol A 29:172–194

Schuijf A (1975) Directional hearing of cod (*Gadus morhu*a) under approximate free field conditions. J Comp Physiol 98:307–332

Schuijf A (1976) The phase model of directional hearing in fish. In: Schuijf A, Hawkins AD (eds) Sound reception in fish. Elsevier, Amsterdam, pp 63–86

Schuijf A, Buwalda R (1975) On the mechanism of directional hearing in cod *(Gadus morhua* L.). J Comp Physiol 98:333–343

Schuijf A, Hawkins A (1983) Acoustic distance discrimination by the cod. Nature 302:143–144

Schuijf A, Baretta J, Wildshut J (1972) A field discrimination on the discrimination of sound direction in *Labrus berggylta* (Pisces: Perciformes). Norwegian J Zool 22:81–105

Schulze FE (1870) Ueber die sinnesorgane der seitenlinie bei fischen und amphibien (About the sense organs of the lateral line in fish and amphibians). Archiv für mikroskopische Anatomie 6:62–88

Stetter H (1929) Untersuchungen über den Gehörsinn der Fische, besonders von *Phoxinds laevis* L. und *Amiurus nebulosus* Raf (Studies on the auditory sense of the fish, especially of *Phoxinds laevis* L. and *Amiurus nebulosus* Raf). Z Vgl Physiol 9:339–477. https://doi.org/10.1007/bf00340160

Stipetić E (1939) Über das Gehörorgan der Mormyriden. J Comp Physiol 26:740–752

Tavolga WN (1967) Masked auditory thresholds in teleost fishes. In: Tavolga WN (ed) Marine bio acoustics. II Pergamon Press, Oxford, pp 233–245

Tavolga WN (1971) Sound production and detection. In: Hoar WS, Randall DJ (eds) Fish physiology, vol V. Academic Press, New York, pp 135–205

Tavolga WN (1974) Signal-noise ratio and the critical band in fishes. J Acoust Soc Am 55:1323–1333

Tavolga W (1976) Sound reception in fishes: benchmark papers in animal behavior, vol 7. Dowden, Hutchinson & Ross, Stroudsburg

Tavolga W (1977) Sound production in fishes: benchmark papers in animal behavior, vol 9. Dowden, Hutchinson & Ross, Stroudsburg

Tavolga WN (2002) Fish bioacoustics: a personal history. Bioacoustics 12:101–104

Tavolga WN, Wodinsky J (1963) Auditory capacities in fishes: pure tone thresholds in nine species of marine teleosts. Bullet Am Museum Nat History 126:177–240

Tavolga WN, Wodinsky J (1965) Auditory capacities in fishes: threshold variability in the blue-striped grunt, *Haemulon sciurus*. Anim Behav 13:301–311

Tester AL, Kendall JI, Milisen WB (1972) Morphology of the ear of the shark genus *Carcharhinus*, with particular reference to the macula neglecta. Pacific Sci 26:264–274

van Bergeijk WA (1964) Directional and nondirectional hearing in fish. In: Tavolga WA (ed) Marine bio-acoustics. Pergamon, New York, pp 281–299

van Bergeijk WA (1967) The evolution of vertebrate hearing. In: Neff WD (ed) Contributions to sensory physiology, vol 2. Academic Press, New York, pp 1–49

von Frisch K (1923) Ein Zwergwels, der kommt, wenn man ihm pfeift (A catfish who comes when one whistles to it). Biol Zentralbl 43:439–446

von Frisch K (1936) Über den Gehörsinn der Fische. Biol Rev 11:210–246

von Frisch K (1938a) The sense of hearing in fish. Nature 141:8–11

von Frisch K (1938b) Über die Bedeutung des Sacculus und der Lagena für den Gehörsinn der Fische. Z Vgl Physiol 25:703–747. https://doi.org/10.1007/bf00340903

von Frisch K, Dijkgraaf S (1935) Können Fische die Schallrichtung wahrnehmen? (Can fish perceive sound direction)? Z Vgl Physiol 22:641–655. https://doi.org/10.1007/bf00338993

von Frisch K, Stetter H (1932) Unterbuchungen über den Sitz des Géhörsinnes bei der Elritze. Z Vgl Physiol 17:686–801. https://doi.org/10.1007/bf00339067

von Heusinger C (1826) Bemerkungen über das Gehörwerkzeug des Mormyrus cyprinoides, *Gastroblecus compressus* und *Pimelodus synodontis*. Arch Anat Physiol Meckel 1:324–327

Walton I (1653) The compleat angler or the contemplative man's recreation. Richard Marriot, London

Walton I, Cotton C (1861) The complete angler, or the contemplative man's recreation. H. G. Bohn, London

Walton PL, Christensen-Dalsgaard J, Carr CE (2017) Evolution of sound source localization circuits in the nonmammalian vertebrate brainstem. Brain Behav Evol 90:131–153

Weber EH (1820) De aure et auditu hominis et animalium. Pars I. De aure animalium aquatilium. Gerhard Fleischer, Leipzig

Werner CF (1926) Die Cupula im Labyrinth der Fische. Z Zellforsch Mikrosk Anat 4:459–474. https://doi.org/10.1007/bf00391214

Wersäll J, Flock Å (1965) Functional anatomy of the vestibular and lateral line organs. Contrib Sens Physiol 14:39–61

Westerfield F (1922) The ability of mudminnows to form associations with sounds. J Comp Psychol 2:187–190. https://doi.org/10.1037/h0072963

Wodinsky J, Behrend ER, Bitterman ME (1962) Avoidance-conditioning in two species of fish. Anim Behav 10:76–78. https://doi.org/10.1016/0003-3472(62)90134-3

Wohlfahrt T (1936a) Ein neuer Versuch über das Tonunterscheidungsvermögen der Fische (A new attempt on the sound discrimination of fish). Verh dtsch zool Ges:135–140

Wohlfahrt TA (1936b) Das Ohrlabyrinth der Sardine (*Clupea pilchardus* Walb.) und seine Beziehungen zur Schwimmblase und Seitenlinie. Zeitschrift für Morphologie und Ökologie der Tiere 31:371–410

Yost W, Popper AN, Hawkins AD (2020) Dick Fay and goldfish. Acoustics Today 16(3):53–60

Chapter 4
A Nasty, Brutish, and Short History of Amphibian Bioacoustics

Peter M. Narins, H. Carl Gerhardt, and Jakob Christensen-Dalsgaard

4.1 Introduction

This chapter is divided into three sections. In Sect. 4.1, we provide a short introduction. Section 4.2 reviews the literature on what was known to the ancients about amphibian auditory anatomy, the function of the ear and its ability to transduce sound and vibrations, and, finally the development of our knowledge of frog calls and their function. Section 4.3 reviews the early research on frog communication from the modern (post-sound spectrograph) era which features the work of three notable pioneers in the field: W. Frank Blair, Murray Littlejohn, and Robert Capranica. Their groundbreaking studies inspired many researchers to enter the field with fresh ideas and modern techniques to bear on the recording of amphibian calls in the wild, electronic analysis and synthesis of calls, and a variety of new methods to make quantitative micromechanical measurements of inner ear function.

With apologies to Thomas Hobbes

P. M. Narins (✉)
Departments of Integrative Biology and Physiology, and Ecology and Evolutionary Biology, University of California Los Angeles, Los Angeles, CA, USA
e-mail: pnarins@ucla.edu

H. C. Gerhardt
Division of Biological Sciences, University of Missouri, Columbia, Columbia, MO, USA
e-mail: GerhardtH@missouri.edu

J. Christensen-Dalsgaard
Department of Biology, University of Southern Denmark, Odense, Denmark
e-mail: jcd@biology.sdu.dk

D. R. Ketten et al. (eds.), *A History of Discoveries on Hearing*, Springer Handbook of Auditory Research 77,
https://doi.org/10.1007/978-3-031-41320-9_4

Fig. 4.1 We dedicate this chapter to Jasper J. Loftus-Hills (1946–1974) (left, Photo: PM Narins) and Phillip Bishop (1957–2021) (right, Photo: U Otago photographer Graham Warman, with permission), both of whom left us far too early. Nobody could ever fill your shoes, but we all follow in your footsteps. Jasper was a brilliant, energetic, and creative scientist with deep insights into amphibian bioacoustics. Phil was a larger-than-life amphibian biologist and conservationist who used bioacoustical tools to understand the animals he loved the most. He was an acknowledged world leader in global amphibian conservation

In no sense is this chapter meant to be exhaustive. For a more complete treatment of the literally hundreds of frog bioacoustics experiments, the reader is referred to several excellent reviews (Rand 2001; Gerhardt and Huber 2002; Narins et al. 2007; Wells 2007). We dedicate this chapter to Jasper Loftus-Hills and Phillip Bishop (Fig. 4.1).

4.2 Amphibian Communication as Known to the Ancients

> Some say the goddess turned him into a frog among the reeds, with haunted frog voice chanting. (Gilgamesh epos, 2100 BCE).

This to our knowledge is the earliest literary reference relevant to frog bioacoustics,[1] but likely the loud calls of frogs and toads have aroused interest since (and before) the origin of humans. In many human societies frogs are an important food source, so attending to and localizing calling aggregations would be advantageous, and the obvious association of frogs' calls with rain (many species call at the onset of rains) and fecundity also generated interest in these animals. Several folk

[1] However, the translation of Akkadian cuneiform is not an exact science. In other translations of this passage the fate of Ishtar's unfortunate gardener is to be transformed into a caterpillar or a dwarf.

instruments capture the sounds of frogs, among them the guiro, a frog-shaped wooden block with a crest that is stroked and produces a very frog-like sound. Ocarino-like frog-shaped instruments were used in the Mesoamerican culture, very likely to call rain or maybe to incite calling in frogs in order to collect them. Another measure of the attention to frog sounds is that many frog names are onomatopoetic, for example, the Puerto Rican coqui (*Eleutherodactylus coqui*), the German Unken (*Bombina bombina* and *Bombina variegata*), or the Central American Túngara frog (*Engystomops pustulosus*), as are the generic names for frogs in some languages (e.g., Thai and Malay); in fact, in some languages like Dalabon (Australian Northern Territory) onomatopoetic names for frogs are apparently the rule (Cutfield 2016).

4.2.1 Antiquity to Middle Ages

The first accurate written description of frog calls is found in the comedy "Frogs" by Aristophanes (BCE 405), where a chorus of frogs accompanies the descent of Dionysus into the underworld. Their (incessant) call is transliterated as "Brekkekekex Koax Koax," and by the sound of it, almost certainly a precise rendering of the call of a European waterfrog (*Pelophylax sp.*). The call of the European edible frog (*Pelophylax esculentus*) is transliterated exactly the same way by Boulenger (1897–98).

Thus, interest in and perception of the call of anurans is well documented from antiquity, but the study of hearing in those animals did not attract much attention. Aristotle, in his impressive biological treatises, did not write extensively on frogs. He notes (Historia Animalium 4, ch 8) that the call of frogs attracts females (which implies that females have a sense of hearing) and that all "blooded animals" (roughly meaning vertebrates) possess the five normal senses, including hearing. However, this and a few anecdotes about the calls of frogs comprise all that was known of frog hearing to that point, and observers in later antiquity (Pliny and others) do not substantially supplement the information in Aristotle. This is in stark contrast to the plethora of information on fish hearing in the classical literature, for example in Aristotle (Historia Animalium, 4, ch. 8) and in Pliny the Elder [Gajus Plinius secundus] (1855, Natural History, ch. 89), reflecting that fish were domesticated, even kept as pets in ancient Rome, and that fish could be conditioned to respond to acoustic stimuli (see Sand, Hawkins, and Popper, Chap. 3).

The lack of information on frogs persisted throughout the Middle ages, during which the focus seems to have been to compile and annotate the earlier literature, including scattered information from the church fathers. A notable exception is the important German scholastic philosopher Albertus Magnus (Albert the Great), who adds his own observations to a large compilation of animal natural history called "De animalibus" ("On animals," ca. 1250). The entry on frogs is short (under the category "Vermes" i.e., "Creeps" (!)), but here is an example of an authentic observation of antiphonal calling in fire-bellied frogs (*B. variegata*): *"There is a particular species of toad called "horned"* [cornutus] *after the sound of its voice. It is a dusky color and is yellow on the belly. They sit in stagnant swamps and call one to the other. It is said that they do not call outside of France. But I have experienced that this is false because they call very shrilly throughout all of Germany.*" (Albertus Magnus, 1250, transl. Kitchell and Resnick 2018)

4.2.2 The Onset of Systematic Zoology—The Sixteenth Century

Neither of the great zoology writers of the Renaissance—the Swiss Conrad Gesner nor the Italian Ulisse Aldrovandi (1663), both working around 1550—adds much to the study of frog hearing. In fact, these works are mostly large compilations of ancient lore about animals with a focus on the utility or disutility for humans, mostly in medicine or magic in the case of amphibians. Both authors stand out, however, in the quality of their animal illustrations (e.g., Fig. 4.2). In this figure, from Gesner's article on the grass frog (*Rana temporaria*), the eardrum is accurately depicted, probably for the first time.

4.2.3 The Structure of Frog Middle and Inner Ear

Figures 4.3a, b show the structures of frog middle and inner ear, based on the tympanic ears of ranid frogs (note that there is considerable variation among amphibian species, especially in middle ear structures). The diagrams are based on two important publications: the first full paper on frog auditory neurophysiology by Frishkopf and Goldstein (1963) and the landmark anatomical study of Gustaf Retzius (1881).

Fig. 4.2 A reasonably accurate drawing of *Rana temporaria*, showing the eardrum (the circle beneath the eye) (Gesner 1586)

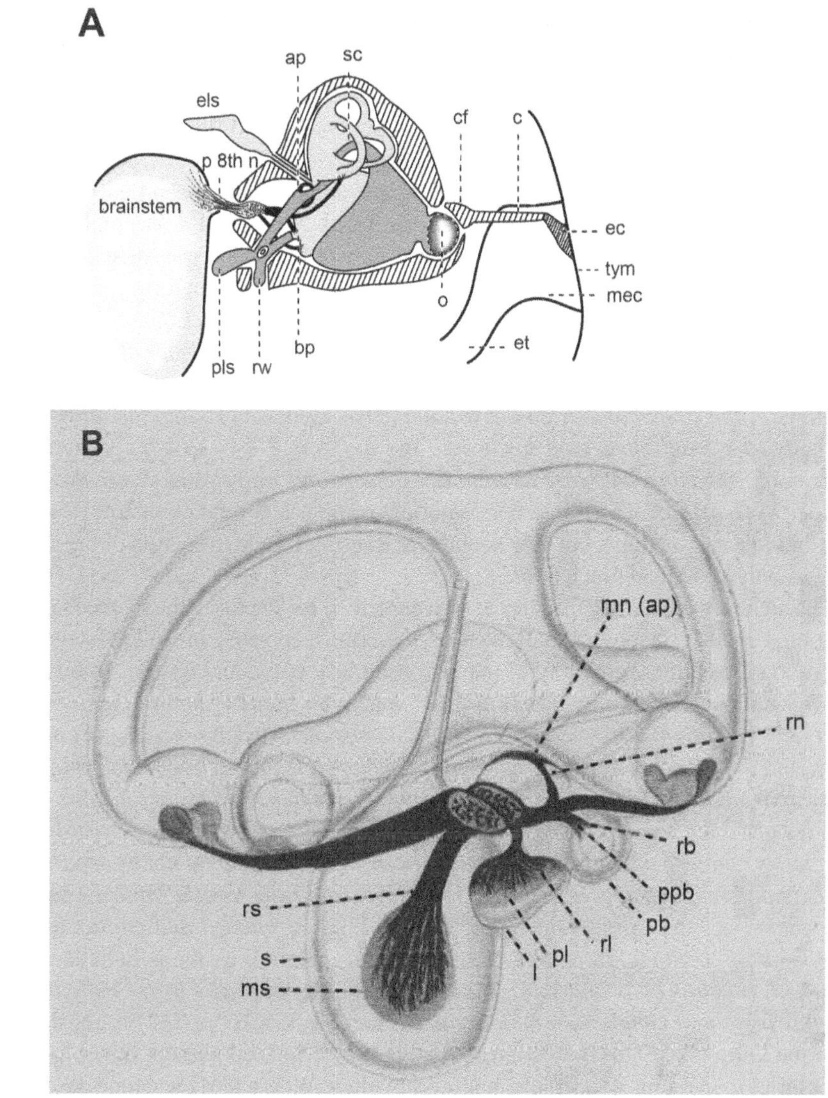

Fig. 4.3 (**a**) Diagram of the frog ear seen in transverse section at the level of the eardrum. (Redrawn and altered from a diagram by Frishkopf and Goldstein (1963), by permission from American Institute of Physics). *pls* perilymphatic sac, *p 8th n.* posterior branch of the eighth nerve, *rw* round window, *bp* basilar papilla, *o* operculum (would not be visible in the transverse section, but is outlined in simulated 3-D extending out of the plane of the figure), *et* Eustachian tube, *mec* middle ear cavity, *tym* tympanum, *ec* extracolumella, *c* columella, *cf* columellar footplate, *sc* semicircular canals, *ap* amphibian papilla, *els* endolymphatic sac. (**b**) Drawing of the inner ear of *P. esculentus*, from Retzius (1881) (slightly retouched for clarity). The drawing shows seven of the eight inner-ear organs in a lateral view: The hearing organs amphibian and basilar papillae (*ap* and *pb*), the three semicircular canals, and the otolithic organs sacculus and lagena (*s* and *l*). *rs* ramus sacculi (saccular nerve), *ms* macula sacculi, *pl* papilla acusticus lagena, *rl* ramus lagenae, *ppb* papilla acusticus basilaris, *rb* ramulus basilaris, *rn* ramulus neglectus, *mn* maculus neglectus (Amphibian papilla macula). The eighth inner ear organ, the otolithic utriculus is not visible in the drawing

4.2.4 Anatomy of the Middle Ear—The Seventeenth and Eighteenth Centuries

Observation of structures in these small animals was very difficult until optically improved lenses were available for magnifying glasses and early microscopes during the seventeenth century. Human middle ear anatomy had been described by sixteenth century Italian anatomists (review in Gouk 2004), and it was understood that the middle ear apparatus was an integral part of the sound transmission pathway. One early account, by Helkiah Crooke (1615, in Gouk 2004, page 139) describes the action of the human ear: *"The membrane being stricken doth move the three little bones and in a moment maketh impression of the character of the sound. This sound is presently received of the inbred Ayre, which it carryeth through the windows of the stony bone…into the winding burroughs, and so into the labyrinth, after into the Snaileshell, and lastly into the Auditory Nerve which conveyeth it thence unto the common Sense as unto his Censor and Judge. And this is the true manner of Hearing."* The following centuries would see several comparative studies of middle ear anatomy, but how sound was transduced and the action of the inner ear was only described much later.

The anatomy of the frog middle ear was described for the first time by the Danish anatomist Holger Jacobsen (Latinized as Olgerus Jacobæus) in 1676. Jacobæus reports on dissections, probably of the European waterfrog, and writes: "*The hearing organ shows the following structure: on both sides of the cranium is an ossified or at least cartilaginous structure extending skin as the skin covering the rest of the body. And this skin-substance forms a circle, that covers an internal membrane like the tympanum, leaving an opening and a small canal like the auditory meatus. The ossicles are connected to the above-mentioned tympanum, two are just visible, so seamlessly connected that they would appear as one, if they could not be separated by their different substances. Thus, it is apparent that the ossicle attached to the tympanum* [i.e., the extracolumella, ec in Fig. 4.3a] *is rounded and convex like a little knob of cartilage. Here attaches the second ossicle* [i.e., the columella, c in Fig. 4.3a] *which is bony and solid. It bends towards its base and forms there a triangular plate that connects to the cranium*" (Jacobæus 1676, p. 78–79, translated from the Latin text). This dissection was probably performed without other optical accessories than (at most) a simple magnifying glass and is a fairly accurate description of the ear (Fig. 4.4). Modern accounts of the frog tympanum usually state that the tympanum is formed from normal skin. However, when the skin is peeled off from the head of ranid frogs a membrane actually remains (personal observation, Simmons and Ketten 2020), so there are two layers of the tympanic membrane, and Jacobæus' account is accurate. However, in contrast to Jacobæus' description, the internal membrane does not contain any opening, and this observation likely reflects damage during the dissection.

The French anatomist Étienne-Louis Geoffroy published the first comparative study of the ear in lizards, snakes, and amphibians in 1755. Geoffroy's dissections were performed through a magnifying glass, and he corroborated most of the

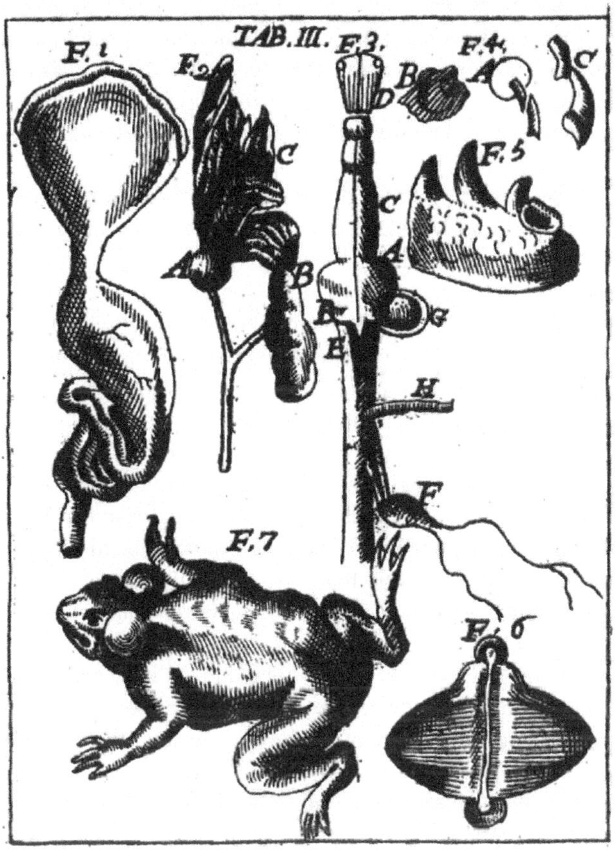

Fig. 4.4 The first depiction of a frog dissection showing details of the middle ear, from Jacobæus (1676). The eardrum and middle ear are shown in F.4; fig. F.4.C shows columella and extracolumella. The figure shows various other body parts such as pharynx and alimentary canal (F.1), and nervous system (F.3). The frog species name is not mentioned, but the drawing shows a frog with two extended vocal sacs (F.7). This feature is only found in one Danish species: the European waterfrog

findings of Jacobæus regarding frogs and added many new observations of middle and inner ear structures. He states that the tympanum does not contain any openings and also comments on the structure of the permanently open Eustachian tubes. Geoffroy also describes the oval window, and parts of the membranous labyrinth. He states that "*Apart from the parts ….that look like those found in many other reptiles* [i.e., reptiles and amphibians, according to the contemporary classification], *there is one that I have only found in this animal* [the frog]*; it is a large opening that can be seen easily through the pharynx of the animal, also in the live animal, that connects to the middle ear cavity; this opening, that I call the ear trumpet, since it has the same function as the Eustachian trumpet, is placed beside the upper jaw, a bit in front of its border….It cannot be doubted that* ***this opening serves to transmit***

sound to the inside of the ear [our highlighting]; *the cartilaginous tympanum of frogs attenuates vibrations considerably, and it* [the extra opening] *is therefore necessary to ensure that this organ is sensitive in an animal that lives both in water and in air, and only moves by jumps, so it cannot risk a tympanum formed by a simple membrane: Nature has solved it by making a second opening, that sound can pass through*."

Geoffroy is here describing a phenomenon seen in most amphibians, lizards, and archosaurs (crocodiles and birds): that sound can reach both surfaces of the eardrums through wide and open Eustachian tubes or interaural canals, which makes the eardrums inherently directional (see van Hemmen et al. 2016). Geoffroy also described the ear of "a toad" (the midwife toad *Alytes obstetricans*) having a thinner eardrum than the frog. He did not see Eustachian tubes in this animal and explained that it has no need for a second sound access to the eardrum.

Two large-scale, comparative studies of ears (in mammals, birds, reptiles, amphibians, fish, and invertebrates) were both published in 1789—by Antonio Scarpa and by Andrea Comparetti. Of amphibians, Scarpa mainly reports on the ear of salamanders (the "Caecilia" described by Scarpa, and also by Geoffroy, is not a caecilian amphibian, but a slow worm, i.e., the limbless lizard *Anguis fragilis*) but also states that the ears of frogs and toads are generally similar to those of lizards, turtles, and crocodiles. Most information on anurans apparently is based on Geoffroy's decription, but Scarpa adds the correct observation that toads also have Eustachian tubes. Scarpa describes the inner ear and auditory nerve of salamanders. Both Scarpa and Comparetti add to the earlier description of the middle ear by describing the other movable element in the otic capsule, the operculum, in frogs and salamanders. Comparetti also gave the first account of the frog inner ear, describing the structure of the semicircular canals and the otoconial organs with anatomical drawings. Structures in frogs are drawn rather crudely; the drawings of the larger lizard, turtle, and bird ears show more detail.

4.2.5 Anatomy of the Inner Ear—The Nineteenth Century

The nineteenth century was a golden age of frog ear anatomy. Probably due to advances in microscopic technique, histology, and histochemistry, the century saw many careful studies of inner ear anatomy. However, it is clear from the rich anatomical literature that a functional interpretation of the inner ear remained a challenge. Hasse (1868) wrote: *The study of the hearing organ of frogs is probably one of the most difficult tasks to undertake, one could even say the most difficult chapter in the chapter of all hearing organs, not so much because of its minuteness as because of the complexity of its parts..... A long and thorough study, a sharpening of the eyes for the smallest microscopical structures is necessary, to achieve a clear overview over this chaos...*" (a familiar sentiment for anyone who has struggled to make sense of the anuran auditory system!).

Likely, a principal stumbling block for the anatomists of the time was that it was unclear which of the eight inner-ear end organs (see Fig. 4.3b) subserved hearing. Du Verney (1683) had associated the sense of human hearing with both the cochlea and the semicircular canals, mostly based on the very conspicuous shape of these organs, and Du Verney had even assumed tonotopy (however, reversed compared to recent knowledge) of the cochlea based on its tapering.

However, the auditory inner ear organs of frogs (the amphibian and basilar papilla, see Fig. 4.5) are rather inconspicuous in comparison. Du Verney and later Scarpa also associated hearing sensitivity with the semicircular canals since these are prominent structures and quite similar in most vertebrates. In any case, this was a moot point, since detailed ablations, let alone neurophysiological recordings, were not possible before the twentieth century. Bioelectricity had been discovered in 1780 by Luigi Galvani by accidentally stimulating the frog spinal nerves, but clear ideas of nerve function emerged only fifty years later with Johannes Müller's important Law of Specific Nerve Energies (1835)— that sensation depends on **which** nerves are stimulated, rather than on **how** they are stimulated—and Müller's student

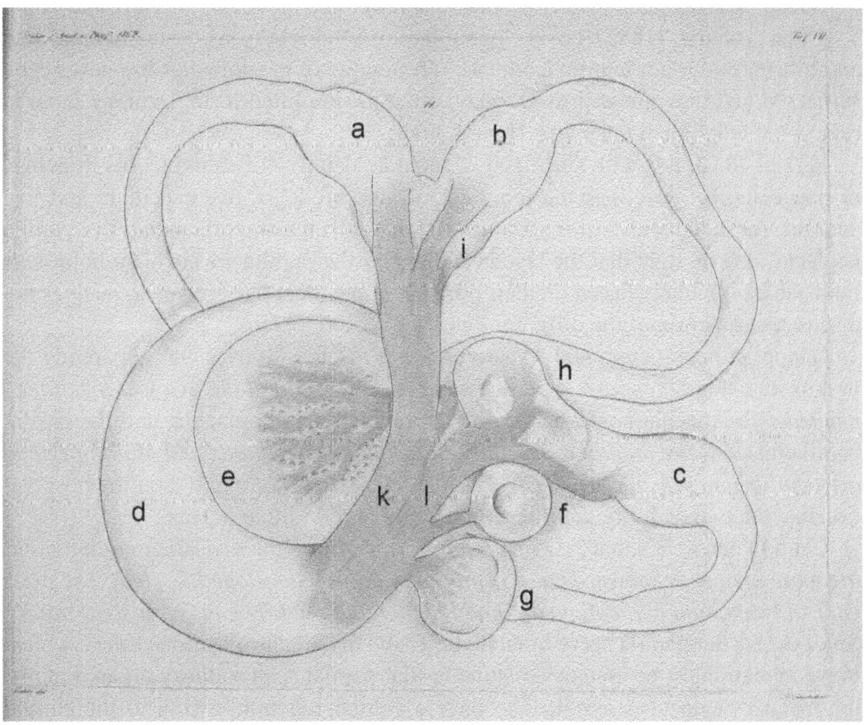

Fig. 4.5 Frog inner ear, from Deiters 1862 (slightly retouched for clarity). **a, b, c** semicircular canals, **d** saccule, **e** saccular macula, **f** basilar papilla, **g** lagena, **h** amphibian papilla, **i** recess joining the semicircular canals a and **b**, **k** anterior and **l** posterior branch of the eighth nerve. This figure conveys a realistic impression of a dissection of the ear, probably from the ventral side, anterior left, and posterior right. Compare the figure to the idealized drawing of Retzius (Fig. 4.3b)

Hermann von Helmholtz's measurements (1850) of nerve conduction velocity in frog spinal nerves.

The growing importance of the frog as experimental animal in physiological studies (Holmes 1993) led to increased interest in frog anatomy, and another student of Müller, Carl Windischmann (1831), made the first comprehensive study of inner ears of "Amphibians," meaning amphibians (here called "naked amphibians") and reptiles ("scaly amphibians"), according to Linnean classification at the time. Windischmann describes middle and inner ears in several urodeles, a caecilian, and several species of frogs and notes the absence of a functional middle ear in urodeles, a caecilian, and one species of frog, *Bufo igneus* (now *Bombina* sp., the fire-bellied toad). He noted the presence of an operculum in all the (naked) amphibians.

Otto Deiters (1862) was the first to suggest that the papilla neglecta, now known as the amphibian papilla, had a similar function as the cochlea, and used the term "Schnecke" (snail) for it. This was not universally accepted, and the Swedish anatomist Gustaf Retzius (1881) kept the term "macula neglecta" for the amphibian papilla, suggesting a homology with the macula neglecta of other vertebrates. Deiters also published detailed drawings of the frog inner ear (Fig. 4.5).

Gustav Retzius published two books on vertebrate inner ears in 1881 and 1884. The first volume (1881) covers agnathans, chondrichthyans, osteichtyans, and amphibians and is a scientific landmark. The quality of the drawings has never been surpassed and they are still used today, often in simplified form, in many modern reviews of amphibian inner ears (Fig. 4.3b).

Retzius illustrates with surprisingly accurate detail—for example, his drawings of hair cells. He describes the inner ear of nine urodeles, one caecilian, and five anurans. As stated above (this section), Retzius calls the amphibian papilla "papilla neglecta" and thought that the lagena and the basilar papilla are cochlear homologs (and analogs?), likely based on their position in the inner ear. However, he does not discuss the function of the different inner ear sensory organs.

Gaupp, in his monumental "Anatomie des Frosches (1904)" was apparently the first to state that "*there are good reasons for assuming that the two last-mentioned maculae* [the macula neglecta; i.e., the amphibian papilla macula, and the basilar papilla macula] *are auditory, and the rest (semicircular canals and otolith organs) are equilibrium sensors"* (p. 693). These good reasons were based on the observation that the two auditory papillae are the only non-otolithic organs.

Gaupp later writes that "...*Furthermore, concerning the acoustic function nothing more precise is known. That it is present, and well developed, is clear. The presence of tympanum and columella, and the fact that the frog calls, is a solid proof.*" Since the basilar papilla nerve branchlet is found in the same position as the cochlear nerve in mammals, he assumes that the basilar papilla is an auditory organ. Finally, the macula neglecta is assumed to have a similar function, owing to the similar structure of tectorial membrane in the two maculae (p. 751).

4.2.5.1 Darwin's Observations of Acoustic Communication in Amphibia

Darwin did not discuss evolution of the ear and auditory system, but noted the distinctive calls of frogs, such as during his voyage on the *H. M. S. Beagle.* In an entry in his diary from the autumn of 1832, written while living in a small house in a Brazilian forest, Darwin provides one of the first nineteenth-century descriptions of a frog and its vocalizations in musical terms:

> 204. Hyla. On the back, a band of "yellowish-brown." Width of head, sides copper yellow; abdomen silvery-yellowish white slightly tuberculated; beneath the mouth, smooth dark yellow, under sides of legs leaden flesh color. Can adhere to perpendicular surface of glass. The fields resound with the noise which this little animal, as it sits on a blade of grass, about an inch from the water, emits. The note is very musical. I at first thought it must be a bird. When several are together, they chirp in harmony; each beginning a lower note than the other, and then continuing or alternating upon the two (I think these notes are thirds to each other). May, Rio de Janeiro.

Later, Darwin (1871) noted the sexual dimorphism in frog vocal communication as an example of sexual selection:

> Frogs and toads offer one interesting sexual difference, namely, in the musical powers possessed by the males; but to speak of music, when applied to the discordant and overwhelming sounds emitted by male bull-frogs and some other species, seems, according to our taste, a singularly inappropriate expression. Nevertheless, certain frogs sing in a decidedly pleasing manner. ..The various sounds are emitted chiefly by the males during the breeding-season, as in the case of the croaking of our common frog. In accordance with this fact the vocal organs of the males are more highly developed than those of the females. In some genera the males alone are provided with sacs which open into the larynx. For instance, in the edible frog (*Rana esculenta*) "the sacs are peculiar to the males, and become, when filled with air in the act of croaking, large globular bladders, standing out one on each side of the head, near the corners of the mouth." The croak of the male is thus rendered exceedingly powerful; whilst that of the female is only a slight groaning noise. In the several genera of the family the vocal organs differ considerably in structure, and their development in all cases may be attributed to sexual selection.

The studies of Reichert (1837) suggested that the middle ear ossicles in mammals were derived from the embryonic branchial skeleton, and that the mammalian stapes and the amphibian columella were homologous. Apparently, Darwin was unacquainted with Reichert's work, but later Gaupp (1913) revised Reichert's findings and suggested that common ontogeny implied homology of parts of the middle ear apparatus in the tetrapods. Gaupp's observations also suggested, however, that "the eardrum of anurans, sauropsids and mammals originated in parallel from different ancestral forms that had a middle ear cavity, but where the tissue between it and the skin surface was not yet thinned to a membrane that could vibrate" (Gaupp 1913, translated from Maier and Ruf 2016). The hypothesis of independent origins of the tympanic ears of these groups was contested during the twentieth century, but was subsequently reestablished by Lombard and Bolt (1979) and Clack (1997).

4.3 The Modern Age—The Twentieth Century

The twentieth century marked the onset of auditory physiological studies of amphibians. Robert M. Yerkes (1905) published an important behavioral study in which tactile stimuli could be delivered to awake green frogs (*Rana clamitans*). This produced a reflex that could be modified by sound—usually increased if sound and tactile stimuli were simultaneous. Interestingly, this is one of the very few studies of auditory behavior not related to communication calls in frogs. Yerkes demonstrated hearing in air and water and showed that the response was strongly reduced by lesioning the auditory nerve (Fig. 4.6). Moreover, cutting the tympanum or the columella did not result in a reduced response.

As mentioned in Sect 4.2.5, anatomical studies of salamander ears had been undertaken in the eighteenth and nineteenth centuries, but the first definitive study was not published until 1909 by Kingsbury and Reed. These workers investigated development of the ear in various species of salamanders and developed an influential hypothesis for the reception of substrate vibrations via the forelimbs and operculum scapula and opercularis muscle (Fig. 4.7). However, the function of the operculum in salamanders and frogs is still debated, and various other functions have been proposed (Hetherington 1992; Mason and Narins 2002).

The function of the inner ear organs was conclusively demonstrated in the early twentieth century. Tait and McNally (1934) showed by ablation studies that the utricle and semicircular canals were implicated in equilibrium sensing. Recordings

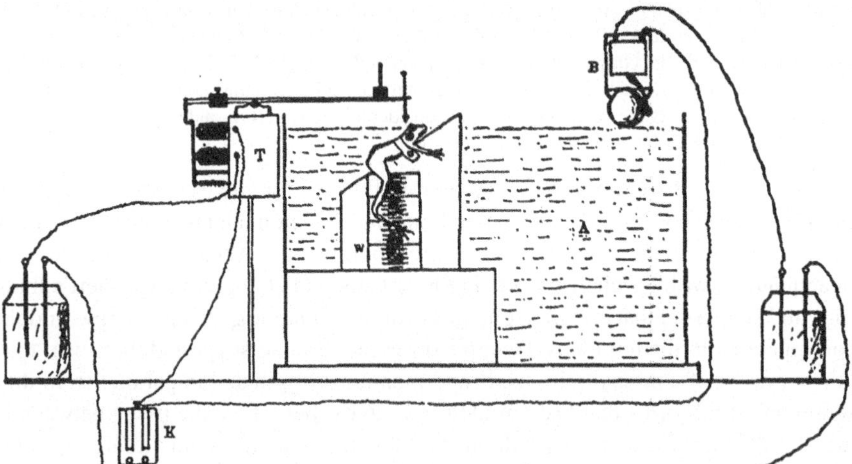

Fig. 7. Auditory apparatus for testing hearing in air and in water. *A*, aquarium; *B*, electric bell; *T*, tactual stimulus apparatus; *K*, hand-key for giving stimuli; *w*, weight to hold leg.

Fig. 4.6 The setup used by Yerkes (1905) to study frog hearing. The frog receives tactile stimulation from the apparatus above its head that is paired with auditory stimulation delivered by a bell (B)

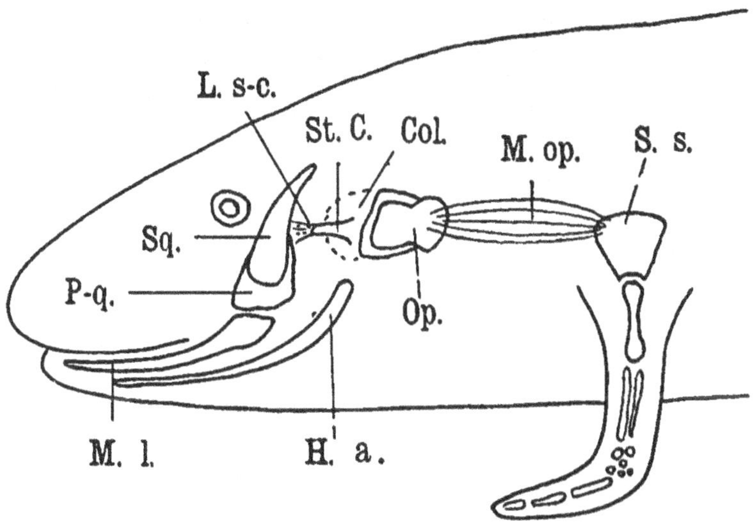

Fig. 4.7 Acoustic detection in salamanders. Diagram of a salamander with the proposed pathway of substrate vibrations through the forelimb, suprascapular (S.s), and operculum (Op) to the inner ear. This mechanism was also thought to apply to anuran amphibians. Col., columella; H.a., hyoid arch; L.s-c, ligamentum squamoso-columellare; M.op., musculus opercularis; M.l., skeleton of lower jaw; Op., operculum; Pq., palatoquadratum; Sq., Os squamosum; S.s., suprascapula; St.C, stilus columellae (From Kingsbury and Reed (1909), fig. 21b)

from nerve branchlets performed by Ashcroft and Hallpike (1934) showed that the saccule was responding to substrate vibrations but was not involved in equilibrium sensing. At the same time, Adrian et al. (1938) recorded the cut auditory nerve in European green frogs (*P. esculentus*) and grassfrogs (*R. temporaria*). They found that the nerve fibers could be stimulated only by intense sound, probably due to lack of blood supply.

The first single-unit responses from the frog auditory nerve in intact frogs were recorded about 1960 in the leopard frog (*Rana pipiens*) by a then undergraduate student, Franklin S. Axelrod (Axelrod 1960, reported in Lettvin and Maturana 1960, p.167–168). Axelrod found three types of nerve fibers: "*The first is a curious auditory element. It responds to frequencies in the 600–700 cps band, but adapts very rapidly so that there is very little response, except initially, to the turning on of a sine wave of constant amplitude in that band.*" The second type is described as sensitive to lower frequencies, and its spontaneous activity is inhibited by tones above its characteristic frequency. The third type is sensitive to seismic vibrations.

Axelrod's supervisors, Lettvin and Maturana had, together with McCullough and Pitts, just published one of the most influential and classic neurophysiological papers entitled: "What the frog's eye tells the frog's brain" (Lettvin et al. 1959). This impressive work is one of the foundational papers in neuroethology, since it uses behavior as a tool to understand neuronal processing. They showed that the response of the visual system in the bullfrog (*Rana catesbeiana* = [*Lithobates catesbeianus*]) was heavily biased toward biologically relevant stimuli.

Certainly, this paper and contacts with Lettvin influenced a promising graduate student at the Massachusetts Institute of Technology (MIT), Robert R. Capranica, in his view of the bullfrog auditory periphery as a "matched filter," that could extract conspecific calls from background noise. A publication by Frishkopf et al. (1968) acknowledged that "...The success of Lettvin's teleologically oriented approach was impressive to us, and ... had considerable influence on our own approach to studies of the neural coding of acoustic stimuli" and "Our story is in many ways a sequel to the earlier paper by Lettvin et al.").

Lawrence Frishkopf and Moise Goldstein (Capranica's thesis advisor) had recorded from single fibers in the bullfrog auditory nerve (Frishkopf and Goldstein 1963) and described two populations of nerve fibers, with thresholds as low as 25 dB SPL. The two populations were termed simple units, mostly with characteristic frequencies (CFs) above 1000 Hz, and complex units, which were stimulated by low-frequency sound, and could be inhibited only by frequencies above CF. Also, and more important, Frishkopf and Goldstein apparently were the first to investigate auditory responses to the mating call, which would be central to future investigations by Capranica.

The focus on neuroethology as an important research strategy initiated by Lettvin et al. and later by Capranica (see Sect. 4.3.3) made it essential to integrate the study of auditory physiology in frogs with studies of their communication behavior. Technological advances of the twentieth century made it possible to record and play back sounds, while developments in electronics and computers created new possibilities of analyzing and modifying sounds for behavioral experiments.

One such advance was the invention of the sound spectrograph. This instrument was originally devised to analyze and provide a permanent visual record of the energy distribution of human speech sounds in both frequency and time. Conceived and built by three engineers working at the Bell Telephone Laboratories in New York, Koenig et al. (1946) had the foresight to include in their seminal paper the very first sound spectrogram of a frog's call (Fig. 4.8). Although the frog species was not identified, both amplitude and temporal modulation are clearly evident in its vocalization.

The true dawn of amphibian bioacoustical studies in the USA was the publication of the first widely available set of frog-call recordings by Arthur Allen and Peter Kellogg at Cornell University (Allen and Kellogg 1948). Charles Bogert at the

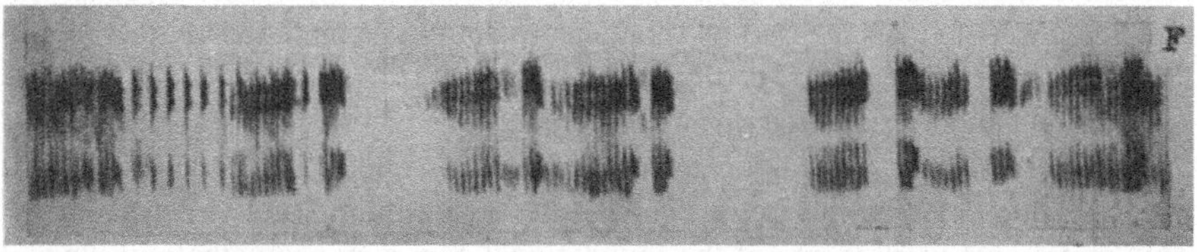

Fig. 4.8 First spectrogram of a frog call. Earliest known published sound spectrogram of frog vocalizations (Koenig et al. 1946, Fig. 33F, with permission). The species represented herein are unknown as are the time and frequency scales

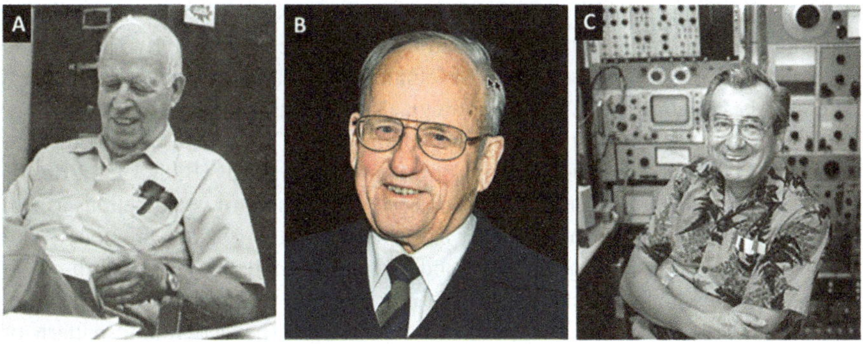

Fig. 4.9 (**a**) Frank Blair, U Texas; (**b**) Murray Littlejohn, U Melbourne; (**c**) Bob Capranica in his lab at Cornell University

American Museum of Natural History also produced an important album of frog calls and a comprehensive review of acoustic communication in amphibians and reptiles (Bogert 1960).

Unlike the history of bat echolocation which can trace its beginnings to *two* research giants—Spallanzani and Griffin (Grinnell et al. 2016), it was not until the mid-twentieth century that W. Frank Blair, Murray Littlejohn, and Robert Capranica (henceforth "*The Holy Trinity,*" Fig. 4.9) introduced a rigorous scientific approach to the study of anuran (frog and toad) vocalizations and their biological significance. These researchers were among the first to use the sound spectrograph not only for description and analysis of anuran vocalizations, but also to consider the behavioral, evolutionary, and mechanistic consequences of these vocalizations. Although we recognize the valuable contributions from bioacousticians of other areas of the world (many mentioned later in the chapter), we nevertheless consider the work of the *Holy Trinity* as the beginning of modern anuran bioacoustics both because of their own personal and technical contributions to the field but also, by extension, the contributions of the many students, postdoctoral fellows, and colleagues who they inspired worldwide.

4.3.1 W. Frank Blair

A Professor of Zoology at the University of Texas at Austin, Blair was among the first to use the sound spectrograph to analyze and describe the vocalizations of anurans. One of Blair's most influential contributions was to show that these calls varied not only within and between populations but also throughout the geographic range of widely distributed species (Blair 1958a). He also combined analyses of vocalizations of different pairs of species with the results of artificial crosses between them to assess what he termed the "stage of speciation" (Blair 1955; Blair and Littlejohn 1960). The crosses estimated the genetic compatibility of different species: the lack of viability or fertility of crosses was considered a post-mating isolating mechanism. Differences in mating calls, were a presumptive pre-mating isolating mechanism because studies of female selectivity were not yet available.

Using these data along with morphological criteria, Blair also proposed species groups of toads and hylid (treefrog) frogs (Blair 1958b), and he was among the first biologists to describe the vocalizations of interspecific hybrids (Blair 1956). Blair's review also considered potential evolutionary consequences of vocalizations as isolating mechanisms. He pointed out for example that if two species previously isolated by geography came into contact and had similar calls that resulted in costly mating mistakes, then selection over time would result in sufficient divergence in call structure to reduce or eliminate those mistakes. The geographic pattern of greater call differences in sympatry than in allopatry is termed reproductive character displacement (RCD) and the process, reinforcement. There are a handful of examples in anuran vocalizations, and a study of a pair of narrow-mouthed toad species by Blair (1955) was the first in anurans (Blair 1974). One of the most robust and convincing examples of southeastern US chorus frogs was published by Blair's student Fouquette (1975). Lemmon et al. have confirmed this and several other examples using massive phylogenomic analyses and extensive samples of vocalizations and female-preference tests (Lemmon 2009).

Another widely accepted example focused on two species of hylid frogs in Australia (Littlejohn 1965; Loftus-Hills and Littlejohn 1992). The paucity of robust examples of RCD probably stems from concentrating on call differences alone. If call differences in allopatry are adequate for discrimination by females, then selection might act to increase female selectivity in sympatry in the face of masking interference, differences in intensity, and variation in other properties that affect mate choice but are not species-specific. Studies of gray and green treefrogs have provided evidence supporting this hypothesis (Gerhardt 1994; Höbel and Gerhardt 2003).

4.3.1.1 Vocal Production by Amphibians

Two outstanding Blair graduate students were W. F. Martin and James Bogart. Martin published important papers on the vocalization system of toads and collaborated with Carl Gans, another pioneer in this field (Martin 1971, 1972; Martin and Gans 1972; Gans 1973). Other early contributors to our knowledge of vocal mechanisms were R. S. Schmidt in the USA and H. Schneider in Germany (Schmidt 1973; Schneider 1988). Since then great progress has been made in both the morphological and physiological mechanisms underlying frog vocal production (reviewed in Wells 2007 and see below). Bogart contributed to the discovery that call differences often lead to the discovery of cryptic polyploid species of frogs (Bogart and Wasserman 1972), and inspired studies of the role of cellular changes such as larger cell size in polyploids that can affect calls (e.g., Keller and Gerhardt 2001)

4.3.2 Murray J. Littlejohn

Blair was a major source of inspiration to the second member of the Trinity, Murray J. Littlejohn, who studied frog vocalizations as an undergraduate and graduate student of A. R. Main at the University of Western Australia. They published the first papers on acoustic communication in Australian frogs (Littlejohn and Main 1959) about the same time that Littlejohn became a post-doc in Blair's lab. During a subsequent study leave in the USA in the late 1960s, many of Blair's graduate students, including Carl Gerhardt, were greatly inspired by Littlejohn.

Whereas Blair's principal personal focus was on differences in vocalizations between species and hybrids, Littlejohn together with Blair's students were also curious about the receiver side of communication systems. The first evidence that sound alone was sufficient to attract females was provided by Martof and Thompson Jr (1958), who reported that gravid females of the Southern chorus frog *Pseudacris nigrita* were attracted to earphones that emitted recorded calls of conspecific males. Inspired by these results, Littlejohn returned to the Blair lab and used a pair of speakers to create a "discrimination task" (Fig. 4.11). Littlejohn and Michaud (1959) showed that gravid females of the chorus frog *Pseudacris streckeri* selectively responded to a speaker broadcasting recordings of a conspecific male rather than to alternating broadcasts of a sympatric species *P. clarkii*. This was the first evidence supporting Blair's assumption that call differences alone could function as pre-mating isolating mechanisms.

After joining the faculty at the University of Melbourne in 1959, Littlejohn and his students studied many questions derived from his experiences in the Blair lab. Their research endeavored to understand and explore questions about speciation (Littlejohn 1981, 1993), the significance of hybrid zones and hybridization, and their role in reinforcement (Littlejohn 1965; Littlejohn and Loftus-Hills 1968; Littlejohn and Watson 1973; Littlejohn and Roberts 1975). Besides addressing evolutionary questions, Littlejohn and his students, especially Jasper Loftus-Hills (Fig. 4.1), conducted the first two-speaker tests with females using electronically synthesized calls (Loftus-Hills and Littlejohn 1971a), made measurements of the sound levels of frog vocalizations (Loftus-Hills and Littlejohn 1971b), and began research on the mechanisms underlying frog-call production (Loftus-Hills 1974) and central auditory mechanisms (Loftus-Hills and Johnstone 1970; Loftus-Hills 1973). Loftus-Hill's premature and tragic death during fieldwork in Texas was a significant loss for the field. A paper that supported Blair's conclusions about the importance of RCD was based in large part on data collected during this fieldwork (Loftus-Hills and Littlejohn 1992).

Littlejohn was also influenced by interactions with R.D. Alexander at the University of Michigan at a time when the topic of sexual selection was beginning to command widespread attention. His change in perspective is exemplified by this quote from an important review: "The main purpose of mate choice is to find the best mate, not to find the 'correct species' …" (Littlejohn 1999). Clearly, the best mate will nearly always also be a member of the same species, but vocalizations can also indicate the current physical fitness and/or genetic fitness of males and to function in male-male competition (Wells and Schwartz 1984; Wells and Taigen 1986; Welch et al. 1998).

The results of behavioral experiments by Blair and Littlejohn have also been important for guiding studies of the proximate mechanisms underlying female auditory selectivity and testing neurobiological hypotheses (Rose and Capranica 1984; Rose et al. 1985; Ryan et al. 1992; Wilczynski et al. 2001; Gerhardt and Huber 2002). In particular, another important result reported by Blair serves to illustrate the interplay between studies of behavior and the underlying neural mechanisms, which is the main focus of neuroethology. Blair (1958a) showed that intra-individual variation in many kinds of frogs was largely a result of the temperature dependence of some of the physical properties of their calls. This effect is especially strong in the eastern gray treefrog (*Hyla versicolor*) and in sibling species, Cope's gray treefrog (*Hyla chrysoscelis*, Fig. 4.10a). We now know that this property and the duration of pulses in the trills of these frogs are not only temperature-dependent but also that female preferences change in parallel, a phenomenon termed "temperature coupling" (Gerhardt 1978; Fig. 4.10b). Note too that the most effective pulse-repetition rate of a midbrain, bandpass neuron changed dramatically with temperature in the same way that the female preferences changed (Fig. 4.10c). This phenomenon can be characterized as the maintenance of parallel tuning of temporal filters in the face of significant changes in temperature—a kind of dynamic matched filtering.

The emphasized frequencies in the calls of many frogs and toads are inversely correlated with body size, and other physiological changes can also affect the spectral tuning of the peripheral auditory system to create a match. As mentioned above, Capranica likened this form of biological coordination between sender and receiver to the electronic communications concept of a "matched filter." Long-term changes in body size are almost certainly the main driver of matched filtering among species and even within species in some situations. For example, in the Coqui frog (*E. coqui*) in Puerto Rico, male body size increases along an altitudinal gradient (Narins and Smith 1986) resulting in decreases in call frequency with altitude. The frequencies to which the ear is tuned also decrease with altitude in part because of body size and decreases in temperature with altitude. Taken together, these factors largely account for matched filtering over an extensive range of altitudes (Meenderink et al. 2010).

Fig. 4.10 (continued) playback experiments with synthetic calls with pulse-repetition rates typical of males calling at 16 °C and 20 °C, respectively "(see arrows in Part A)". Lines connecting open squares show the proportions of females tested at female preferences for synthetic calls with pulse-repetition rates of males calling at 16 °C and choosing an alternative of 40 pulses/s to alternatives with lower or higher values; lines connecting closed squares show proportions tested at 20 °C choosing alternatives of 50 pulses/s, choosing an alternative of 50 pulses/s to alternatives of lower and higher values. (From Gerhardt (2015), 73–85. Originally published by and used with permission from Dove Medical Press Ltd). (**c**) Spike numbers of the bandpass neurons at two different temperatures in the midbrain of *H. versicolor* (From Brenowitz et al. (1985). Originally published by and used with permission from Dove Medical Press, Ltd. in Gerhardt (2015) cited above)

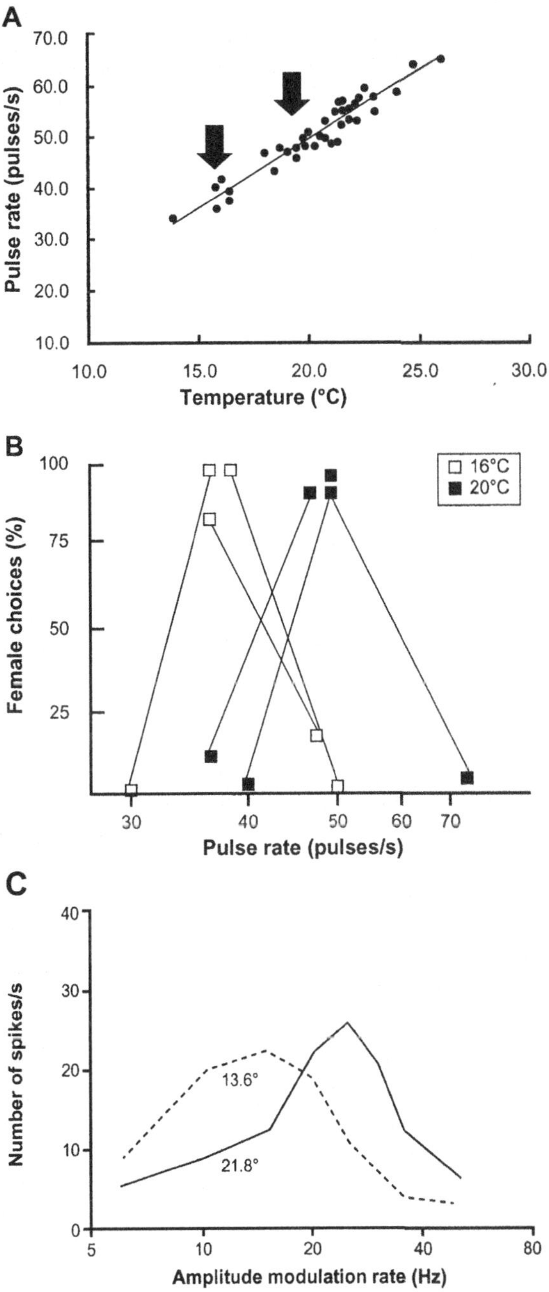

Fig. 4.10 Temperature effects on calls and preferences in gray treefrogs: correlates with temporal tuning in the auditory system. (**a**) Scatter diagram showing positive effect of temperature on pulse-repetition rate in male gray treefrogs, *Hyla chrysoscelis*. (**b**) Results of two-stimulus, forced-choice

However, the difference in the time course over which call frequency and auditory tuning occur can also weaken or destroy the matched filtering. For example, green treefrog females tested at much lower than normal breeding temperatures actually preferred synthetic calls with a low-frequency peak that was nearly an octave lower than that of conspecific males (Gerhardt and Mudry 1980). At this low temperature, auditory tuning but not call frequency changed significantly. This drop in tuning was noted by Mudry and Capranica (1987) and was the impetus for the study showing the mismatch in female preferences at low frequency. We now turn to a discussion of the main source of inspiration and guidance of neuroethological research on anuran bioacoustics.

4.3.3 Robert R. Capranica

The third of our *Holy Trinity* was trained as an electrical engineer and received his Sc.D. from MIT in 1964. Capranica had been working at the famed Bell Laboratories (Murray Hill, New Jersey) since 1958 where he developed a research program dedicated to the understanding of the frog auditory system. This work formed the basis of his doctoral dissertation, which has been referred to as "… arguably the most elegant dissertation research ever completed in animal bioacoustics" (Adler et al. 2013), and was published by MIT Press as a Research Monograph (Capranica 1965). In this seminal work, Capranica developed the one-male acoustic playback paradigm, in which an isolated male bullfrog in a laboratory tank is presented with an acoustic stimulus through a loudspeaker and the vocal responses of the male to this playback are recorded with a directional microphone.

Using this apparatus, Capranica broadcast the recorded mating calls of 34 representative species of frogs and toads to an isolated male bullfrog. He found that only the playback of the conspecific call resulted in an evoked vocal response (EVR) from the male under test, thus demonstrating the bullfrog's advertisement call is species-specific (Capranica 1966).

In a second experiment using the one-male acoustic playback paradigm, he synthesized a full-spectrum bullfrog advertisement call consisting of a narrowly-tuned spectral peak centered at 200 Hz and a broadly-tuned spectral peak centered at 1400 Hz (Fig. 4.11a–c). He then presented a synthetic advertisement call to an isolated male bullfrog in a laboratory tank and recorded the frog's EVRs in response to this stimulus (the one-male, acoustic playback paradigm; Fig. 4.11d). Remarkably, every presentation of this full-spectrum advertisement call (Fig. 4.11c) resulted in an EVR; i.e., a 100% EVR rate from the male tested (Fig. 4.11e). Then in a novel twist, Capranica electronically manipulated the full-spectrum stimulus; first by low-pass filtering (Fig. 4.11f) and then by high-pass filtering (Fig. 4.11g), but neither filtered version of the call resulted in EVRs. Capranica thus concluded that both the low-frequency and high-frequency spectral peaks must be present simultaneously in the stimulus to obtain the EVR from the male bullfrog (Capranica 1966).

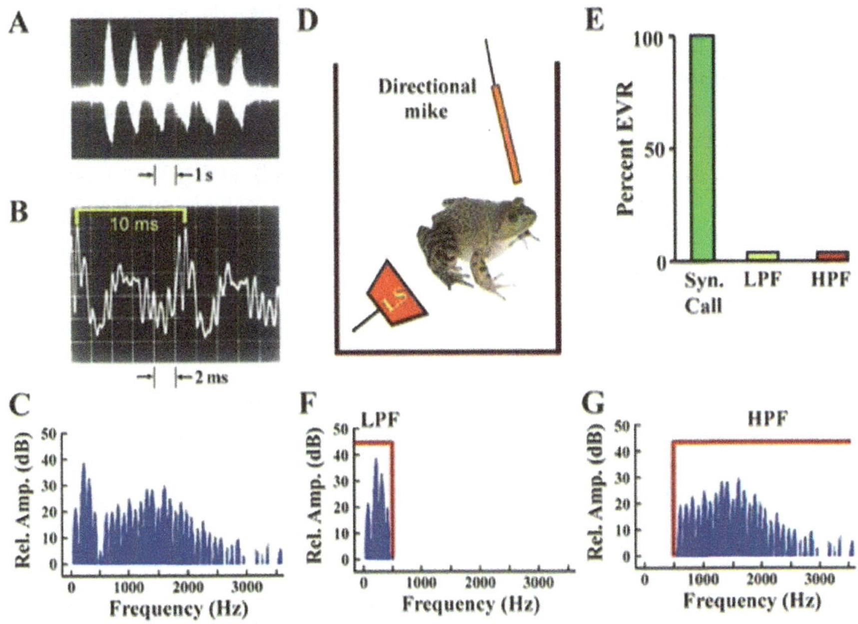

Fig. 4.11 Determining the spectral components required to elicit the bullfrog EVR. (**a**) Waveform of a synthetic pulsatile bullfrog call; (**b**) Same as A with expanded time axis revealing an internal call periodicity of 10 ms; (**c**) Amplitude spectrum of the synthetic bullfrog mating call exhibiting a narrowly-tuned spectral peak centered at 200 Hz and a broadly-tuned spectral peak centered at 1400 Hz. (**d**) This call was broadcast through a loudspeaker (LS) to an isolated male bullfrog. All calls emitted by the bullfrog were recorded with a directional microphone and each presentation of the full-spectrum synthetic call resulted in an evoked vocal response (100% EVR) from the male under test (**e**). Low-pass filtered (**f**) or high-pass filtered (**g**) versions of the call were completely ineffective at evoking an EVR (**e**). (Modified from Capranica (1966) with permission)

Capranica was a giant in the field of amphibian bioacoustics. His contributions include, among others: he developed the one-male acoustic playback paradigm and used it to measure the EVR of the isolated male bullfrog; he established the anuran auditory system as a convenient vertebrate model for studying communication (Capranica 1965); he pioneered a signal processing approach for quantifying sender-receiver dynamics (Capranica 1965, 1966), and he first applied the engineering concept of the matched filter to amphibian communication systems to provide a framework in which to approach the analysis of the neural processing of complex vocal signals in noise (Capranica and Moffat 1983).

4.3.3.1 Work Inspired by the Holy Trinity

More recently, the Capranica paradigm was applied to bullfrogs in the field and this work revealed that males learn the calls of territorial neighbors with which they fight to establish boundaries early in the season (Davis 1987). By habituating them to playbacks of synthetic calls, which are initially treated as a dangerous rival, Mark Bee showed that territorial males use the spectral structure of the call for such individual recognition (Bee and Gerhardt 2001a, b).

Capranica's matched filter hypothesis inspired many other researchers, and the widespread nature of this phenomenon is clearly illustrated by the high correlation between the carrier frequency of the advertisement calls and the estimated frequency at the lowest threshold measured electrophysiologically in many amphibian species (Fig. 4.12). The high coefficient of determination shown in this figure is remarkable given the wide variation in methods, species, and recording temperatures. At the same time, we caution that such matched filtering is seldom very sharp and only partially effective in reducing masking interference in mixed-species choruses.

In the fledgling Capranica lab of 1970, the cohort of Albert Feng and Peter Narins (graduate students) and Carl Gerhardt (postdoctoral fellow) would go on to pursue careers in amphibian bioacoustics. As mentioned earlier, the studies of Frishkopf and Goldstein (1963) placed all frog auditory nerve fibers into two populations:

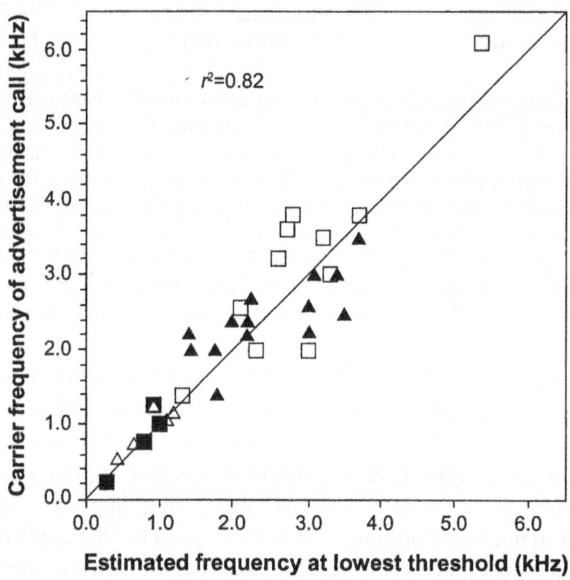

Fig. 4.12 Match between calls and hearing thresholds. Scatter diagram of spectral peaks in the advertisement calls of 24 species of anurans (plus three populations of one species) against estimates of minimum threshold (BEF). Solid symbols show data from studies in which closed system stimulation was used; open symbols show data from studies in which free-field stimulation was used. Squares indicate low-frequency sensitivity attributed to the amphibian papilla, and triangles indicate high-frequency sensitivity attributed to the basilar papilla. (Adapted from Gerhardt and Schwartz (2001) Auditory tuning and frequency preferences in anurans. In Michael J. Ryan, editor; Anuran Communication. Washington, DC: Smithsonian Press (Fig. 7A). Adapted by Gerhardt (2015), 73–85 and used with permission from Dove Medical Press Ltd)

simple and complex. One of the first publications to come out of this new Capranica cohort was one that critically revisited those results. Single-unit recordings from the eighth nerve of the bullfrog in response to tones revealed *three* populations of auditory fibers: low-frequency inhibitable fibers, mid-frequency non-inhibitable fibers, and high-frequency non-inhibitable fibers. By selectively lesioning individual nerve branchlets within the inner ear, it was shown that the amphibian papilla gives rise to low- and mid-frequency sensitive units whereas the basilar papilla gives rise to high-frequency sensitive units (Feng et al. 1975).

One of us (Peter Narins), inspired by Capranica's experiments in which he showed that both the low-frequency and high-frequency spectral peaks must be present simultaneously in the stimulus to obtain the EVR from the male bullfrog (Fig. 4.11), asked the question if this result might also obtain in the time domain. That is, by choosing another species producing a sequence of notes, we could ask if the natural note order is critical to evoke the frog's behavioral response. Thus, the ideal species for this study was as follows: (a) one in which males produced a sequence of notes (and for simplicity—the shortest possible call-note sequence: two-notes), (b) preferably a treefrog (which as a group, are often more accessible to the researcher than fully aquatic species), and (c) one with a head size large enough to facilitate electrophysiological recordings. Taking into account these considerations (i.e., applying the The Krogh Principle) it was decided that the most appropriate species for this study (Krogh 1929) would be the Puerto Rican Coqui (*E. coqui*). Males produce a two-note advertisement call (Co-Qui) but when a synthetic call is broadcast to calling males in the field, they drop their Qui note 42% of the time on average, producing just the Co note (the "One-note response," Narins and Capranica 1976). This then was the metric used to quantify the specificity of a frog's behavioral response in the time domain. The results were disappointing at first. The frogs dropped their Qui note equally frequently in response to the Co-Qui and Qui-Co stimuli; thus, the forward- and the reverse-note sequences were equally effective at evoking the one-note response from males of *E. coqui*. Further experiments revealed, however, that the Co note alone was equally effective as either of the two-note sequences. Thus, if the Co note were presented alone, or whether it was followed (natural sequence) or preceded by (reverse-sequence) a Qui note, the one-note response was evoked in 42% of the trials on average. Curiously, the Qui note alone was almost totally ineffective in evoking the one-note response from calling males (Narins and Capranica 1976).

It had been observed that males physically and vigorously defend their arboreal territories and that fighting is preceded by bouts of Co notes exchanged between a resident and an intruder (Reyes Campos 1971). If the Co note is used in male-male territorial interactions, what about the Qui note? Using the Littlejohn Paradigm, a series of two-choice discrimination experiments revealed that given a choice of a Co note or a Qui note, female Coquis showed strong preferences for the Qui-note speaker (Narins and Capranica 1976, 1978; Lopez and Narins 1991). A study of auditory nerve fiber tuning properties of males and females of *E. coqui* revealed that males have a larger population of fibers tuned to the Co-note frequency than the females, whereas females have more fibers tuned to the Qui-note frequency than the males. This sensory sexual dimorphism was the first to be reported for vertebrates and reflects the importance of the Co note for males and the Qui note for females (Narins and Capranica 1976). Subsequently, playbacks of advertisement calls or different segments of such calls that have acoustically distinct motifs were found to elicit immediate changes in those calls or the production of one or more different calls in the species repertoire. Kentwood Wells and Michael J. Ryan, and their students also contributed to dissecting the functional significance of complex calls (Wells and Schwartz 1984; Ryan 1985).

Wells and Ryan were both graduate students at Cornell in the 1970s, and although not studying with Capranica, they were obviously inspired by him and went on to produce a multitude of important papers and influential books about frog bioacoustics. For example, Wells's (1977) review of social behavior and his landmark book on the ecology and behavior of amphibians (Wells 2007) include detailed and lucid discussions of research using the Littlejohn and Capranica paradigms in field experiments (Wells 2007). His studies with Taigen on calling energetics are also classics (Taigen and Wells 1985; Wells and Taigen 1986) that inspired detailed studies of neuromuscular mechanisms (e.g., Girgenrath and Marsh 1997; Marsh 1999).

Michael Ryan's many significant contributions to the understanding of anuran vocal communication include multiple studies of sexual selection in the túngara frog (*Physalaemus pustulosus*) summarized in his influential book, *The Túngara Frog: A Study of Sexual Selection and Communication* (Ryan 1985). His and Stanley Rand's studies of closely related species in Central and South America provided support for the concept of sensory exploitation (Ryan 1990; Ryan and Rand 1990). The idea is that hidden sensory biases can be revealed when some acoustic feature that is currently absent in the repertoire of a species is added. For example, the túngara frog has a two-part or multiple-part call consisting of a "whine" and one or more "chuck" elements. A closely related species (*Physalaemus coloradorum*) produces only whines, but females of this species preferred a complex call consisting of the whine to which three chuck elements were added. In another species with a whine-like call, females did not prefer calls with added chucks (Tárano and Ryan 2002), but adding extra elements or increasing the rate or duration of calls beyond the usual range of variation increased attractiveness to females in many species (reviewed in Wells 2007). Ryan et al. have done extensive explorations of the evolution of calls and preferences using the comparative approach (e.g., Cocroft and Ryan 1995; Canatella et al. 1998) and have explored some "cognitive" aspects of frog behavior (e.g., Phelps and Ryan 1998).

Another area of research that flourished in the Capranica lab was the study of sound localization in anurans. To this end, Albert Feng conducted behavioral studies with green treefrogs along with Carl Gerhardt, showing that disrupting the input from one of the tympanic membranes adversely affected localization (Feng et al. 1976). Feng performed a series of detailed neurophysiological studies of the frog central auditory pathways which added a great deal to our understanding of the neural processing underlying sound localization (Feng and Capranica 1976, 1978; Feng 1981).

When Jürgen Rheinlaender joined the lab as a postdoc, he and Carl Gerhardt quantified localization accuracy in green treefrogs (*Hyla cinerea*) and found it more acute than could be accounted for by interaural time or intensity differences (Rheinlaender et al. 1979). This and the fact that small treefrogs without pinnae can also localize sound sources accurately in three dimensions suggested that frogs, like many insects, birds, and lizards use a pressure difference system for sound localization rather than relying on external differences between the two ears (Gerhardt and Rheinlaender 1982; Passmore et al. 1984; Jørgensen and Gerhardt 1991; Christensen-Dalsgaard 2011; van Hemmen et al. 2016). In parallel with studies by Feng, Rose, and Capranica, Bibikov et al. carried out an extensive series of neurophysiological studies on auditory processing by the frog midbrain (Bachtin and Bibikov 1974; Bibikov 1990; Bibikov and Nizamov 1996; Bibikov 2002). It is unfortunate that much of this interesting research is less accessible to Western scientists than it might have been had it been published in English.

Today, the frog ear can be accurately conceptualized as a multi-port (multi-input) system (Narins 2016). In addition to the tympanic inputs, the tympanic membrane can be stimulated by sound entering through the mouth floor (Aertsen et al. 1986) and/or through the "extratympanic" pathways that presumably account for low-frequency sensitivity and directionality (Wilczynski et al. 1987; Christensen-Dalsgaard 2005). Moreover, direct experimental evidence supports a significant path for low-frequency sound from the environment to the inner ear through the frog's body wall (Narins et al. 1988; Ehret et al. 1990; Jørgensen and Gerhardt 1991; Ehret et al. 1994). By this route, sound enters the two lungs and the larynx and then travels via the Eustachian tubes where it can act on the inner surfaces of both tympanic membranes. A recent study convincingly revealed one function of this lung-ear connection in frogs. In the green treefrog (*H. cinerea*), the lung input to the ear appears to cancel the noise entering the ear in a frequency band adjacent to the spectral peaks of the call, thus increasing the received signal-to-noise ratio (Lee et al. 2021).

In the late 1990s, Kraig Adler at Cornell University sent one of us (Peter Narins) a letter describing a frog he saw in China that had an unusual ear morphology—ear canals with sunken tympanic membranes. To explore this further, Albert Feng and Peter Narins traveled to Huangshan, China in 1998 to record the vocalizations of the Concave-eared torrent frog, *Odorrana tormota*. The result of that and many subsequent trips to China were the discoveries that: (a) males of *O. tormota* (= *Amolops tormotus*) produce both audible and ultrasonic call components (Feng et al. 2002; Narins et al. 2004; Suthers et al. 2006); (b) This frog can detect the ultrasonic components of the call, and (c) this remarkable upward extension of both the call and the hearing sensitivity likely to have co-evolved in response to the wideband ambient noise from the torrential streams in their environment (Narins et al. 2004; Feng et al. 2006). Subsequent studies have revealed that *Huia cavitympanum* in Malaysian Borneo produces pure ultrasonic advertisement calls (Arch et al. 2008, 2009). The quest continues to locate other ultrasonic species such as *Odorrana graminea* (Shen et al. 2011) to determine the phylogenetic extent of this unique amphibian trait.

4.3.4 E.R. (Ted) Lewis

No discussion of amphibian bioacoustics would be complete without the inclusion of the wide-ranging contributions to the field made by Ted Lewis et al. Ted spent his academic career at the University of California Berkeley where he and his students pioneered the use of the scanning electron microscope (SEM) to examine the eight frog inner ear organs. Their work established the basilar papilla as auditory; the amphibian papilla and the sacculus as auditory and seismic; the lagena as seismic; and the utricle, sacculus, lagena, and semicircular canals as vestibular.

Then, in a landmark study, functionally identified, dye-filled fibers of the eighth cranial nerve in the American bullfrog (*R. catesbeiana*) were traced to their peripheral origins providing the first precise functional overlays for the micro-structural maps of inner-ear sensory surfaces (Lewis et al. 1982a). In another classic study, Lewis et al. (1982b) used intracellular dye-injection studies to reveal tonotopic organization of the bullfrog amphibian papilla, an auditory organ lacking a basilar membrane or its equivalent. In addition, Lewis et al. pioneered the quantitative study of the response of frogs to substrate-borne (seismic) stimuli (Fig. 4.13). They carried out studies of vibratory sensitivity of the frog ear with an elaborate setup to reduce ambient substrate vibrations and thus allow low-noise recordings of single-fiber sensitivity to vibrations (Koyama et al. 1982; Narins and Lewis 1984; Yu et al. 1991; Fig. 4.13a). Moreover, in a series of tour-de-force experiments, Lewis et al. transported a reconstructable version of the seismic

Fig. 4.13 (continued) steel plates and enclosing it inside a plywood box sitting on a series of bricks, plywood, foam, and rubber inner tubes. The result was a highly effective mechanical filter for isolating ground vibrations from the animal under test, achieving a noise floor below 2×10^{-6} m/s^2. (**b**) Setup for producing seismic "thumps" mimicking those made by calling male white-lipped frogs in Puerto Rico. The thumper is enclosed in two nested plywood boxes for airborne sound attenuation. The geophone records the substrate-borne stimulus and the microphone records the frog calls. (**c**) Control setup for the experiment shown in (**b**). The thumper with its plywood boxes is now placed on a vibration-isolation platform re-assembled in the field for each frog tested. This platform ensured any vibrational signal from the thumper being transmitted to the ground, as measured with the geophone, was below the detection threshold of a frog under test. (Modified from Lewis et al. (2001) with permission)

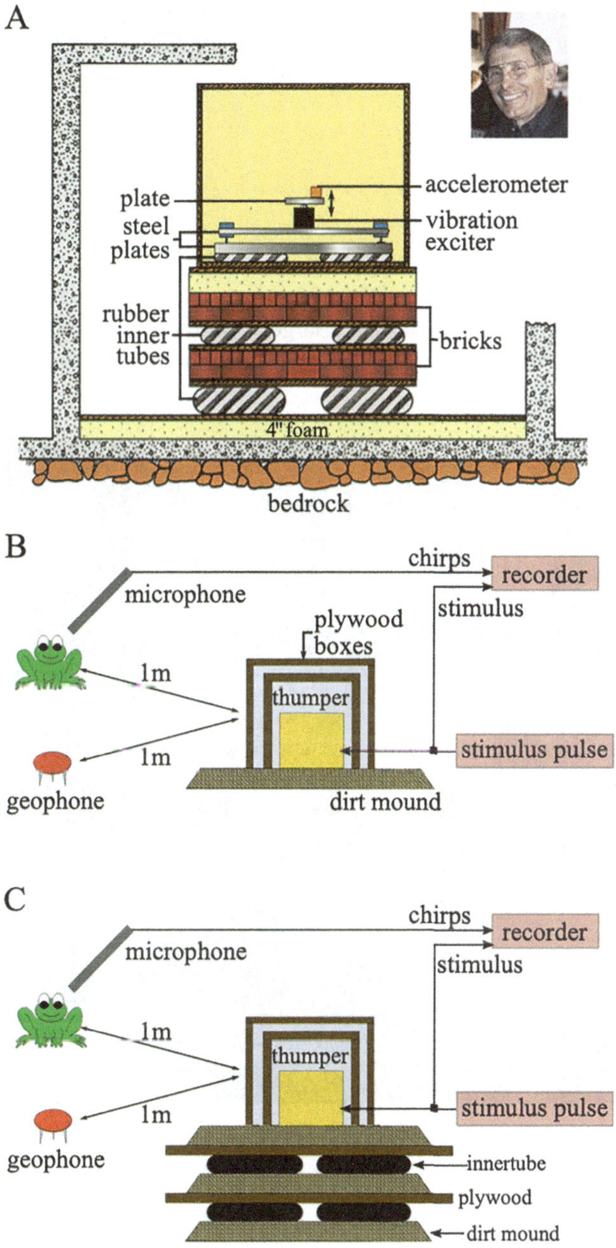

Fig. 4.13 Seismic sensitivity of amphibians. (**a**) Schematic diagram of the setup in a concrete-walled basement room at UC Berkeley used by Lewis et al. to study seismic sensitivity of the frog ear. A piezoelectric accelerometer was mounted on a plate adjacent to the frog to be studied. The plate was then affixed to a vibration exciter that provided a dorso-ventral vibrational stimulus to the animal. Acoustic isolation was achieved by mounting the aforementioned apparatus on massive

isolation apparatus to the field to demonstrate that pure seismic stimuli in the absence of airborne sound were capable of altering the calling patterns of males of the white-lipped frog, *Leptodactlylus albilabris*, in Puerto Rico (Lewis et al. 2001; Fig. 4.13b, c).

4.4 Conclusion

Shown in Figs. 4.14, 4.15, 4.16, and 4.17 is a selection of researchers (and some of their equipment) who have dedicated a large portion of their scientific lives to the study of acoustic communication in amphibians. These include some workers who

Fig. 4.14 Frog Bioacousticians I. (**a**) Anthony Arak, Frog Bioacoustics (FB) leader in the UK; (**b**) Carl Gerhardt, neuroethology and evolution; (**c**) Neville Passmore, South Africa FB; (**d**) Gary Rose, neuroethology, with emphasis on temporal tuning in the frog brain; (**e**) Michael Ryan introduced the túngara frog as a model of auditory communication and sexual selection; (**f**) Kentwood Wells, social behavior in Anura; (**g**) Andrea Megela Simmons, leader in frog auditory development; (**h**) Darcy Kelley, leader in frog vocal production; (**i**) Walter Hödl, Austrian FB leader; (**j**) Peter Narins in French Guiana, amphibian neuroethology; (**k**) Albert Feng in Sichuan Province, China, neuroethology of sound localization in frogs. Not shown: Hans Schneider. German FB; Alan Dubois, French FB; Nikolai Bibikov, auditory central nervous system mechanisms; Patricia Burrowes and Ignacio de la Riva, FB for conservation

Fig. 4.15 Frog Bioacousticians II. (**a**) Jürgen Rheinlaender with apparatus for testing 3-D sound localization in frogs; (**b**) Arne Schiøtz, African FB; (**c**) Kim Hoke, comparative neuroethology; (**d**) Miguel Vences, use of FB in integrative taxonomy; (**e**) Walt Wilczynski, hormonal effects on FB; (**f**) Margaret Stewart, FB of Puerto Rican and Malawian frogs; (**g**) Frank Glaw, Madagascar FB; (**h**) Sebastiaan Meenderink, frog neuroethology; (**i**) Heike Pröhl, neotropical FB; (**j**) T. Mitchell Aide, FB for conservation; (**k**) Jun-Xian Shen, Chinese leader in ultrasonic FB; (**l**) Matt Mason, inner and middle ear anatomy and physiology; (**m**) Wolfgang Walkowiak, frog acoustic neuroanatomy and physiology; (**n**) Andreas Elepfandt, water surface wave communication in frogs; (**o**) The Audio Utility Trailer developed by Neville Passmore and students in South Africa. It could be towed into place, making it a lion-proof frog lab

have concentrated on investigating these animals in their natural habitat, others who prefer laboratory work and still others who combine studies in both venues to discover the secrets of frog call function, their generation, their transmission through the natural habitat and the mechanisms that have evolved for their detection and extraction from noise by the amphibian auditory system. In some sense, they are arranged in chronological order, without intentional offense.

In this chapter, we have tried to present a brief history of frog bioacoustics. Starting with the earliest literary reference relevant to frog bioacoustics (Gilgamesh epos, 2100 BCE), we are taking it to the present (Lee et al. 2021). Obviously, it is impossible to cover all amphibian bioacoustics studies over this more than

Fig. 4.16 Frog Bioacousticians III. (**a**) Jakob Christensen-Dalsgaard, Danish FB leader; (**b**) Ximena E. Bernal, roles of eavesdropping natural enemies on amphibians, Photo: Mark Simmons; (**c**) Karen Warkentin, leader in frog biotremology; (**d**) Jaime Bosch, FB of EU frogs; (**e**) Gerlinde Höbel, evolution and function of anuran communication; (**f**) Mario Penna, Chilean FB leader; (**g**) Adolfo Amézquita, Colombian frog communication expert; (**h**) Ulmar Grafe, authority on SE Asian and Bornean frog calls and multimodal communication; (**i**) Rafael Márquez, Spanish FB leader; (**j**) Pamela Lopez, neotropical frog biology; (**k**) Luís Felipe Toledo, Brazilian leader in amphibian communication; (**l**) Joshua Schwartz, developed multi-speaker frog playback paradigm; (**m**) Pim van Dijk, Dutch leader in mechanisms underlying sound processing in the frog inner ear; (**n**) J. Mark Bee, neuroethology of the cocktail party effect in frogs

4000-year period and we apologize to those whose work was not included in this short review. What is clear is that since the advent of electronics and computers in the 1940s, the field has burgeoned. Sophisticated call synthesis and analysis programs have become widely available. Autonomous recordings of frog vocalizations have now been used to quantify acoustic biodiversity in a variety of habitats for taxonomic use (Köhler et al. 2017), for guiding conservation policy (Moreno-Gómez et al. 2019), and new experimental tools from biophysics and molecular biology (Cobo-Cuan et al. 2021) continue to make the study of amphibian bioacoustics a vibrant and exciting field.

Fig. 4.17 2022 generation of frog bioacoustic researchers and their current home base. (**a**) Felipe Moreno-Gómez, Chilean FB for conservation; (**b**) Ryan Taylor, FB, US; (**c**) Ariadna Cobo-Cuan, frog neuroethology, US; (**d**) Eva Ringler, neotropical frog biology, Switzerland; (**e**) Marcos Gridi-Papp, frog vocal production, US; (**f**) Ikkyu Aihara, chorus dynamics, Japan; (**g**) Rigoberto Solís, FB, Chile; (**h**) Matias Muñoz Sandoval, FB, Chile; (**i**) Fernando Vargas-Salinas, FB, Colombia; (**j**) Bibiana Rojas, FB, Austria; (**k**) Norman Lee, neuroethology of FB, US; (**l**) Jianguo Cui, FB, China; (**m**) Alejandro Vélez, evolution of FB, US; (**n**) Diego Llusia, FB, Spain; (**o**) Wouter Halfwerk, frog communication, The Netherlands. (Photo: Jussi Puikkonen); (**p**) Doris Preininger, SE Asian FB, Austria; (**q**) Sandra Goutte, evolution of FB, UAE; (**r**) Nelson Velásquez, FB, Chile; (**s**) Max Ringler, neotropical frog biology, Switzerland/Austria; (**t**) Valentina Zaffaroni-Caorsi, frog biotremology, Italy; (**u**) Luis Manuel Diaz Beltran, FB, Cuba; (**v**) Patricia (Yuki) Quiñones, frog inner ear, US; (**w**) Johana Goyes Vallejos, FB for phylogenetics, US; (**x**) Karen Beard, FB and invasive species, US; (**y**) Ariadne Angulo, FB and conservation, Canada. Not pictured: Jodi Rowley, using bioacoustics to help delineate species boundaries in cryptic groups of frogs, Australia; Rodrigo Alonso, FB, Cuba

Acknowledgments We thank Ariadna Cobo-Cuan for help with formatting Figs. 4.11 and 4.13. We thank Iris Adam for help with Figs. 4.3b and 4.5, and Karin Dreyer Jørgensen for correcting the translation of Jacobæus' Latin text.

Compliance with Ethics Requirements Peter Narins declares that he has no conflict of interest. Jakob Christensen-Dalsgaard declares that he has no conflict of interest. Carl Gerhardt declares that he has no conflict of interest.

References

Adler K, Narins PM, Ryan MJ (2013) Obituary Robert R Capranica (1931-2012) and the Science of Anuran Communication. Herpetol Rev 44(4):554–556

Adrian ED, Craik KJW, Sturdy RS (1938) The electrical response of the auditory mechanism in cold-blooded vertebrates. Proc R Soc London B 841:435–455

Aertsen AMHJ, Vlaming MSMG, Eggermont JJ, Johannesma PIM (1986) Directional hearing in the grassfrog (*Rana temporaria* L) II Acoustics and modeling of the auditory periphery. Hear Res 21:17–40

Albertus Magnus (c 1250) De animalibus, ch 26. Translated by KF Kitchell and IM Resnick (2018) On Animals, Ohio State University Press, Vol 2, p 1745

Aldrovandi U (1663) De quadrapedibus digitatis oviparis, book 2. Giambattista Ferroni, Bologna

Allen AA, Kellogg PP (1948) Voices of the Night (phonograph record). Cornell Laboratory of Ornithology, Ithaca

Arch VS, Grafe TU, Narins PM (2008) Ultrasonic signaling by a Bornean frog. Biol Lett 4:19–22

Arch VS, Grafe TU, Gridi-Papp M, Narins PM (2009) Pure ultrasonic communication in an endemic Bornean frog. *PloS One* 4(4):e5413. https://doi.org/10.1371/journal.pone.0005413

Aristophanes (405 BCE) Frogs, Translated by R Lattimore (1962) University of Michigan Press, Ann Arbor, MI

Aristotle (c 400 BCE) Historia Animalium, translated as History of Animals by D'Arcy Wenthworth Thompson Clarendon Press, Oxford

Ashcroft DW, Hallpike CS (1934) On the function of the saccule. J Laryngol 49:450–460

Axelrod FS (1960) Prolechthesomena to frog hearing. B.Sc. thesis, Dept. Biol., MIT

Bachtin GA, Bibikov NG (1974) Variation of sensitivity to interruption of a sound signal in adaptation of frog auditory-system. Soviet Phys Acoust USSR 19:387–388

Bee MA, Gerhardt HC (2001a) Neighbor-stranger discrimination by territorial male bullfrogs (*Rana catesbeiana*): I. Acoustic basis. Animal Behaviour 62:1129–1140

Bee MA, Gerhardt HC (2001b) Habituation as a mechanism of reduced aggression between neighboring territorial male bullfrogs. J Comp Psychol 115:68–82

Bibikov NG (1990) Response of frog midbrain neurons to tones amplitude-modulated by pseudorandom noise. Neurophysiology 22:180–186

Bibikov NG (2002) Addition of noise enhanced neural synchrony to amplitude-modulated sounds in the frog's midbrain. Hear Res 173:21–28

Bibikov NG, Nizamov SV (1996) Temporal coding of low-frequency amplitude modulation in the torus semicircularis of the grass frog. Hear Res 101:23–44

Blair WF (1955) Mating call and stage of speciation in the *Microhyla olivacea- M. carolinensis* complex. Evolution 9:469–480

Blair WF (1956) The mating call of hybrid toads (genus *Bufo*). Texas J Sci 8:350–355

Blair WF (1958a) Mating call in the speciation of anuran amphibians. *Am Nat* 92:27–51

Blair WF (1958b) Call structure and species groups in US treefrogs (*Hyla*). Southwestern Nat 5:129–135

Blair WF (1974) Character displacement in frogs. Am Zool 14:1119–1125

Blair WF, Littlejohn MJ (1960) Stage of speciation of two allopatric populations of chorus frogs (*Pseudacris*). Evolution 14:82–87

Bogart, JP, Wasserman AO (1972) Diploid-polyploid cryptic species pairs: a possible clue to evolution by polyploidization in anuran amphibians. Cytogenetics 11:7–24

Bogert CM (1960) The influence of sound on the behavior of amphibians and reptiles. In: Lanyon WE, Tavolga WN (eds) Animal sounds and communication. AIBS, Washington, DC, pp 137–320

Boulenger GA (1897–98) The tailless batrachians of Europe, vol 1-2. The Ray Society, London

Brenowitz EA, Rose GJ, Capranica RR (1985) Neural correlates of temperature coupling in the vocal communication system of the gray treefrog (*Hyla versicolor)*. Brain Res 359:364–367

Canatella DC, Hillis DM, Chippindale PT, Weigt L, Rand AS, Ryan MJ (1998) Phylogeny of frogs of the *Physalaemus pustulosus* species group, with an examination of data incongruence. Syst Biol 47:311–335

Capranica RR (1965) The evoked vocal response of the bullfrog: a study of communication by sound. MIT Press, Cambridge, MA

Capranica RR (1966) Vocal response of the bullfrog to natural and synthetic mating calls. J Acoust Soc Am 40:1131–1139

Capranica RR, Moffat AJM (1983) Neurobehavioral correlates of sound communication in anurans. In: Ewert J-P, Capranica RR, Ingle DJ (eds) Advances in vertebrate neuroethology. Plenum Press, London/New York, pp 701–730

Christensen-Dalsgaard J (2005) Directional hearing in non-mammalian tetrapods. In: Popper AN, Fay RR (eds) Sound source localization. Springer, New York, pp 67–123

Christensen-Dalsgaard J (2011) Vertebrate pressure-gradient receivers. Hear Res 273:37–45

Clack JA (1997) The evolution of tetrapod ears and the fossil record. Brain Behav Evol 50:198–212

Cobo-Cuan A, Quiñones PM, Macias-Escriva F, Oghalai JS, Narins PM (2021) Vibrometry of the frog inner ear: a comparative approach using optical coherence tomography. Abstr. 44th ARO Res Mtg.

Cocroft RB, Ryan MJ (1995) Patterns of mating call evolution in toads and chorus frogs. Anim Behav 49:283–303

Comparetti A (1789) In gymnasio patavino observationes anatomicæ de aure interna comparata. Patavii

Cutfield S (2016) Common lexical semantics in Dalabon ethnobiological classification. In: Austin PK, Koch H, Simpson J (eds) Language, land & song: studies in honour of Luise Hercus. EL Publishing, London, pp 209–227

Darwin C (1871) The descent of man and selection in relation to sex. John Murray, London

Davis MS (1987) Acoustically mediated neighbor recognition in the North American bullfrog, *Rana catesbeiana*. Behav Ecol Sociobiol 21:185–190

Deiters O (1862) Über das innere Gehörorgan der Amphibien. Archiv für Anatomie, Physiologie und wissenschaftliche Medicin 1862, pp 1–310

Du Verney JD (1683) Traité de l'organe de l'ouie: Contenant la structure, les usages, et les maladies de toutes les parties de l'oreille. E Michallet, Paris, France, 1683

Ehret G, Tautz J, Schmitz B, Narins PM (1990) Hearing through the lungs: lung eardrum transmission of sound in the frog *Eleutherodactylus coqui*. Naturwissenschaften 77:192–194

Ehret G, Keilwerth E, Kamada T (1994) The lung-eardrum pathway in three treefrog and four dendrobatid frog species: some properties of sound transmission. J Exp Biol 195:329–343

Feng AS (1981) Directional response characteristics of single neurons in the torus semicircularis of the leopard frog (*Rana pipiens*). J Comp Physiol 144:419–428

Feng AS, Capranica RR (1976) Sound localization in anurans I. Evidence of binaural interaction in dorsal medullary nucleus of bullfrog (*Rana catesbeiana*). J Neurophysiol 39:871–881

Feng AS, Capranica RR (1978) Sound localization in anurans II. Binaural interaction in superior olivary nucleus of the green tree frog (*Hyla cinerea*). J Neurophysiol 41:43–54

Feng AS, Narins PM, Capranica RR (1975) Three populations of primary auditory fibers in the bullfrog (*Rana catesbeiana*): their peripheral origins and frequency sensitivities. J Comp Physiol 100:221–229

Feng AS, Gerhardt HC, Capranica RR (1976) Sound localization behavior in the green treefrog (*Hyla cinerea*) and the barking treefrog (*Hyla gratiosa*). J Comp Physiol 107:241–252

Feng AS, Narins PM, Xu C-H (2002) Vocal acrobatics in a Chinese frog *Amolops tormotus*. Naturwissenschaften 89:352–356

Feng AS, Narins PM, Xu C-H, Lin W-Y, Yu Z-L, Qiu Q, Xu Z-M, Shen J-X (2006) Ultrasonic communication in frogs. Nature 440:333–336

Fouquette MJ (1975) Speciation in chorus frogs I. Reproductive character displacement in the *Pseudacris nigrita* complex. Syst Zool 24:16–22

Frishkopf LS, Goldstein M (1963) Responses to acoustic stimuli from single units in the eighth nerve of the bullfrog. J Acoust Soc Am 35:1219–1228

Frishkopf LS, Capranica RR, Goldstein M (1968) Neural coding in the bullfrog's auditory system A teleological approach. Proc IEEE 56:969–980

Gans C (1973) Sound production in the Salientia: mechanisms and evolution of the emitter. Am Zool 15:179–194

Gaupp E (1904) A Ecker's und R Wiedersheim's Anatomie des Frosches, part 3 F. Vieweg, Braunschweig

Gaupp E (1913) Die Reichertsche Theorie. Archiv für Anatomie und Entwicklungsgeschichte (Suppl) 1912:1–416

Geoffroy E-L (1755) Premiér Mémoire sur l' Organe de l'Ouie des Reptiles. Mémoires de Mathématique et de Physique. Présentés a l'Académie Royale de Sciences, par divers Savans, et lûs dans ses Assemblées. Tome 2. Paris

Gerhardt HC (1978) Temperature coupling in the vocal communication system of the gray treefrog *Hyla versicolor*. Science 199:992–994

Gerhardt HC (1994) Reproductive character displacement of female mate choice in the grey treefrog *Hyla chrysoscelis*. Anim Behav 47:959–969

Gerhardt HC (2015) Auditory selectivity for acoustic properties of conspecific mate-attracting signals in lower vertebrates and songbirds. Open Access Anim Physiol 2015:1–13

Gerhardt HC, Huber F (2002) Acoustic communication in insects and anurans: common problems and diverse solutions. University of Chicago Press, Chicago

Gerhardt HC, Mudry KM (1980) Temperature effects on frequency preferences and mating call frequencies in the green treefrog, *Hyla cinerea* (Anura: Hylidae). J Comp Physiol A 137:1–6

Gerhardt HC, Rheinlaender J (1982) Localization of an elevated sound source by the green tree frog. Science 217:663–664

Gerhardt HC, Schwartz JJ (2001) Auditory tuning and frequency preferences in anurans. In: Ryan MJ (ed) Anuran communication. Smithsonian Inst Press, Washington, pp 73–85

Gesner C (1586) Historia Animalium, libri 2: De quadripedibus oviparis Johann Wechel, Frankfurt [History of animals, book 2: Four-legged, egg-laying animals]

Gilgamesh (Anon, ca. 2100 BCE) Translation: Ferry, D. (1992). Gilgamesh: a new rendering in English verse. (p. 31) New York: Farrar, Straus and Giroux.

Girgenrath M, Marsh RL (1997) In vivo performance of trunk muscles in tree frog during calling. J Exp Biol 200:3101–3108

Gouk P (2004) English theories of hearing in the seventeenth century. In: Smith MM (ed) Hearing history: a reader. The University of Georgia Press, Athens/London

Grinnell AD, Gould E, Fenton MB (2016) A history of the study of echolocation. In: Fenton MB, Grinnell AD, Popper AN, Fay RR (eds) Bat bioacoustics. Springer, New York, pp 1–24

Hasse C (1868) Das Gehörorgan der Frösche. W Engelmann, Leipzig

Helmholtz H (1850) Vorläufiger Bericht über die Fortpflanzungs-Geschwindigkeit der Nervenreizung Archiv für Anatomie. Physiologie und wissenschaftliche Medicin 1850:71–73

Hetherington TE (1992) The effects of body size on the evolution of the amphibian middle ear. In: Webster DB, Popper AN, Fay RR (eds) The evolutionary biology of hearing. Springer, New York, pp 421–438

Höbel G, Gerhardt HC (2003) Reproductive character displacement in the acoustic communication system of green tree frogs (*Hyla cinerea*). Evolution 57(4):894–904

Holmes FL (1993) The Old Martyr of science: the frog in experimental physiology. J Hist Biol 26:311–328

Jacobæus O (1676) De Ranis Paris, France (in Latin)

Jørgensen MB, Gerhardt HC (1991) Directional hearing in the gray tree frog *Hyla versicolor*: eardrum vibrations and phonotaxis. J Comp Physiol A 169:341–347

Keller MJ, Gerhardt HC (2001) Polyploidy alters advertisement call structure in gray treefrogs. Proc R Soc London 268:341–345

Kingsbury BF, Reed HD (1909) The columella auris in amphibians. J Morphol 20:549–628
Koenig W, Dunn HK, Lacy LY (1946) The sound spectrograph. J Acoust Soc Am 18:19–49
Köhler J, Jansen M, Rodríguez A, Kok PJR, Toledo LF, Emmrich M, Glaw F, Haddad CFB, Rödel MO, Vences M (2017) The use of bioacoustics in anuran taxonomy: theory, terminology, methods and recommendations for best practice. Zootaxa 4251(1):001–124. https://doi.org/10.11646/zootaxa.4251.1.1
Koyama H, Lewis ER, Leverenz EL, Baird RA (1982) Acute seismic sensitivity in the bullfrog ear. Brain Res 250:168–172
Krogh A (1929) Progress of physiology. Am J Physiol 90:243–251
Lee N, Christensen-Dalsgaard J, White LA, Schrode KM, Bee MA (2021) Lung mediated auditory contrast enhancement improves the signal-to-noise ratio for communication in frogs. Curr Biol 31:1–11
Lemmon EM (2009) Diversification of conspecific signals in sympatry: geographic overlap drives multidimensional reproductive character displacement in frogs. Evolution 63:1155–1170
Lettvin JY, Maturana HR (1960) Hearing senses in the frog. Mass Inst TechnolResLab ElectronQuartProgrRept 57:167–168
Lettvin JY, Maturana HR, McCullouch WS, Pitts WH (1959) What the frog's eye tells the frog's brain. Proc Inst Radio Eng (now IEEE) 47:1940–1951
Lewis ER, Baird RA, Leverenz EL, Koyama H (1982a) Inner ear: dye injection reveals peripheral origins of specific sensitivities. Science 215:1641–1643
Lewis ER, Leverenz EL, Koyama H (1982b) The tonotopic organization of the bullfrog amphibian papilla, an auditory organ lacking a basilar membrane. J Comp Physiol A 145:437–445
Lewis ER, Narins PM, Cortopassi KA, Yamada WM, Poinar EH, Moore SW, Yu X-L (2001) Do male white-lipped frogs use seismic signals for intraspecific communication? Am Zool 41:1185–1199
Littlejohn MJ (1965) Premating isolation in the *Hyla ewingi* complex (Anura: Hylidae). Evolution 19:234–243
Littlejohn MJ (1981) Reproductive isolation: a critical review. In: Atchley WR, Woodruff DS (eds) Evolution and speciation essays in honor of MJD White. Cambridge University Press, New York, pp 613–634
Littlejohn MJ (1993) Homogamy and speciation: a reappraisal. Oxford Surv Evol Biol 9:135–165
Littlejohn MJ (1999) Variation in advertisement calls of anurans across zonal interactions: the evolution and breakdown of homogamy. In: Foster SA, Endler JA (eds) Geographic Variation in Behavior. Oxford University Press, New York, pp 209–233
Littlejohn MJ, Loftus-Hills JJ (1968) An experimental evaluation of premating isolation in the *Hyla ewingi* complex (Anura: Hylidae). Evolution 22:659–663
Littlejohn MJ, Main AR (1959) Call structure in two genera of Australian burrowing frogs. Copeia 1959:266–270
Littlejohn MJ, Michaud TC (1959) Mating call discrimination by females of Strecker's chorus frog (*Pseudacris streckeri*). Texas J Sci 11:86–92
Littlejohn MJ, Roberts JD (1975) Acoustic analysis of an intergrade zone between two call races of the *Limnodynastes tasmaniensis* complex (Anura: Leptodactylidae). Aust J Zool 23:113–122
Littlejohn MJ, Watson GF (1973) Mating-call variation across a narrow hybrid zone between *Crinia laevis* and *C. victoriana* (Anura: Leptodactylidae). Aust J Zool 21:277–284
Loftus-Hills JJ (1973) Comparative aspects of auditory function in Australian anurans. Aust J Zool 21:253–267
Loftus-Hills JJ (1974) Analysis of an acoustic pacemaker in Strecker's chorus frog *Pseudacris streckeri* (Anura: Hylidae). J Comp Physiol 90:75–87
Loftus-Hills JJ, Johnstone BM (1970) Auditory function, communication, and the brain-evoked response in anuran amphibians. J Acoust Soc Am 42:1131–1138
Loftus-Hills JJ, Littlejohn MJ (1971a) Pulse-repetition rate as the basis for mating call discrimination by two sympatric species of *Hyla*. Copeia 1971:154–156

Loftus-Hills JJ, Littlejohn MJ (1971b) Mating call sound intensities of anuran amphibians. J Acoust Soc Am 39:1327–1329

Loftus-Hills JJ, Littlejohn MJ (1992) Reinforcement and reproductive character displacement in *Gastrophryne carolinensis* and *G. olivacea* (Anura: Microhylidae): a reexamination. Evolution 46:896–906

Lombard RE, Bolt JR (1979) Evolution of the tetrapod ear: an analysis and reinterpretation. Biol J Linn Soc 11:19–76

Lopez PT, Narins PM (1991) Mate choice in the neotropical frog *Eleutherodactylus coqui*. Anim Behav 41:757–772

Maier W, Ruf I (2016) The anterior process of the malleus in Cetartiodactyla. J Anat 228:313–323

Marsh RL (1999) Contractile properties of muscles used in sound production and locomotion in two species of gray treefrog. J Exp Biol 202:3215–3223

Martin WF (1971) Mechanics of sound production in the genus *Bufo*: passive elements. *J Exp Zool* 176:273–294

Martin WF (1972) Evolution of vocalizations in the genus *Bufo*. In: Blair WF (ed) Evolution in the Genus *Bufo*. University of Texas Press, Austin, pp 279–308

Martin WF, Gans C (1972) Muscular control of the vocal tract during release signaling in the toad *Bufo valliceps*. J Morphol 137:1–18

Martof BS, Thompson EF Jr (1958) Reproductive behavior of the chorus frog *Pseudacris nigrita*. Behaviour 13:243–257

Mason MJ, Narins PM (2002) Vibrometric studies of the middle ear of the bullfrog (*Rana catesbeiana*) II: the operculum. J Exp Biol 205:3167–3176

Meenderink SWF, Kits M, Narins PM (2010) Frequency matching of vocalizations to inner-ear sensitivity along an altitudinal gradient in the coqui frog. Biol Lett 6:278–281

Moreno-Gómez FN, Bartheld J, Silva-Escobar AA, Briones R, Márquez R, Penna M (2019) Evaluating acoustic indices in the Valdivian rainforest, a biodiversity hotspot in South America. Ecol Indic 103:1–8

Mudry KM, Capranica RR (1987) Correlation between auditory thalamic area evoked responses and species-specific call characteristics II. *Hyla cinerea* (Anura: Hylidae). J Comp Physiol A 161:407–416

Müller J (1835) Handbuch der Physiologie des Menschen für Vorlesungen, Bd 1, 2nd edn. Hölscher, Bonn

Narins PM (2016) ICE on the road to auditory sensitivity reduction and sound localization in the frog. Biol Cybern 110:263–270

Narins PM, Capranica RR (1976) Sexual differences in the auditory system of the treefrog, *Eleutherodactylus coqui*. Science 192:378–380

Narins PM, Capranica RR (1978) Communicative significance of the two-note call of the treefrog *Eleutherodactylus coqui*. J Comp Physiol 127:1–9

Narins PM, Lewis ER (1984) The vertebrate ear as an exquisite seismic sensor. J Acoust Soc Am 76:1384–1387

Narins PM, Smith SL (1986) Clinal variation in anuran advertisement calls: basis for acoustic isolation? Behav Ecol Sociobiol 19:135–141

Narins PM, Ehret G, Tautz J (1988) Accessory pathway for sound transfer in a neotropical frog. Proc Natl Acad Sci 85:1508–1512

Narins PM, Feng AS, Schnitzler H-U, Denzinger A, Suthers RA, Lin W, Xu C-H (2004) Old World frog and bird vocalizations contain prominent ultrasonic harmonics. J Acoust Soc Am 115:910–913

Narins PM, Feng AS, Fay RR, Popper AN (2007) Hearing and sound communication in amphibians. Springer, New York

Passmore NI, Capranica RR, Telford SR, Bishop PJ (1984) Phonotaxis in the painted reed frog (*Hyperolius marmoratus*): the localization of elevated sound sources. J Comp Physiol A 154:189–197

Phelps SM, Ryan MJ (1998) Neural networks predict response biases in female túngara frogs. Proc R Soc London Ser B 265:279–285

Pliny the Elder [Gajus Plinius secundus] (1855) Natural history, translation J Bostock. Taylor and Francis, Red Lion Court, Fleet Street

Rand AS (2001) A history of frog call studies 405 B C to 1980. In: Ryan MJ (ed) Anuran communication. Smithsonian Inst Press, Washington, pp 8–19

Reichert C (1837) Über die Visceralbogen der Wirbelthiere im Allgemeinen und deren Metamorphosen bei den Vögeln und Säugethieren. Archiv Anat Physiol Wiss Med 1837:120–222

Retzius G (1881) Das Gehörorgan der Fische und Amphibien. In: Das Gehörorgan der Wirbelthiere, vol 1. Samson & Wallin, Stockholm

Reyes Campos NB (1971) Observaciones sobre la conducta del coqui, *Eleutherodactylus coqui*. Caribbean J Sci 11:209

Rheinlaender J, Gerhardt HC, Yager D, Capranica RR (1979) Accuracy of phonotaxis by the green treefrog (*Hyla cinerea*). J Comp Physiol 133:247–255

Rose GJ, Capranica RR (1984) Processing of amplitude modulated sounds by the auditory midbrain of two species of toads: matched temporal filters. J Comp Physiol A 154:211–219

Rose GJ, Brenowitz EA, Capranica RR (1985) Species specificity and temperature dependency of temporal processing by the auditory midbrain of two species of treefrog. J Comp Physiol A 157:763–769

Ryan MJ (1985) The Túngara frog: a study in sexual selection and communication. University of Chicago Press, Chicago

Ryan MJ (1990) Sensory systems, sexual selection, and sensory exploitation. Oxford Surv Evol Biol 7:157–195

Ryan MJ, Rand AS (1990) The sensory basis of sexual selection for complex calls in the túngara frog, *Physalaemus pustulosus* (sexual selection for sensory exploitation). Evolution 44:370–383

Ryan MJ, Perrill SA, Wilczynski W (1992) Auditory tuning and call frequency predict population based mating preferences in the cricket frog, *Acris crepitans*. Am Nat 139:1370–1383

Scarpa A (1789) Disquisitiones anatomicae de auditu et olfactu. Pietro Galeati, Ticino

Schmidt RS (1973) Central mechanisms of frog calling. Am Zool 13:1169–1177

Schneider H (1988) Peripheral and central mechanisms of vocalization. In: Fritzsch B, Ryan MJ, Wilczynski W, Hetherington TE, Walkowiak W (eds) The evolution of the amphibian auditory system. Wiley, New York, pp 537–558

Shen J-X, Xu Z-M, Feng A, Narins PM (2011) Large odorous frogs (*Odorrana graminea*) produce ultrasonic calls. J Comp Physiol A 197:1027–1030

Simmons AM, Ketten DR (2020) How a frog hears. Acoustics Today 16 (issue 4):67–74

Suthers RA, Narins PM, Lin W, Schnitzler H-U, Denzinger A, Xu C-H, Feng AS (2006) Voices of the dead: complex nonlinear vocal signals from the larynx of an ultrasonic frog. J Exp Biol 209:4984–4993

Taigen TL, Wells KD (1985) Energetics of vocalization by an anuran amphibian (*Hyla versicolor*). J Comp Physiol B 155:163–170

Tait J, McNally WJ (1934) Some features of the action of the utricular maculæ (and of the associated action of the semicircular canals) of the frog. Philos Trans R Soc London B 224:241–286

Tárano Z, Ryan MJ (2002) No pre-existing biases for heterospecific call traits in the frog *Physalaemus enesefae*. Anim Behav 64:599–607

van Hemmen JL, Christensen-Dalsgaard J, Carr CE, Narins PM (2016) Animals and ICE: meaning origin and diversity. Biol Cybern 110:237–246

Welch AM, Semlitsch RD, Gerhardt HC (1998) Call duration as an indicator of genetic quality in male gray treefrogs. Science 280:1928–1930

Wells KD (1977) The social behaviour of anuran amphibians. Anim Behav 25:666–693

Wells KD (2007) The ecology and behavior of amphibians. University of Chicago Press, Chicago/London

Wells KD, Schwartz JJ (1984) Vocal communication in a neotropical treefrog, *Hyla ebraccata*: advertisement calls. Anim Behav 32:128–145

Wells KD, Taigen TL (1986) The effect of social interactions on calling energetics in the gray treefrog (*Hyla versicolor*). Behav Ecol Sociobiol 19:9–18

Wilczynski W, Resler C, Capranica RR (1987) Tympanic and extratympanic sound transmission in the leopard frog. J Comp Physiol 161:659–669

Wilczynski W, Rand AS, Ryan MJ (2001) Evolution of calls and auditory tuning in the *Physalaemus pustulosus* species group. Brain Behav Evol 58:137–151

Windischmann CJ (1831) De penitiori Auris in Amphibiis structura Dissertatio inauguralis anatomica. Leopold Voss, Leipzig

Yerkes RM (1905) The sense of hearing in frogs. J Compar Neurol Psychol 15:279–304

Yu X-L, Lewis ER, Feld D (1991) Seismic and auditory tuning curves from bullfrog saccular and amphibian papillar axons. J Comp Physiol 169:241–248

Chapter 5
The History of Auditory Research in Lizards

Geoffrey A. Manley

5.1 Introduction

Until the second half of the nineteenth century, authors supported the "ladder" hypothesis of Aristotle: Vertebrate groups were arranged in the supposedly historical sequence of: amphibians, lizards, crocodilians, birds, mammals (reading "which gave rise to" for each comma!). Thus, the ears of lizards were viewed as the forerunners of the ears of crocodilians and these of the ears of birds, etc. Today, remnants of this long defunct hypothesis on vertebrate systematics remain, in the tendency to incorrectly and misleadingly refer to anything other than birds and mammals as "lower vertebrates," where birds and mammals are "higher vertebrates." It has been known for decades that mammalian ancestors were the *first* to diverge from "stem reptiles," so they arose before the ancestors of lizards, crocodilians, turtles, and birds (Kemp 2005). Far from sequentially, each modern land vertebrate group diverged separately from stem ancestors and evolved in parallel to others for 250 million years. This enormously influences how the evolution of middle and inner ears is viewed (Manley 2016).

Squamate vertebrates (lizards or Sauria, and snakes or Serpentes; the name means "having scales"), with more than 11,000 described species (Utz et al. 2022), form the second largest group of vertebrates (after fishes). Snakes contribute about one-third of squamate species. Research on the ears of lizards began in the second half of the eighteenth century. At that time, and until the first quarter of the twentieth century, most anatomists published their work in German. Howard Ayers (1892) is a major exception and worked in the USA. Interestingly, his "paper" (360 pages

G. A. Manley (✉)
Cochlear and Auditory Brainstem Physiology, Department of Neuroscience, School of Medicine and Health Sciences, Cluster of Excellence "Hearing4all", Carl von Ossietzky University Oldenburg, Oldenburg, Germany
e-mail: geoffrey.manley@uol.de

D. R. Ketten et al. (eds.), *A History of Discoveries on Hearing*, Springer Handbook of Auditory Research 77,
https://doi.org/10.1007/978-3-031-41320-9_5

long, not untypical for the time!) contained 295 references, of which about 90% were in German, the other 10% in Danish, Dutch, French, Italian, Latin, and Swedish. By the mid-twentieth century, partly as a result of the loss of university personnel through two world wars, the language of publications had changed dramatically in favor of English. Two examples, which also show shortened manuscripts, are: Grey (1955; 31 pages, 26 references, of which 20% were in German); Shute and d'A Bellairs (1953; 19 pages, 13 references, 15% in German). The great length of early papers was partly due to excessive—and often disrespectful—discussion of previous work.

In the nineteenth century, when vertebrate systematics was still in its infancy, lizards were often included under the group "Amphibians." Even today, the term "Herpetology" is used for the combined study of Amphibians and "Reptiles" (the latter is a group no longer recognized as closely related). In modern systematics, squamates are a group separated not only from amphibians but also from other groups previously known as "reptiles" (e.g., crocodilians and turtles). Squamate ancestors separated from all other land vertebrates in the early Mesozoic, before the evolution of middle ears, and this had profound consequences for the (separate) evolution of their unique inner ears (Manley 2016). Snakes (Serpentes) evolved about 150 million years ago. They partially lost their middle ears, which resulted in a simplified inner-ear hearing organ.

Like many other fields of science, the history of approaches to studying hearing research is mainly the history of technical advances. Since 1850, anatomical studies progressed from the rough, magnifying-glass level to cellular ultrastructure. Since about 1950, biochemical studies moved from the crude to the molecular, and physiological studies moved from the use of mm-sized electrodes to microelectrode studies of responses of single neurons or ion channels using computer-controlled stimuli in real time.

5.2 Authors Before 1900

Early authors had broad interests and discussed both hearing and vestibular systems of all vertebrate groups. Clason (1873) and Jan Versluys (1898), who concentrated on lizards, were exceptional in that respect. According to Edward Clason (1873), Andreae Camparetti (1789) was the first to draw the middle ear of lizards. Considering that in the mid-nineteenth century, Schleiden and Schwann's cell theory was quite new, it is not surprising that older anatomical descriptions showed no cellular details of organs such as the auditory papilla. Chemical fixation was either not used or inadequate to preserve cellular details, and magnification was mostly via hand lenses. There was also no useful laboratory micro-photography, until it became practical on a small scale in the late nineteenth century, and early anatomical papers are illustrated only by—often quite beautiful and detailed—hand drawings (Fig. 5.1). In 1851, Corti described the mammalian inner-ear organ that soon came to bear his name.

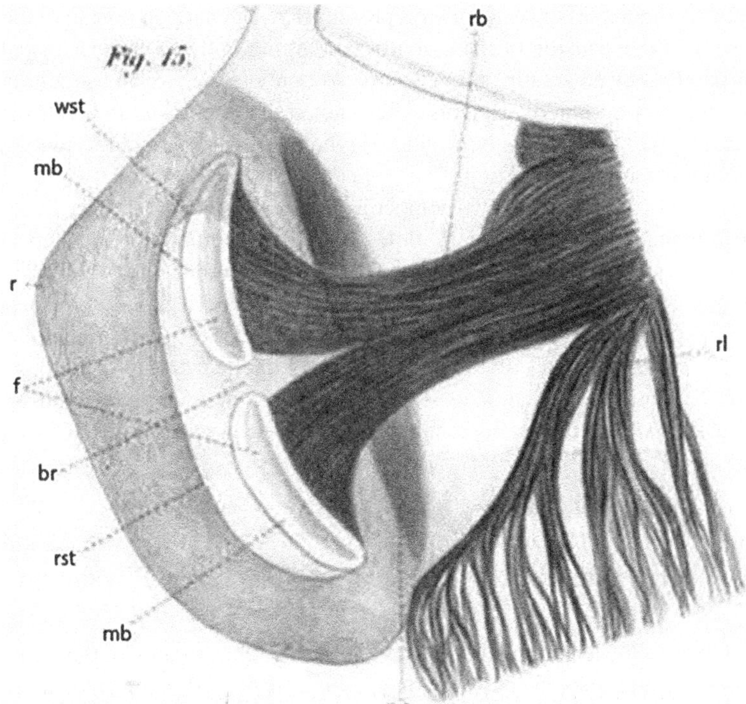

Fig. 5.1 A drawing from Retzius (1884, plate VIII, his Fig. 15) of the cochlear duct of the European green lizard *Lacerta viridis*, showing both that the papilla basilaris is divided into two sub-papillae and that the auditory nerve splits into two branches. *Br*—bridge (hiatus) between sub-papillae; *f*—sub-papillae; *mb*—basilar membranes; *r*—edge of inner ear; *rb*—auditory nerve; *rl*—branches of the lagenar nerve; *rst, wst*—edges of scala tympani

According to Gustav Retzius (1884), some earlier authors had studied lizard ears, but most of these works are not, or not easily, available; further details can be found in Retzius' famous work (1884, pp. 55–70). After discussing the ears of chelonians and snakes, Retzius (1884) described previous work on lizard ears by at least 20 authors of the 18th and 19th centuries. Of these, only Camparetti (1789), Clason (1873), and Retzius (1884) were available to me. On modern standards, such work on lizard ears is of limited use in a detailed comparative discussion.

By Retzius' (1884) time, cellular detail was observable. His drawings illustrate that he was able to see significant height differences between the tall and short hair cells of Crocodilians as compared to the more gradual changes in birds. Many terms were already established, such as papilla basilaris and tegmentum vasculosum, but little was known about lizard ears. It was clear that there is a basic correspondence among the inner ears of most vertebrates, especially in the structures of the vestibular system. Each vestibular component had a recognizable structure in all species, although size differences in for example, the saccular macula, and differences in diameters and curvatures of the semicircular canals were found, and some maculae (e.g., macula neglecta) were not always present. The lagenar component of the inner ear, a sack-like expansion of the sacculus containing both the lagenar macula and basilar papilla, varied greatly among lizards, but only in the size and structure of the basilar papilla. Importantly, Retzius (1884) noted that there was so much variation in lizard papilla basilaris that he wondered whether it was even fair to place lizards together in one systematic group!

Versluys (1898) discussed two competing hypotheses concerning the histories of non-mammalian and mammalian middle ears. The hypothesis of Reichert (1837), who saw the incus and malleus of mammals as new components of the middle ear but homologous with the stem reptile bones articular and squamosal, respectively, was less explanatory to Versluys than Ernst Gaupp's (1898) idea. Gaupp believed that the non-mammalian columella and extracolumella were homologous to the chain of three ossicles in mammals. Today, it is the Reichert hypothesis that receives universal support from developmental and paleontological studies (Clack 1997; Luo 2007).

5.3 The Transition to Modern Times and Early Physiology

Following Retzius' lead, De Burlet (1929) noted that the structure of the lizard inner ear varies considerably, the cochlear duct being especially long in geckos. In 1936, Evans also studied the inner ears of geckos and concluded that the size differences of the lagenar macula and the basilar papilla were indicative of evolutionary "progress" in hearing. Some lizards—including geckos—crossed what he called the *vocality threshold* among amniotes, which suggests that when the basilar papilla is longer than the lagenar macula, vocality is common due to an improved sense of hearing. In the absence of physiological studies, however, it was not even known whether the lagena or the basilar papilla were the main sound receptors (it is the latter alone—see Sect. 5.5.2).

5.3.1 Modern Anatomical Studies (1950 Onward)

Hamilton (1964) states (p. 255): "From the time of de Burlet ('29) until Shute and Bellairs ('53) published their paper on the cochlear apparatus in Geckonidae and Pygopodidae (geckos and legless geckos), no major papers were published on the inner ear in lizards." This curious 20-year lapse "… is best explained … by the fact that the amphibian ear is more amenable to experimentation than that of lizards." He did not, however, explain why lizard ears were less amenable.

Shute and d'A Bellairs (1953) noted that the variability of the basilar papilla in geckos could not easily be related to differences in species' way of life and, as other authors, suggested using lizard inner ears as key features in species' classification. They wrote (p. 695): "…we have observed certain striking features of the cochlear limbus which appear to be characteristic of two families of lizards, the Gekkonidae and the Pygopodidae .... Although the differences ... may well have some relationship to their powers of hearing, it is seldom possible to explain them in terms of adaptive modification.... Observations on ... the inner ear are therefore likely to have a special significance in a consideration of the phylogenetic history and affinities of the groups studied." They used the term "limbus" for the basilar membrane support in lizards and noted that it is "ring-like." They also noted a width change along the basilar membrane and asked whether this is associated with changes in frequency perception, as in mammals. They missed, however, the unique hair cell arrangements in both geckos and their subfamily, the pygopods. Underwood (1957) also worked on geckos, but referred to: "The 'class' Reptilia (as) that motley assortment of all the amniotes which are neither mammals nor birds." Obviously, at that time, the term "Reptile" was already falling into disrepute among zoologists.

Hamilton (1964, p. 264) wrote of lizard species, "The most apparent differences in the otic system involve the cochlear duct." He noted that (p. 255): "... lizards tend to present a rag-bag of characters, some ... considered primitive and others specialized (or highly evolved).... This has obfuscated phylogenetic relationships here much more than in other animal groups where the paleontological record is more complete." Hamilton (1960) also emphasized the use of inner-ear characteristics in the systematics of lizards. Interestingly, he, like Shute and d'A Bellairs (1953), noted that the inner ear of xantusids, which are "generally regarded as having affinities with Gekkonidae," lacks the massive, unique limbic lip of geckos. Instead "... Scincidae and Xantusiidae ... inner ears are remarkably alike" (Hamilton 1964). Malcolm Miller (1992) noted the same. Systematics sometimes put the xantusiids closer to skinks (Estes et al. 1988; Vidal and Hedges 2009), but mitochondrial phylogeny (Lee 1998) correctly placed them closer to the geckos.

It was not until Schmidt's paper (1964) that formal recognition was given to the usefulness of lizard cochlear anatomy in systematics and phylogenetic studies. Although within most major groups the cochlea shows only slight variation, the lizards form a major exception. "The ... histology of the cochlea of a number of lizards are compared. There are conspicuous variations in cochlear anatomy... *The histological characteristics of the cochlea are of great potential value and have only begun to be utilized for phylogenetic studies*" (Schmidt 1964, p. 542, my italics). It is the recognition of the systematic variety of otic anatomy within the lizards that led to the decision to concentrate this chapter on lizards, rather than the so-called Reptiles.

Lizard ear studies received a great boost during the 1960s, as Glen Wever, Irwin Baird, and Malcolm Miller (all in the USA) followed Schmidt and carried out broadly based anatomical work. Wever's group began anatomical and physiological studies on lizards. Wever et al. (1964) used cochlear microphonic (CM) measurements and found a correlation between the number of hair cells and the maximum CM output. They claimed that lizards having larger papillae: "...have reached a level of development in which there is both intensity and frequency differentiation of considerable degree" (p. 1174). The CM recording method, however, has one general drawback: An arbitrary voltage amplitude must be chosen as a "threshold," assuming both that it is comparable across all frequencies and that it is related to the animal's behavioral threshold. As discussed in Sect. 5.5.1 in more detail, both assumptions are unjustified and recording the CM was especially unsuitable for studying lizard ears. Thus, although Wever's extensive anatomical work was really useful, his physiological data (summarized in Wever 1978) were not.

Baird (1966) reported (p. 433) that "Preliminary investigation ... of the reptilian cochlear duct indicates considerable specialization at the cellular level and the need for continuing anatomical and physiological study" since "detailed information ... is extremely limited." Even as he wrote, both Miller and Wever were working on the anatomical details necessary to understand future physiological data. Fortunately, Miller, who used transmission and scanning electron microscopy, and Wever, who used light microscopy, had chosen complementary anatomical techniques that facilitated later comparisons.

5.3.2 Anatomical Studies of Miller and Wever

Before entering a discussion of the differences between lizard papillae as revealed by Miller and Wever's studies, it is appropriate to describe the structure of a generalized lizard papilla. The auditory organ of lizards is within a set of three tubes that form part of the labyrinth, including the vestibular system. In these organs, the sensory epithelia are covered on their upper side—where the hair cells have their stereovillar bundles—by a unique body fluid called endolymph. Outside the endolymph tube containing the sensory organs are two further tubes connected at the cochlear apex and filled with perilymph. The latter is a "normal" body lymph, and in that, it contains much sodium chloride, whereas endolymph contains little sodium and much potassium chloride. The auditory organ is, on average, only a couple of millimeters in length, and sits upon the basilar membrane that covers the area between the arms of the limbic ring. The ring is donut-shaped or much more elongated than that, and the basilar membrane stretches between its walls like a drum membrane. On the brain side of the organ, the nerve fibers of the auditory nerve are organized into very flat bands of fibers that traverse the wall of the "donut" limbus and run through the neural limbus and basilar membrane to the sensory hair cells. The hair cell structure varies systematically from one end of the papilla to the other. Apically, the hair cells are always covered by tectorial material and the hair cell stereovillar bundles are short. These hair cells respond to frequencies below 1 kHz. Within the cochlear duct at this apical end, there is also a second, non-auditory vestibular organ arching over the basilar papilla, the lagena macula. After a transition area, the hair cells change their shape, and the stereovillar bundles more reflect the response frequencies of the hair cells (up to more than 10 kHz); the highest-frequency hair cells have the shortest bundles. This high-frequency area may or may not (depending on the lizard family) be covered by tectorial material. If tectorial material is lacking, the hair cell bundles are much taller. A scanning electron microscope picture illustrating the typical lizard papillar structure is shown in Fig. 5.2.

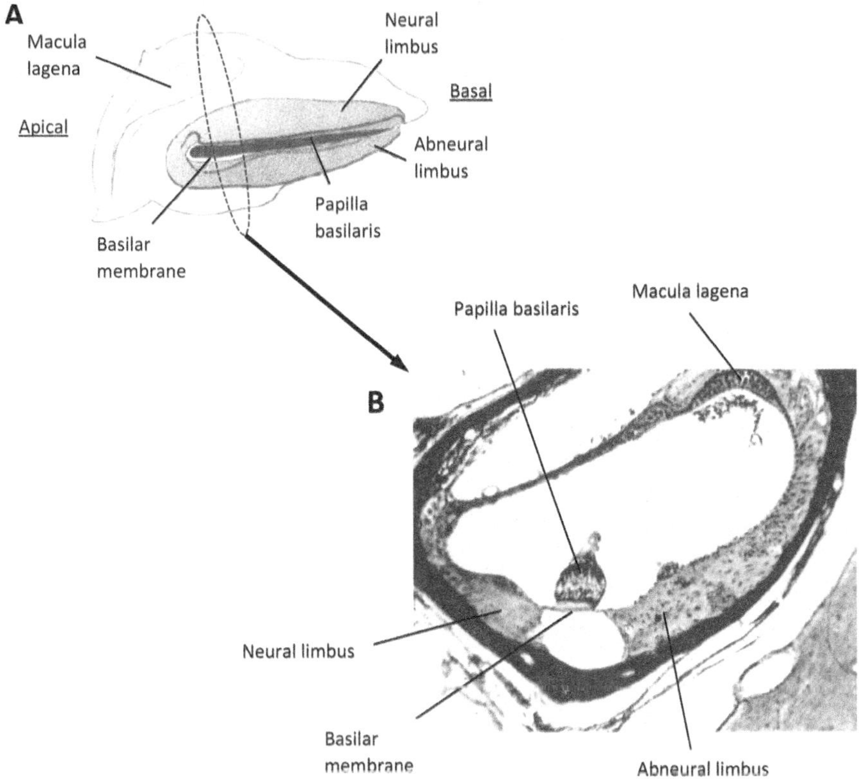

Fig. 5.2 (**a**) The upper surface of a relatively elongated lizard papilla (>1 mm) embedded in its limbus. (Modified from: Miller (1992)). (**b**) The transverse section taken at the position of the dotted circle in (**a**), showing the basilar papilla seated on the basilar membrane that is supported on each side by a thick limbus. (From Wever (1974))

In an early work, Miller (1966) noted that "The differences in the structure of the cochlear duct are related to both the acoustical capacities and the taxonomic relationships of certain lizards and snakes. … *Each family of lizards has a morphologically characteristic cochlear duct*"… (p. 421) (my italics). Miller studied 150 genera of lizards and 130 genera of snakes. His 1966 paper dealt with the gross morphology of the basilar papilla and provided initial impressions of each. Between lizard families, he described differences in the shape and size of the cochlear duct, in the development of the neural limbus into a ridge, bar, or lip-like process, and in the length and shape of the papilla basilaris. He noted that the cochlear duct is distinctively different for each lizard family, and generic differences, when they exist, never exceed the limits of a characteristic family type. There are, nevertheless, similarities in the cochlear ducts of related families and these aid in understanding taxonomic relationships.

In contrast to lizard inner ears, Miller showed that variations in snake cochlear ducts are more related to habitat than to family. The limbus and papillae basilaris of snakes are—regardless of family relationship—most elongated in burrowing species, moderately elongated or ovoid in terrestrial species, and small or reduced in arboreal and aquatic species. Since we know nothing regarding the physiological significance of these anatomical differences, there is no useful basis to discuss snake hearing organs.

Miller's many anatomical studies of lizard basilar papillae were summarized in 1980 and 1992. He showed that (Fig. 5.3): "The papillae (of lizards) vary greatly in length and number of hair cells, type of tectorial cover, the specific disposition of unidirectional and bidirectional hair cell groups, the ratio of the papillar nerve-fiber number to hair cell number (from 0.8/1 to 12/1), and the pattern of afferent nerve-fiber innervation of the hair cells. The structural details of the duct and papilla … differentiate one lizard family from another. However, similarities in duct and papilla structure of certain lizard families indicate close taxonomic relationship (1992, p. 485)."

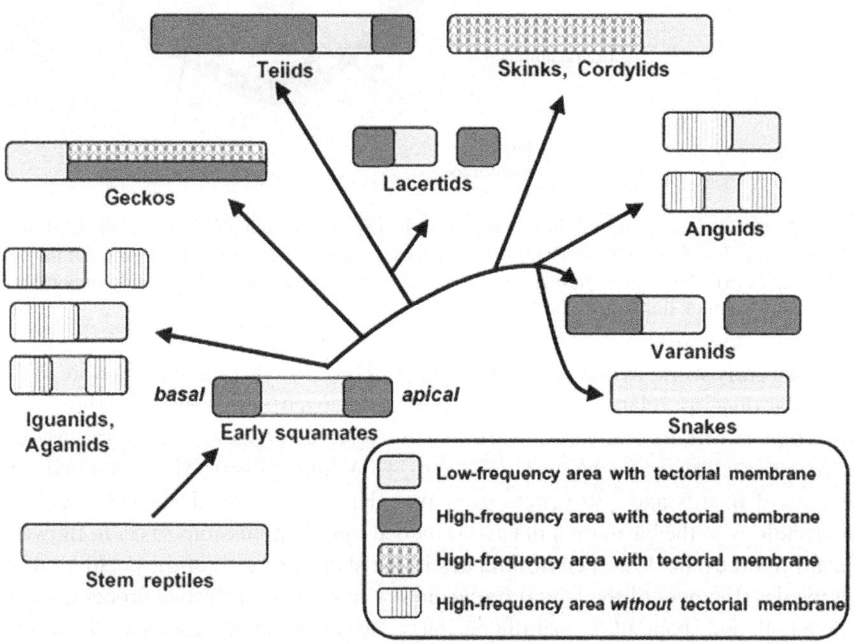

Fig. 5.3 Schematic illustration of the morphological patterns of hair cell and tectorial types in major lizard families and the putative sequences of their evolution. From stem reptiles of the early Mesozoic (*Early squamates*), a tripartite papilla arose and represents the proto-papilla of lizards. From this, various lineages arose in which the hair cell and tectorial patterns and distributions became typical for individual families or family groups. In lizards, "high" frequencies are those above ~1 kHz. (After Fig. 7.8 of Manley (2004a) and used with permission)

Miller (1973a, b, 1974, 1980, 1985, 1992) characterized major anatomical features of lizard papillae:

1. The existence of two types of hair cells distinguished by cytological features, one in the (later shown to be) low-frequency area of the papilla and the other in the higher-frequency area(s).
2. In general, the stereovillar bundles of low-frequency hair cells are all oriented in one direction, that is, on each hair cell surface, the wedge-shaped stereovillar bundle slopes down toward the side of the papilla with the entrance of the nerve fibers (Fig. 5.4; unidirectionally abneural—his "UHC" type). In the high-frequency hair cell areas, the bundles of different cells are oriented in this or in the opposite direction (bidirectional—"BHC" type). The directional sensitivity of hair cell bundles was confirmed by study of their transduction tip links (Pickles et al. 1989). In some families, hair cell groups are systematically arranged on the papilla (e.g., geckos); in other families, the arrangement is much less orderly (e.g., teiids; Fig. 5.4). Exceptions occur and Miller later placed less emphasis on bundle orientations and more on the cells' overall cytology. Thus in skinks, the low-frequency hair cell areas are bidirectionally oriented, although clearly of the UHC cytological type (thus he changed his designating term to "UDT or unidirectional type"). Other such exceptions exist, and BHC became "BDT type."
3. There are either two or three hair cell areas along lizard papillae, only one of which responds to low frequencies. The other areas are of the high-frequency type and, when there are three areas in total, the two high-frequency areas flank the low-frequency area apically and basally.
4. In some families (e.g., lacertids, see also Fig. 5.1), sub-papillae exist that are separated by limbic tissue. In varanids, the sub-papillae are less clearly separated. Independently evolved sub-papillae are also found in basiliscs (an iguanid subfamily).
5. By comparing "reptilian" groups outside of lizards (e.g., papillae of turtles), Miller concluded that ancestral papillae consisted only of UDT hair cells. He

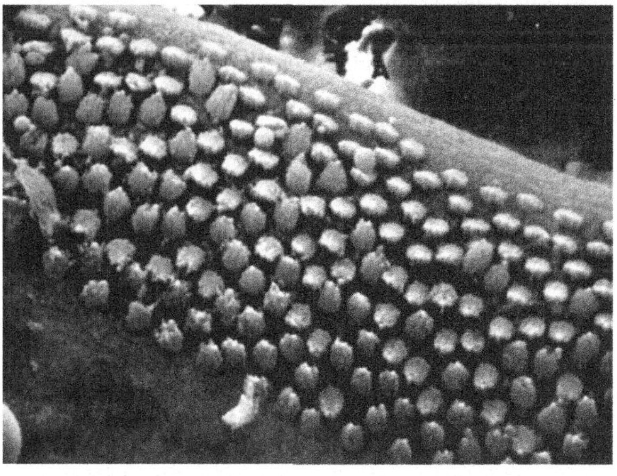

Fig. 5.4 A scanning EM photograph of a small section of the hair cell epithelium of the apical region of the papilla of the Golden (or Black) Tegu, *Tupinambis teguixin*, from which the tectorial membrane has been removed. The opposing orientation of the hair cell stereovillar bundles can clearly be seen and quantified. (From Miller (1973a) and used with permission)

(1992, p. 485) noted that "The probable ancestral lizard auditory papilla was intermediate in length (ca. 1 mm) and consisted of ca. 800 to 1,000 unidirectionally oriented hair cells covered by a substantial, relatively unspecialized tectorial structure. Each hair cell was innervated by relatively few afferent nerve fibers (two to three) and by efferent nerve fibers … some hair cells were exclusively innervated (each nerve fiber innervates only one hair cell) and others were non-exclusively innervated (a nerve fiber innervates more than one hair cell)."

Miller thus suggested that bidirectionality is not ancestral, even though it is generally found in vestibular receptor organs. Bidirectionality presumably arose at both ends of the evolving papilla, forming a "proto-papilla" with three hair cell areas. Miller proposed that varanid and teiid papillae, with three poorly defined hair cell regions, were the most ancient in morphology and represent the ancestral state of lizard ear evolution (Fig. 5.4). From these, various lineages arose whose papillar morphology evolved toward unique configurations (Manley 2004a, 2011). One lineage lost the *apical* bidirectional hair cell region and achieved the papilla typical of skinks and their relatives (cordylids, gerrhosaurids, xantusiids). A second lineage lost the *basal* bidirectional hair cell region and formed a geckonid-type papilla. A third lineage led to the iguanid-agamid type, in which the papillae are small and mostly retain all three hair cell areas, but bidirectionally oriented areas lost their tectorial membranes. Modern lacertids are a continuation of the varanid–teeid type, but evolved a hiatus between sub-papillae (Fig. 5.1). Convergent evolution led to anguid papillae strongly resembling those of distantly related iguanids and agamids.

The functional consequences of many of these anatomical variations were discussed by Köppl and Manley (1992) and Manley (2004a). Many lineages lost one of the two high-frequency regions, as they covered redundant frequency ranges. In geckos, unusually, the basal region was lost, which left the highest frequencies at the apical end. In all other land vertebrates, the highest frequencies lie basal. The evolution of sub-papillae enabled the differentiation of the two high-frequency papillar regions, so that their frequency response ranges are no longer the same. The loss of the tectorial membrane in small papillae (thus reducing coupling between hair cells) enabled the maintenance of frequency selectivity, albeit with a loss of sensitivity and tuning sharpness. Larger papillae often have what Wever (1978) termed "tectorial sallets"—a term that refers to their similarity in shape to a nineteenth-century French military helmet (Figs. 5.5 and 5.6). Later studies showed that sallets enable a sharper frequency tuning compared to hair cells that either have no tectorial cover or are covered by a continuous tectorial membrane (see Sect. 5.3.4).

Wever (1978), and his colleagues studied the anatomy and CM responses in a huge variety of lizards (and chelonians and crocodilians). Before discussing their work, a comment is necessary: Wever embedded whole, fixed heads in celloidin, sectioned these to assess both middle- and inner-ear anatomy, and made drawings of the histological sections (Fig. 5.6). The section thickness was rarely the same as the diameter of a single hair cell and a given hair cell may appear in more than one cross section. This can lead to cells being counted twice and requires correction. From comparisons to unambiguous hair cell counts in the same species made by Miller

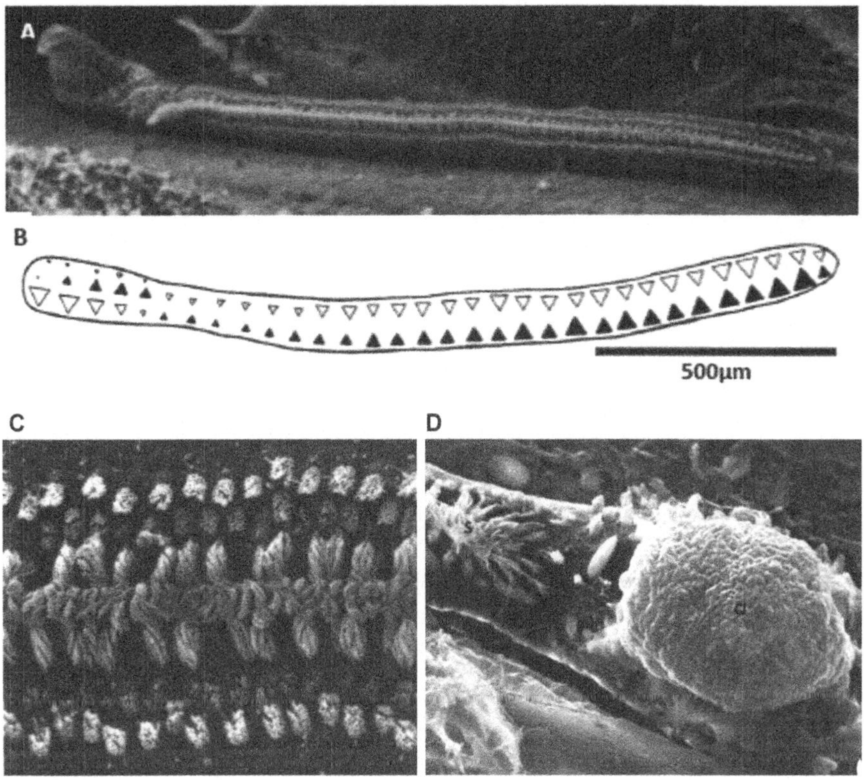

Fig. 5.5 The basilar papilla of the Bobtail skink, *Tiliqua rugosa*. (**a**) As seen in the scanning electron microscope at low magnification. Apical is to the left. (**b**) The polarity of the hair cell bundles along the papilla, apical to the left. The scale applies to (**a**) and (**b**). The apical hair cell area is, unusually, not unidirectionally oriented. Both kinds of tectorial structure have shrunk, pulling on the connections to hair cells at the papilla's edges. (From Köppl (1988) and used with permission). (**c**, **d**) The two types of tectorial membrane in *Tiliqua rugosa*. (**c**) A chain of tectorial sallets overlying hair cell bundles in the basal area. In the living state, the sallets cover the entire width of the hair cell epithelium. (From Köppl (1988) and used with permission). (**d**) A huge culmen (**c**) overlying hair cells at the apical end of the papilla of Gilbert's skink, *Eumeces gilbertii*. To the left is the beginning of the chain of sallets (s). (From Miller (1974) and used with permission)

using scanning EM, it is clear that Wever used a very conservative correction and that his counts were up to 30% too low.

In his classic 1978 book, Wever offered few summaries and the present review cannot discuss findings in each family of lizards, let alone each species. His data mostly conform to findings later summarized by Miller (1992), except in some respects. One difficult area is the structure of the tectorial membrane, which would have been strongly affected by both Wever's and Miller's fixation techniques. By combining studies, however, it is possible to conclude that the lizard tectorial membrane differs from those of birds and mammals and, indeed, those of other "reptiles." The most notable feature is that low-frequency areas always have a tectorial

Fig. 5.6 The same parts of the skink papilla as shown in Fig. 5.5, but from the species *Chalcides chalcides* (Italian three-toed skink), a typical drawing from light-microscopical cross sections by Glen Wever. The basal, high-frequency area is shown on the left, the apical, low-frequency area on the right. (From Wever (1970) and used with permission)

cover that is more-or-less large and continuous and may even be huge—Wever called this a "culmen" (Figs. 5.5 and 5.6).

In contrast, there are three fundamental forms of tectorial membrane covering the high-frequency hair cell areas in lizards:

1. A continuous, often substantial, structure (e.g., in the large papillae of teiids, varanids).
2. A discontinuous, string-of-beads-like tectorium connected by a thin thread along its length (a chain of units that Wever termed "sallets"), in the large papillae of skinks, geckos and their relatives (Figs. 5.5 and 5.6).
3. No tectorial cover. These have been called "freestanding" hair cells (but of course only their bundles are freestanding in the endolymph). Such only occur in small papillae.

5.3.3 Further Anatomical Studies

Our current knowledge of lizard papillae is still largely based on Miller's and Wever's major publications (Miller 1973a, b, 1974, 1978a, b, 1980, 1981, 1985; Miller and Beck 1988; Wever 1978). Curiously, for many years after 1973, other authors only studied lizard families with very small papillae. Baird (1970) concentrated on *Anolis* and *Iguana*, both iguanid species. Bagger-Sjoback (1976) examined *Calotes*, an agamid lizard, and other authors studied the anguid Alligator lizard *Elgaria multicarinata* (earlier known as *Gherrhonotus multicarinata*; Mulroy 1974; Mulroy and Williams 1987). Iguanids and agamids are closely related but have evolutionary radiations in different geographical regions. Anguids are not closely related to them, but the papilla evolved similarly; a small, low-frequency hair cell area has short stereovillar bundles and a tectorial plate covering it. There are also one or two high-frequency areas with no tectorial covering, whose hair cell bundles vary systematically in height along the papilla. The hair cell innervation is only afferent. Later, detailed physiological studies of the *Elgaria* papilla were published (see Sect. 5.5.3).

Information on the innervation patterns of lizard hair cells came from extensive comparative studies by Miller (1985) and Miller and Beck (1988), who showed hair cells were mainly innervated by afferent fibers with an innervation density (ratio of nerve fibers to hair cells) that was greater in smaller papillae. In all species, afferent fibers to UDT hair cells were larger in diameter than those to the BDT cells. Whether an afferent fiber only innervated one hair cell (exclusive) or more than one (non-exclusive) was not uniform in families or papillar types, except that in small papillae, the innervation was generally exclusive. Efferent innervation was generally limited to low-frequency hair cells, and these had more afferent synapses than high-frequency hair cells. The nerve fiber-to-hair cell numbers ratio was up to 11.1 to 1 in small papillae of the iguanid-agamid-anguid type, up to 3.2 to 1 in the teiid type, and between 0.6 to 1 and 1.5 to 1 in the larger, specialized papillae of the scincid and gekkonid types, respectively.

Two groups of hair cells raised the question as to how their afferent input is distributed in the brain. Szpir et al. (1990) studied this in *Elgaria* by following characterized, stained fibers to their peripheral and central contacts. They wrote, referring to the low-frequency hair cells as "tectorial" and the high-frequency hair cells as "freestanding" (p. 530): "Neurons whose peripheral processes contacted tectorial hair cells in the cochlea projected to three divisions of the cochlear nucleus… Neurons whose peripheral processes contacted freestanding hair cells projected primarily to (one division)…we conclude that different divisions of the cochlear nucleus are associated with separate frequency ranges…" (Szpir et al. 1990). The peripheral separation of frequency responses is thus continued into the cochlear nuclei. In a study of cochlear nucleus neurons in the brainstem of *Varanus*, Manley (1976) reported: "Almost all units with CF (characteristic frequency) above 1.2 kHz were found in the nucleus angularis, mainly in the lateral region. Low-frequency response units were found mainly in the nucleus magnocellularis and the medial region of the nucleus angularis. There thus appears to be a crude tonotopic organization of units in the cochlear nuclei." These results do not fully correspond to Szpir et al.'s (1990) findings for nucleus angularis, but there are presently too little data to reach conclusions on any common tonotopy in lizard brainstems.

Subsequently, Köppl (1988) described in detail the morphology of the auditory papilla in the Australian Bobtail skink (*Tiliqua rugosa*), for which a little physiological data was already available (Johnstone and Johnstone 1969). The papilla is large (about 2.1 mm long) and has more than 1900 hair cells. The apical papilla contains 15% of the hair cells covered by a very large tectorial culmen (Fig. 5.5). Hair cells of the long basal segment are covered by about 80 tectorial sallets, with 5 to 11 hair cells in any cross section. Each sallet connects between 9 and 24 hair cells, with this number increasing toward the basal end. The details in Köppl (1988; Fig. 5.5), which correspond to the typical anatomical pattern for skinks (Fig. 5.6), were later used to create a quantitative model of this papilla to compare to the physiology (Manley et al. 1988).

A different class of large lizard papilla was studied quantitatively by Köppl and Authier (1995), that of the Tokay gecko *Gekko gecko* (see also Miller 1973b). The Köppl and Authier study provided quantitative data for modeling papillar frequency responses, to compare to physiological data (Eatock et al. 1981). Miller had clearly shown that, unusually, the hair cells of the BDT area were separated *across* the papilla. In Gekko, there is a hair cell-free hiatus along the center of the BDT area and, on both sides of it, hair cell bundles are in highly ordered groups of opposite orientation (i.e., each side is bidirectional). Neural hair cells (Miller's preaxial) are covered by a tectorial plate that is connected to the over-arching limbic lip. Abneural hair cells (Miller's postaxial) carry small sallets (smaller than in *Tiliqua*), each covering one row of hair cells across the papilla. The sallets have no connections to the tectorial plate. Also unusually, the UDT area is not apical, as in most lizard species, but basal. Köppl and Authier (1995) provided details on hair cell density, the number of stereovilli per bundle, the height of the stereovillar bundles, and the tectorial membrane volumes (Fig. 5.7). These data were used in a model calculation of frequency responses (Authier and Manley 1995; see Sect. 5.5.3).

5.3.4 Types of Tectorial Membrane and Their Physiological Consequences

Neurophysiological studies since the 1970s have made it possible to interpret the structural patterns in tectorial material. The most unusual case for vertebrate ears—no tectorial membrane—is only seen in the higher-frequency areas of very small lizard papillae. Since the frequency ranges covered by small and large lizard auditory papillae are very similar, in small papillae the frequency changes more rapidly with distance along the papilla. Thus, neighboring hair cells along small papillae differ in their best frequency by a larger amount. In the iguanid Bahama anolis, *Anolis sagrei*, for example, there are 140 hair cells in the high-frequency area. With an average of five hair cells across the papilla, there are about 28 hair cell rows between 1 and 8 kHz (Manley and Gallo 1997) and thus 28 specific center frequencies on a logarithmic scale. Between 2 and 4 kHz, this works out to a 220 Hz difference between rows, whereas between 4 and 8 kHz, the difference is 440 Hz between neighboring rows. If these neighbors were coupled by tectorial material, the frequency selectivity of the associated auditory nerve fibers would be much poorer than without coupling. In such papillae, the rapid change in frequency (from 1 kHz to between 4 and 8 kHz) is determined by the steep height gradient in the bundles, from 30 μm near 1 kHz to 7 μm at high frequencies (Turner et al. 1981; Mulroy and Williams 1987). In the low-frequency area itself, the tectorial mass strongly reduces the best frequencies of the hair cells; here, the bundles are relatively short, but coupled.

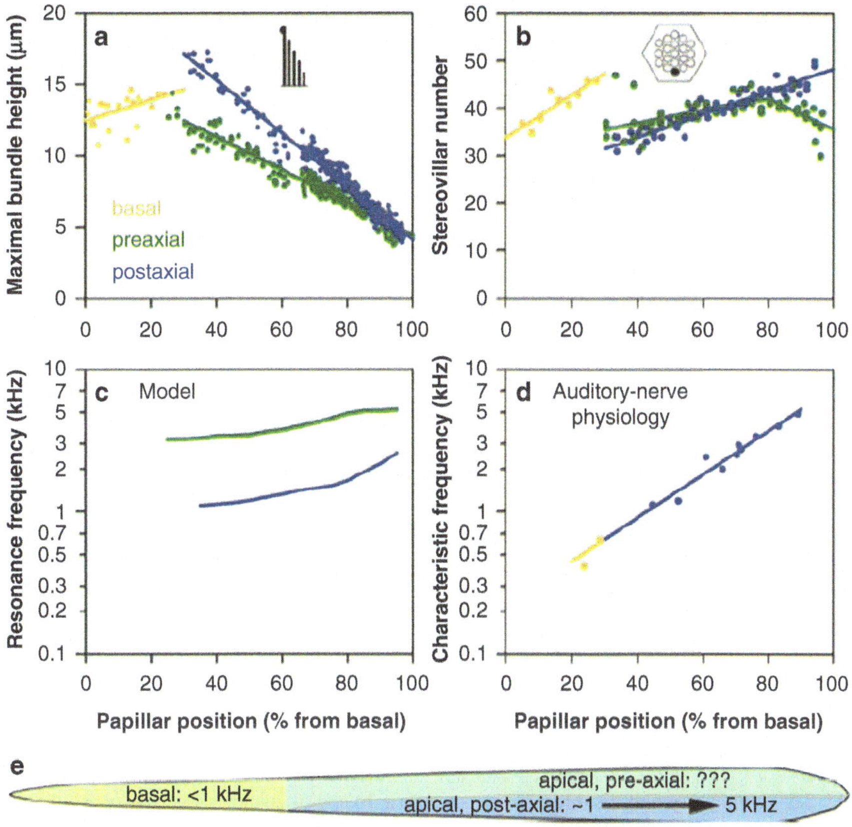

Fig. 5.7 Reversed tonotopic organization of the Tokay gecko, *Gekko gecko* basilar papilla. Yellow codes for the basal region, green for pre-axial and blue for post-axial regions of the apical, bidirectionally oriented papillar areas. (**a, b**) Hair bundle morphology suggests a gradient of increasing characteristic frequency from basal to apical: Hair bundle height (**a**) decreases, and the number of stereovilli per bundle (**b**) increases along this dimension. Lines are linear regressions to emphasize trends. (**c**) Resonance frequencies predicted by a model based on the quantitative anatomy of hair cells and tectorial structures in *Gekko* (Model B of Authier and Manley (1995); assuming 50% shrinkage of tectorial structures during histological processing). Note the prediction of different frequency ranges for the apical pre- and postaxial hair cell areas (basal hair cells were not included in the model). (**d**) Physiological frequency map of *Gekko* basilar papilla, derived by labeling auditory nerve fibers of known characteristic frequency and fitting a linear regression to the data. (**e**) Papillar areas showing the distribution of frequency responses. There are no pre-axial nerve fibers. (From Manley et al. (2014) and used with permission)

Salletal tectorial chains are unique to lizards and mainly found in skink-group families scincids, xantusids, gerrhosaurids, and cordylids, but also in geckos. These all possess moderate to large papillae, up to 2.2 mm length and up to 2000 hair cells (Wever 1978; Miller 1992). The coupling between hair cell rows in salletal papillae is both higher than without a tectorial membrane, but smaller than if the tectorial membrane were thicker and continuous. In large papillae, the frequency distances

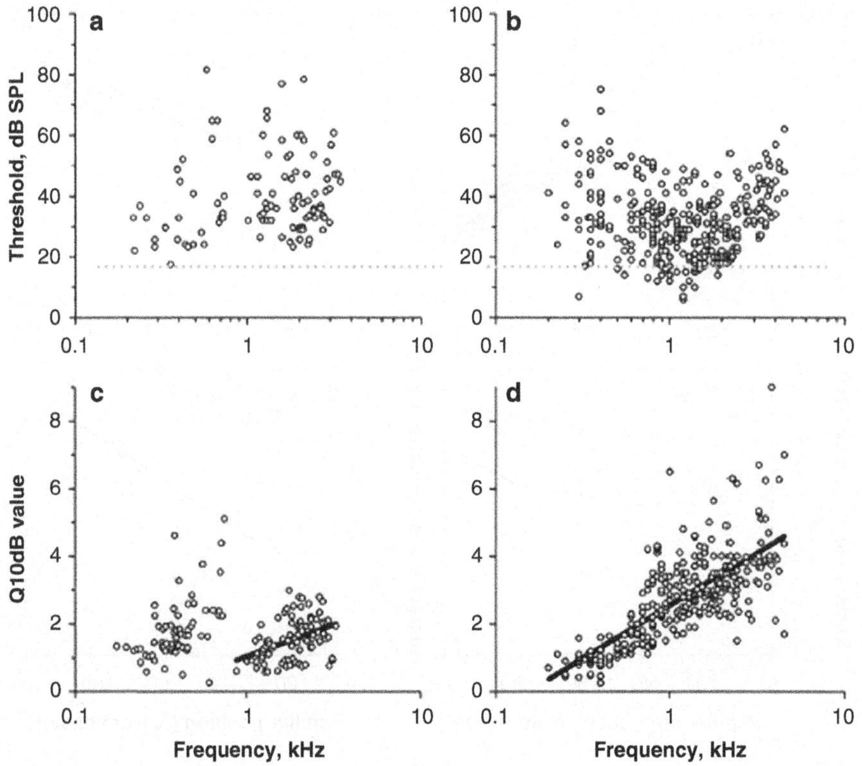

Fig. 5.8 A comparison of the threshold sensitivity (**a**, **b**) and frequency selectivity (**c**, **d**) of auditory nerve fibers in (**a**, **c**), *Elgaria multicarinata* that has no tectorium in its high-frequency region (above ~1 kHz), and the more sensitive and more selective fibers of (**b**, **d**), *Tiliqua rugosa*, which has sallets over the high-frequency hair cells. (From Manley and Köppl (2008), and used with permission)

between hair cell rows are much smaller than in, say, *Anolis*, suggesting better frequency selectivity. Model calculations indicate that the coupling of hair cells with similar frequency responses through small sallets *across* the papilla both improves sensitivity (frequency tuning curves are deeper) and selectivity (tuning curves are sharper; Manley et al. 1988). Thus, the frequency selectivity of auditory nerve fibers of skinks and geckos is twice or three times as sharp as in papillae lacking a tectorial membrane (Fig. 5.8; Köppl and Manley 1992; Manley and Köppl 2008).

5.4 Middle-Ear Studies

Like the "single-ossicle" middle ears of other non-mammalian amniotes (chelonians and archosaurs), the middle ear of squamates (Lepidosauria) evolved independently during the Triassic (Manley and Clack 2004). This "simple" kind of middle ear is surrounded by the same space that formed the first gill slit of aquatic ancestors, the spiracle, but is now sealed at the outside by an eardrum. These spaces are continuous from both sides of the head to the mouth cavity.

The columella is most of the inner portion of the ossicle. The outside portion, the extracolumella, connects the columella to the eardrum through processes that have their tips either near the center of the eardrum or as struts connecting to the eardrum surrounds. In some lizards, there is a middle-ear muscle that is not homologous to mammalian middle-ear muscles, but can presumably create tension in the conduction chain (Wever 1978). This has not yet been experimentally studied in lizards.

According to Miller and Beck (1988), such small papillae always show exclusive innervation that, in combination with the lack of a tectorial coupling, makes frequency selectivity possible in the high-frequency region. The neurons are less sensitive than those from large papillae, both features being consequences of the absence of a tectorial membrane (Köppl and Manley 1992; Manley 2000). Species possessing very small papillae faced evolutionary-selective pressures favoring either hair cell coupling that would improve sensitivity but result in poorer frequency selectivity—or retaining a degree of selectivity and accepting a loss of sensitivity. The latter won out (Manley 2004a).

Using the velocity-sensitive Mössbauer measurement technique, the responses to sound of the middle ears of two gecko species, the small Australian Dtella (*Gehyra variegata*) and larger *G. gecko* were studied (Manley 1972a, b; Fig. 5.9). The amplitude responses of the differently sized eardrums of these two species were very similar, except that, as expected, the corner frequency was lower in the Tokay (near 500 Hz) than in the *Gehyra* (1.25 kHz). There is an interplay between species size, individual age, and species' papillar size that influences the function of the middle ear (Werner et al. 2002).

Compared to the air-particle velocities driving the system (constant across frequencies for a constant sound pressure), gecko middle ears showed a loss of only about 7 dB (i.e., about a half) in its most sensitive area, which is similar to such data from mammalian middle ears (e.g., Manley and Johnstone 1974; Manley 2010). Thus, it is not surprising that many lizards show roughly the same hearing sensitivities as those of mammals and birds (Manley 2016) and the same magnitude of loss when the middle ear is disrupted (Wever 1978).

The inferior process of the extracolumella steadily creates tension on the eardrum and tents the center of the eardrum outward (the opposite of mammals). Compared to mammals, however, the middle-ear system is not very stiff. Air pressure applied to the middle ear achieved an equivalent result to that of mammals with only one-tenth of the pressure difference (van Dijk and Manley 2013). Perhaps the "floppiness" of the lizard middle ear allows it to absorb displacements that presumably can occur when a lizard consumes, for example, a large insect whose extremities may enter the middle ear through the wide connection to the mouth. It would also make externally exposed eardrums (e.g., in iguanid lizards) less sensitive to damage.

Measurements of the lizard middle ear (discussed in Sect. 5.5.1, in connection with cochlear microphonics measurements) show that single-ossicle and

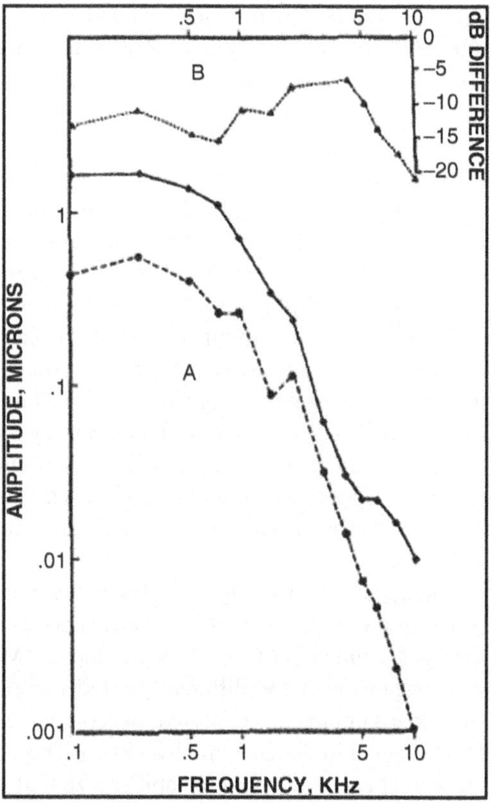

Fig. 5.9 The displacement amplitude of the tip of the inferior process of the extracolumella in the center of the eardrum (*continuous line*), and the footplate of the columella (*dashed line*) in the Tokay gecko, *Gekko gecko*. The smaller panel at the top shows the dB difference between these two functions and suggests that the transmission system is most efficient between 1 and 6 kHz. (From Manley (1972a) and used with permission)

three-ossicle middle ears perform similarly (Manley 2010). This can also be deduced from the fact that hearing thresholds of lizards (at their operating temperature), birds, and mammals are very similar (Manley 2016). This is compatible with paleontological evidence that mammalian middle ears arose *de novo* and are not "improved" single-ossicle middle ears (Manley 2010).

5.4.1 Pressure-Gradient Middle Ears in Lizards

Lizards have open connections between the ears, allowing sound to cross inside the body and interact at each ear. Such *pressure-gradient receivers* contrast with the pressure receivers of, for example, almost all mammals. Middle-ear systems of most non-mammals are widely open to the mouth and may additionally be connected through bony canals across the head. Wever (1978) suggested that in the crocodilian *Caiman* and some lizard species, each eardrum is influenced by sound arriving from both sides and the result depends on head size, frequency, and the arrival direction of the sound. In the auditory nerve (Christensen-Dalsgaard and

Carr 2018) and auditory midbrain neurons of Gekko (Manley 1981), the response differences to sounds presented from different sides of the head are much larger than expected from the head-shadowing effect.

The pressure-gradient effect in lizard middle ears is profound. There are differences between species in that some (thinner) eardrums are protected in an ear canal (as in geckos), and some are exposed to the outside world and thicker (as in iguanids). Using laser vibrometric studies of tympanic motion, a pronounced (species-specific) directionality near the best frequency of hearing was found, caused by the interference of the two sound inputs (Christensen-Dalsgaard and Manley 2005, 2008). Occluding the opposite ear abolished the directionality. In the range of maximal directionality, the head is essentially acoustically transparent, but with resonances of the head spaces. The pressure-difference characteristics generate the highest directionality of any terrestrial vertebrate ear. This provides the two auditory nerves with information about sound direction without the need for brain processing (Christensen-Dalsgaard and Carr 2018).

5.5 A Short History of the Auditory Physiology of Lizards

The present review includes, but also supplements, a previous summary of hearing in lizards and other "reptiles" (Manley 1990). Adrian et al. (1938) first recorded the "electrical response of the auditory mechanism" in various species, including alligators and tortoises, but not lizards. Both their stimuli and recording apparatus were crude and responses were only to low frequencies, although higher temperatures produced higher-frequency responses.

5.5.1 Cochlear Microphonic Studies

Following the discovery of cochlear potentials in mammals (Wever and Bray 1930), the earliest physiological work on lizard ears was by Wever and associates (1963–1975). Together with data from chelonians and crocodilians, these are described in Wever's book "The Reptile Ear" (1978). Through a filter that eliminated auditory nerve responses, potentials called cochlear microphonics (CM) were identified as originating from hair cells.

The problems in using CM "auditory thresholds" in lizards are very profound. Not only are "threshold" criteria arbitrary but lizards also show a unique problem that, remarkably, Wever failed to recognize. Measurements of CM assume that all hair cells respond to a tonal stimulus with the same phase and that individual contributions thus simply sum. As detailed above, however, lizard papillae show a species-dependent proportion of hair cells that is not abneurally oriented (the ancestral state), but at 180° to this. Their CM will thus be out-of-phase and cancel contributions from neurally oriented cells. Differently oriented hair cell groups exist with proportions that vary along the papilla and thus with frequency. If the two orientation populations are roughly equal in size, the resulting summed CM will be very small or could even be zero—although all hair cells may be highly active. In such cases, "thresholds," however defined, will be meaningless. In turtle papillae, where all hair cells have the same orientation, the same stimulus might produce a large signal and thus a "threshold" lower than that of a lizard. Such thresholds are meaningless, whether compared to other animals such as mammals, other lizard species (one of the purposes of Wever's work) or even within a single papilla.

The threshold problem is best illustrated using a species in which CM and auditory nerve data can be directly compared. The response thresholds of auditory nerve fibers of different characteristic frequencies in *G. gecko* are known (Fig. 5.10; Eatock et al. 1981; Manley et al. 1999). A comparison of the sound-pressure levels necessary at different frequencies to produce a "threshold" CM in *Gekko* (they defined 0.1 μV; Wever et al. 1963) are compared to neural thresholds for the auditory nerve in Fig. 5.11. *Gekko* papillae have only abneurally, unidirectionally oriented hair cells in their low-frequency areas (Köppl and Authier 1995), so cells here will respond to a tone with the same phase and add their CM voltages. Above 1 kHz, the hair cells are strictly bidirectional. As can be seen in Fig. 5.11, the low-frequency thresholds of both data sets are similar up to near 1 kHz. Above this frequency, they differ greatly: Between 2 and 4 kHz there is an apparent 30 dB difference in sensitivity. Measurements of CM are thus patently unsuitable for measuring hearing sensitivity in lizards. They are uninterpretable and must therefore be largely ignored in this review.

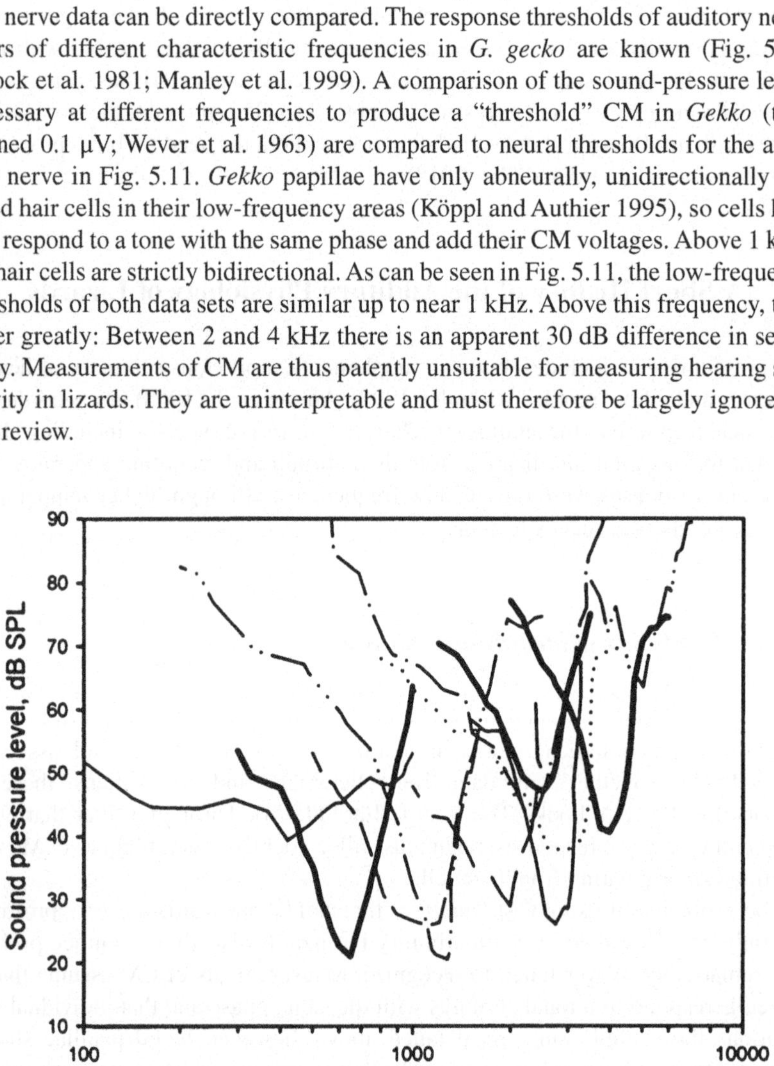

Fig. 5.10 Typical tuning curves recorded from single auditory nerve fibers in *Gekko gecko*. (From Manley et al. (1999) and used with permission)

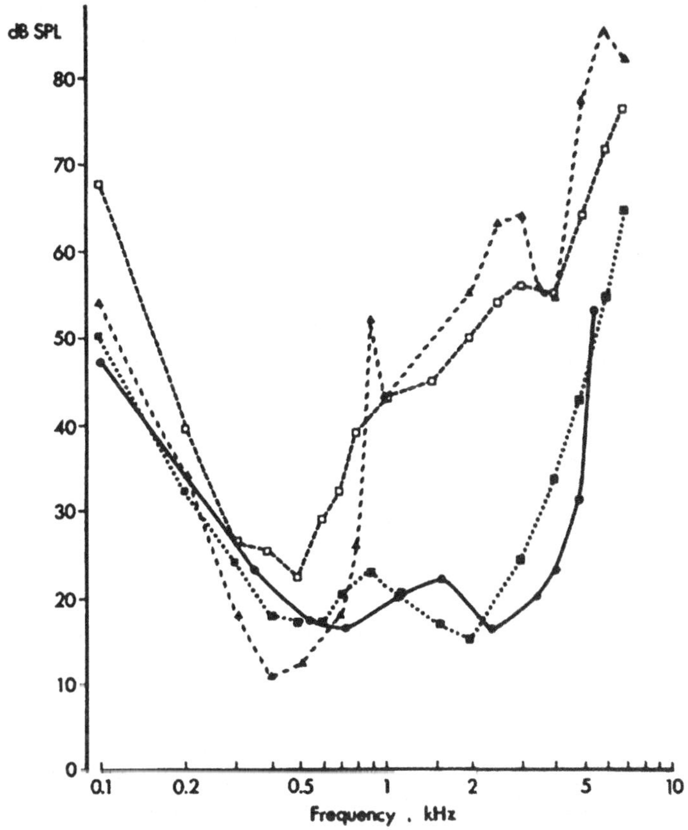

Fig. 5.11 Auditory sensitivity curves or audiograms for *Gekko gecko* derived from neural and microphonic data. Open squares connected by dashed lines and filled triangles connected by dashed lines are cochlear microphonic (CM) data from Hepp-Reymond and Palin (1968) and Werner and Wever (1972) respectively, and represent the sound levels required at different frequencies to produce O.1 μV of CM at the round window. The sensitivity curves derived from the most sensitive points on tuning curves of cochlear nucleus cells (filled squares and dotted lines, Manley 1972a) and of primary nerve fibers (filled circles, continuous line, Eatock et al. 1981) are much more sensitive at frequencies above 1 kHz than are the CM data. (From Eatock et al. (1981) and used with permission)

The CM method did, however, permit middle-ear studies in lizards that simply compared the sound levels needed to achieve a certain CM output using two methods of stimulation. One was via the intact middle ear, the other with the middle ear mostly removed and the sound applied directly to the columellar footplate. In the iguanid Collared lizard, *Crotaphytus* collaris, which has an externalized, thicker eardrum, Wever (1978) showed that the middle ear contributed about 20 dB of sensitivity in the mid-frequency range. In two geckos (*G. gecko* and *Eublepharis macularius*), and in several skink species, by contrast, which have deeper-lying, thinner eardrums, the CM loss was 40–60 dB. These latter data resemble results in mammals, indicating that in lizards, the middle ears can be equally efficient. Similar studies are Wever and Werner (1970) and Werner and Wever (1972).

5.5.2 *The Function of the Lagena*

One of the questions that had plagued nineteenth-century debates about the inner ear was the function of the lagena that, in non-mammalian amniotes and monotreme mammals, lies at the apex of the cochlea and is sometimes as large as the auditory papilla. Anatomically, the lagenar macula resembles macular vestibular structures (e.g., sacculus, utriculus) and not the auditory papilla. Fishes, however, use vestibular maculae to hear, so arguments against the lagena as a hearing organ were not fully convincing. Nerve fibers in the chicken cochlea that were traced to the lagenar macula ($N = 13$) did not respond to sound stimuli (Manley et al. 1991), suggesting a vestibular function. There is no reason to doubt that this is also true for lizard lagenae.

5.5.3 *Studies of Auditory-Neuron Responses and Models of Papillar Function*

Suga and Campbell (1967) first recorded single neural units in lizards, from the brainstem of Western Banded geckos (*Coleonyx variegatus*), with frequency selectivity up to about 4 kHz. Their sample was small, and the best threshold was 27 dB SPL, which is high and suggests that they did not have an optimal preparation.

The first recordings from the lizard auditory nerve were by Johnstone and Johnstone (1969) from the Bobtail skink *T. rugosa* (then called *Trachydosaurus rugosus*). In their small sample, they found frequency-selective units tuned from 0.7 to 3 kHz, with a best sensitivity of 25 dB SPL. They report a "...marked seasonal dependence on the number of active fibers." Later study of this species showed no seasonal dependence of fiber activity, but a clear seasonal change in the required levels of anesthetic—the first indication that lizard auditory responses are sensitive to the level of anesthesia. Under good conditions and appropriate anesthesia, however, this species has (at 30 °C) best frequencies of auditory nerve fibers between 0.2 and 4.5 kHz and sensitivities better than 10 dB SPL (Köppl et al. 1990, Fig. 5.10).

The 1970s and 1980s saw a huge increase in interest in lizard auditory system physiology that has, however, since ebbed away. Three main groups drove this interest, one in Boston (the lab of Tom Weiss), another in Detroit (Robert Turner), and the Manley lab in Montreal and later Munich. The Weiss group investigated the inner ear of *Elgaria multicarinatus* (known then as *Gerrhonotus multicarinatus*). This anguid has a very small auditory papilla with a single, high-frequency hair cell area that lacks a tectorial membrane (Mulroy 1974). They began their studies using intracellular recordings from hair cells (Mulroy et al. 1974), the first such recordings from any vertebrate cochlea. Later, auditory nerve recordings were reported (Fig. 5.8a, c; Weiss et al. 1976; Holton and Weiss 1983a, b; Eatock et al. 1991) showing that (a) the papillar region with a tectorial covering responds below 1 kHz, and the other, lacking a tectorium, responds from 1 kHz to about 4 kHz; (b) only the low-frequency nerve fibers show two-tone suppression and this effect sharpens their tuning to better than that of the high-frequency fibers; (c) hair cell and nerve fiber tuning is similar and reveals a tonotopic organization that in the high-frequency region correlates with the height of the hair cell bundles; (d) rate-level functions and phase-locking capacities of the two fiber populations are also different; and (e) the mean frequency selectivity of high-frequency fibers increases with best frequency.

Concurrently, the similar papilla of an iguanid lizard was studied in the Granite spiny lizard *Sceloporus orcutti* (Peake and Ling 1980; Turner et al. 1981; Frishkopf and DeRosier 1983; Holton and Hudspeth 1983; Nielsen and Turner 1983). Instead of one, the granite spiny lizard has two high-frequency hair cell areas that flank the low-frequency region. Micromechanically, the auditory papillae of *Elgaria* and *Sceloporus* behave similarly. Although the basilar membrane supporting the hearing organ is not locally tuned, the best frequencies are systematically distributed, correlating strongly with the hair cell bundle heights of the high-frequency regions. Direct observation of papillar movements showed greater relative motion of apical hair cell bundles at lower frequencies and of basal hair cell bundles at higher frequencies (Frishkopf and DeRosier 1983; Holton and Hudspeth 1983). The two high-frequency hair cell areas in *Sceloporus* are anatomical and physiological mirror images (Fig. 5.12; Turner 1987). In contrast to hair cells of turtle papillae (review in Fettiplace 1987), electrical resonances at the appropriate frequencies were not found in *Elgaria* "freestanding" hair cells (Eatock et al. 1993), supporting the concept of a primarily micromechanically tuned system in lizard high-frequency hair cells. Such concepts were supported by models assuming that micromechanical features (stiffness, mass, etc.) of the bundles determined the frequency responses of hair cells (Weiss and Leong 1985).

The Manley group initially studied hearing in the Bengal monitor lizard, *Varanus bengalensis* that has a large papilla (1.6 mm and about 1800 hair cells) with two unequally sized sub-papillae separated by a hair cell-free hiatus. The apical subpapillar area is smaller and has only bidirectionally oriented hair cell areas; the basal area has this plus a unidirectionally oriented area, resembling an archetypal "three-hair-cell-areas" papilla. Single neurons in the brainstem cochlear nuclei belonged to two groups; lower-frequency, more sensitive, and more sharply tuned units, and higher-frequency, less-well tuned units with best frequencies only up to 2.8 kHz (Manley 1976). Recordings of auditory nerve fibers recorded peripherally where they "fan out" to innervate hair cells along the papilla's length (Manley 1977) enabled the frequency mapping on part of both sub-papillae. Fibers from the apical area responded above 1.5 kHz, while fibers from the basal area responded to lower frequencies; those innervating the unidirectional area had best frequencies below 0.6 kHz, those innervating the adjacent bidirectional area from 0.65 to 0.95 kHz—an intermediate frequency range. This pattern suggested that unlike in species such

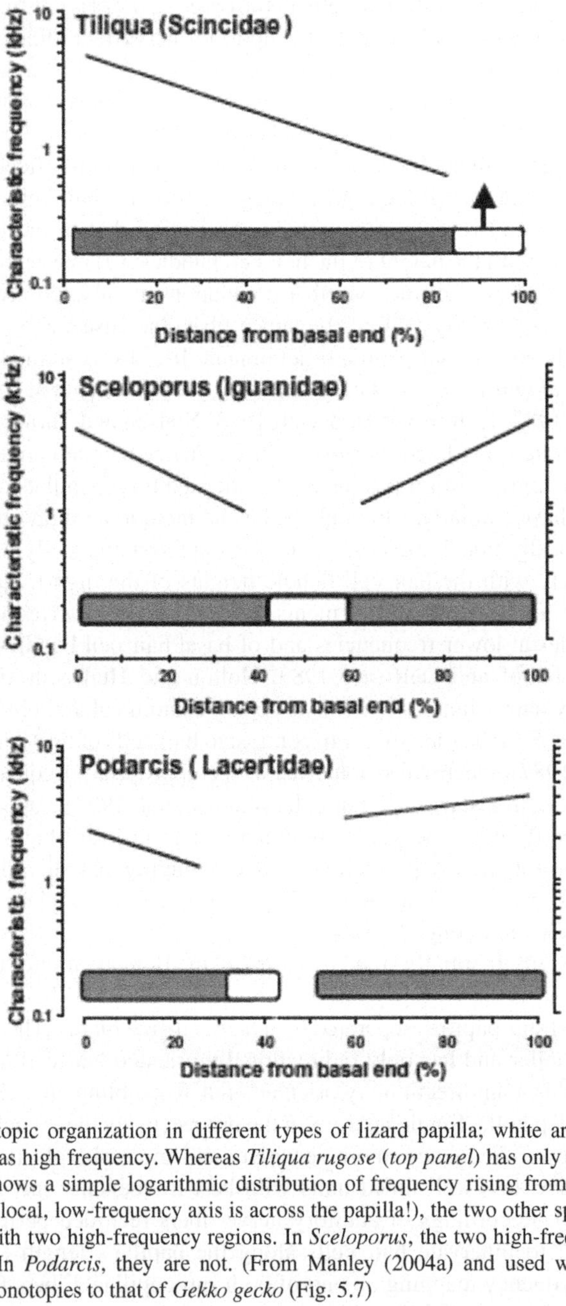

Fig. 5.12 Tonotopic organization in different types of lizard papilla; white areas are low frequency, gray areas high frequency. Whereas *Tiliqua rugose* (*top panel*) has only two papillar hair cell areas and shows a simple logarithmic distribution of frequency rising from the apex on the right (where the local, low-frequency axis is across the papilla!), the two other species have a tripartite papilla with two high-frequency regions. In *Sceloporus*, the two high-frequency areas are mirror images. In *Podarcis*, they are not. (From Manley (2004a) and used with permission). Compare these tonotopies to that of *Gekko gecko* (Fig. 5.7)

as *Sceloporus*, the two bidirectionally oriented areas in *Varanus* are not mirror images.

In 1980, attention turned to lizards with larger papillae that according to Miller (1980) were more highly organized than those of varanids, lacertids, iguanids, or anguids. The most attractive of these were the geckos, whose papillae are anatomically highly ordered (Miller 1973b; Wever 1978). The basal section contains only unidirectionally oriented hair cells, and the larger rest has highly-ordered rows of two sets of hair cells with a unique double bidirectionality; these are separated by a central, hair cell-free strip—the hiatus (Fig. 5.7). The neural area Miller (1973b) called pre-axial, the abneural area post-axial. Rows of hair cells across the papilla on the post-axial side are each covered by tectorial sallets, whereas pre-axial hair cell bundles are connected to a curtain-like tectorial sheet that descends from the gecko-typical, overarching, limbic lip. Eatock et al. (1981) reported the first auditory nerve recordings from the Tokay gecko, *G. gecko*, a species with over 2000 hair cells. There was a set of low-frequency fibers with best frequencies below 1 kHz and a set of high-frequency fibers with best frequencies from 1 to ~5 kHz. Since the latter, that were assumed to come from the bidirectional area, did not reveal two different populations, it was assumed that each fiber innervated both pre- and post-axial hair cells. In contrast to the poorly selective responses from both *Gerrhonotus* and *Sceloporus* fibers, tuning selectivity in *Gekko* rose strongly from low to high frequencies.

As first shown using CM measurements (e.g., Werner 1972), hearing in lizards is temperature sensitive. *Gekko* auditory nerve fibers shifted their best frequencies by 0.05 octaves/°C (up with warming; Eatock and Manley 1981). This is equivalent to a thermal Q10 of 1.4 and thus more than expected from a simple temperature effect on the viscosity of the cochlear fluids. The separation of the hair cell groups across the *Gekko* papilla remained puzzling, since there appeared to be no physiological correlate. If the hair cells were all innervated by the same afferent fibers, what was the function of the two populations? Authier and Manley (1995) modeled the frequency responses of *G. gecko* hair cells using detailed anatomical features of the hair cell groups (Köppl and Authier 1995). Pre- and postaxial populations were modeled separately, as connections between them were not known, with surprising results. The model predicted that the basal papilla (that in all other amniotes responds to high frequencies) covered the lowest frequencies, and the two strips of hair cells in the apical area covered *complementary* high-frequency ranges. The passive model (Authier and Manley 1995) could, however, only produce sensitive, sharply tuned responses if it were assumed that the cochlear fluid viscosity was only one-tenth of its real value. This seemingly inadmissible assumption was later justified by the finding of active processes that enable the hair cell bundles to drive the system and to form the basis of the high sensitivity and selectivity of coupled hair cells in lizards. More recent model simulations have confirmed these ideas (Dierkes et al. 2008; Gelfand et al. 2010; Wit et al. 2012, 2019).

This model prediction made it imperative to map the *Gekko* papilla physiologically. As predicted, staining of characterized nerve fibers showed that low frequencies mapped to the basal end of the papilla; gecko papillae thus have a reversed tonotopy (Fig. 5.7d, e; Manley et al. 1999). This was not too surprising, since during their evolution from early lizard tripartite papillae, either one of the two high-frequency areas could be lost. Most families, like the skinks, lost the basal area, only geckos the apical area. This is of no consequence for neural processing in the brain.

Frequencies above 1 kHz all mapped to a uniform sequence with the highest frequencies apically, instead of the model prediction that the natural frequencies of pre- and post-axial areas were complementary. Thus, although the hair cells of the pre- and postaxial areas are not connected, the neurons indicated that they are in some way coordinated. This conundrum was partially explained by the discovery that in the high-frequency region of *Gekko,* only postaxial (salletal) hair cells are actually innervated (Chiappe et al. 2007). These authors suggested that, as in mammals, two hair cell populations exist in which one set (pre-axial hair cells) acts as an amplifier, while the other (salletal cells) is innervated and communicates the net result to the brain. This does not, of course, explain why the hair cell populations have different "native" (modeled) frequencies.

A field study of a different gecko subfamily, the pygopods (legless geckos native to Australasia) added further mystery (Manley et al. 2014). Using recordings of the compound action potentials (CAP) of the whole auditory nerve, the auditory thresholds of different pygopod species were examined and revealed a correlation to the spectral content of their vocalizations (Manley and Kraus 2010). In pygopods, two genera—*Pygopus* and *Lialis*—are regarded as relatively ancestral, whereas other genera (such as *Delma*) are regarded as derived. The two "ancestral" genera had CAP audiograms similar to those measured in other geckos, with best frequencies between 2.0 and 3.0 kHz and highest response frequencies at or below 6 kHz. *Delma* species, however, showed an extended sensitivity up to 8 kHz and then generally a small improvement in threshold near 10 kHz, and a final loss of responses at 13 or 14 kHz (Fig. 5.13). Masking experiments revealed that the neural elements responsible for this high-frequency extension were the same units producing the "normal" responses up to 8 kHz. This may indicate a mechanical interaction between the two sets of high-frequency hair cells (that pygopods also have), but this is purely speculative. How gecko papillae function is still unclear, but is a fascinating example of complexity among the generally simple hearing organs of lizards.

In her Master's thesis (partially reported in Manley 1990), Köppl studied auditory nerve responses in two species of lacertid lizards, *Podarcis sicula* and *P. muralis*. The papillae of these lizards resemble those of varanids but are much smaller and show complete separation of two sub-papillae (see Fig. 5.1). Tracing auditory nerve fibers showed that the papilla has a low-to-medium-to-high frequency, three-area organization like in *Varanus*. In *Podarcis*, the low best frequency fibers (up to ~0.8 kHz) innervate the apical portion of the basal sub-papilla, and the mid-frequency fibers (up to ~2.7 kHz) the basal part of that sub-papilla. Fibers from the apical sub-papilla had best frequencies between ~2.7 and 5 kHz, the highest in the extreme basal area (Fig. 5.12). It is remarkable how a papilla that contains only about 100 hair cells can be so discretely organized! Perhaps in both varanids and lacertids the hiatus plays an important role in disambiguating the ancestral mirror-imagery of the two high-frequency areas.

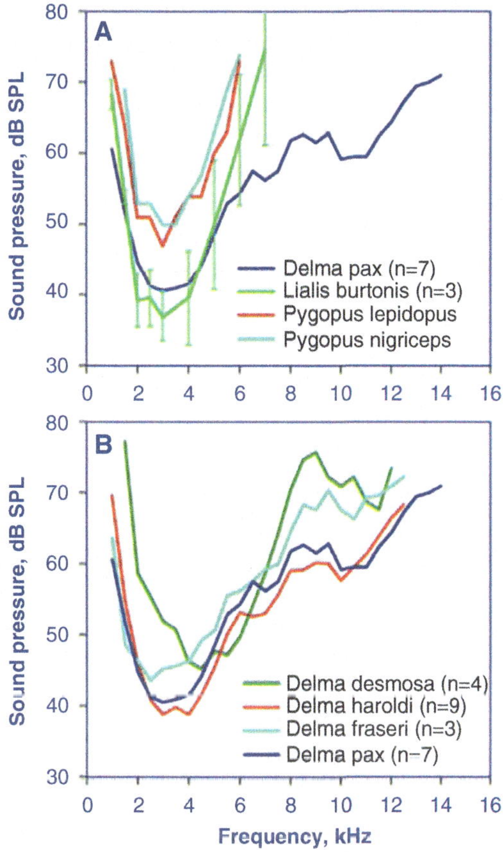

Fig. 5.13 Compound action potential (CAP) audiograms of pygopod geckos. (**a**) From three less-derived species, Burton's Snake Lizard *Lialis burtonis* (*green*, *N* = 3, ±s.d.), the Common Scaly-Foot, *Pygopus lepidopus* (red) and the Western Hooded Scaly-Foot *Pygopus nigriceps* (*light blue*) compared with the mean CAP audiogram for *Delma pax* (*dark blue*, *N* = 7). (**b**) A comparison of mean audiogram data from the four *Delma* species, *Delma desmosa* (*green*, *N* = 4), *Delma haroldi* (red, *N* = 9), *Delma fraseri* (light blue, *N* = 3) and D. pax (dark blue, *N* = 7). (From Manley and Kraus (2010) and used with permission)

Faced with the reversed tonotopicity of the gecko papilla (Fig. 5.7d, e), it was necessary to compare the tonotopy in an equally large papilla of an unrelated family, the skinks. Although the first nerve recordings in lizards had been carried out in Bobtail skinks *T. rugosa* in 1969, this highly robust species was not further investigated until the 1980s. Over the next two decades, more than 20 publications reported the anatomy, but mostly the physiology of the peripheral auditory system of this species. Models of the *Tiliqua* papilla (Dierkes et al. 2008; Gelfand et al. 2010; Wit et al. 2012, 2019) were compared to models of *Gerrhonotus* and established that frequency-response patterns, including tuning-curve shape and sharpness, can be explained in detail by micromechanical tuning through coupling of small groups of hair cells. Local frequency selectivity is entirely the result of the patterns of hair cell bundle and tectorial anatomy and, as in *Gekko*, coupling small groups of hair cells through sallets improves both the sensitivity and the selectivity of the responses.

An initial series of publications (Köppl et al. 1990; Köppl and Manley 1990a, b; Manley et al. 1990a, b) described the responses of *Tiliqua* auditory nerve fibers. A clear difference was seen between the broader, rounded frequency tuning curves of the low-frequency fibers (up to 1 kHz) and the V-shaped higher-frequency tuning curves. The latter showed breaks in the high-frequency slopes that occurred at the transition from broad tuning to local, selective tuning. Tuning selectivity rose in this species, also, from low- to high best frequencies (Manley et al. 1990a). The frequency map of the papilla showed a logarithmic distribution of best frequencies from low to high from apical to basal—typical of amniote papillae (Fig. 5.12; Köppl and Manley 1990a, b). The innervation pattern was consistent with that reported by Miller (1980, 1992), with non-exclusive innervation of groups of hair cells by each auditory nerve afferent (Köppl and Manley 1990a).

5.6 Modern Days: Otoacoustic Emissions and Active Processes in Lizard Ears

The study of distortion-product otoacoustic emissions (DPOAE) in humans and mammals has been common since the 1970s. Distortion products were known in psychoacoustics as audible frequencies that were not in the stimulus (two close frequencies) but were in the sound field of the ear canal and originated in the ear. An earlier study of *Elgaria* (Rosowski et al. 1984) was followed by descriptions of DPOAE in *Tiliqua* (Manley et al. 1993; Köppl and Manley 1993a). Remarkably, the patterns generated by varying the levels and frequency ratios of the two primary tones strongly resembled those of mammals, but there were differences between low- and high-frequency responses (Manley et al. 1993). Like neural tuning, the DPOAE were physiologically sensitive and vanished upon oxygen deprivation. Using third tones to suppress DPOAE, the tuning selectivity was found to show strong similarities to neural tuning curves from the same frequency region (Köppl and Manley 1993a).

While observing the microphone output on the spectrum analyzer during DPOAE measurements, the first spontaneous otoacoustic emissions (SOAE) in non-mammals were discovered as a set of peaks of very much smaller amplitudes initially interpreted as noise. They were present in the absence of stimulation and sensitive to external tones and changes in the animal's temperature. This was unexpected, since SOAE in humans were thought to be due to a uniquely mammalian active hair cell process. Later experimentation showed that lizards are in fact the most robust generators of SOAE of all amniotes—every species examined (more than 30, e.g., Manley 2006) had a clear SOAE spectrum. In mammals and birds, by contrast, SOAE are rare (Manley and Köppl 2008; Manley and van Dijk 2008), but their characteristics across different vertebrate groups are remarkably similar (Bergevin et al. 2015; Manley et al. 2015).

Each Bobtail ear had a unique SOAE spectrum, averaging ten emission peaks per ear, with frequencies above 1 kHz but sensitive to body temperature (Köppl and Manley 1993b, Fig. 5.14d). They originated from the papilla's high-frequency region that, at 30 °C, ends at about 4.6 kHz, as in most other lizards (Manley 2004b,

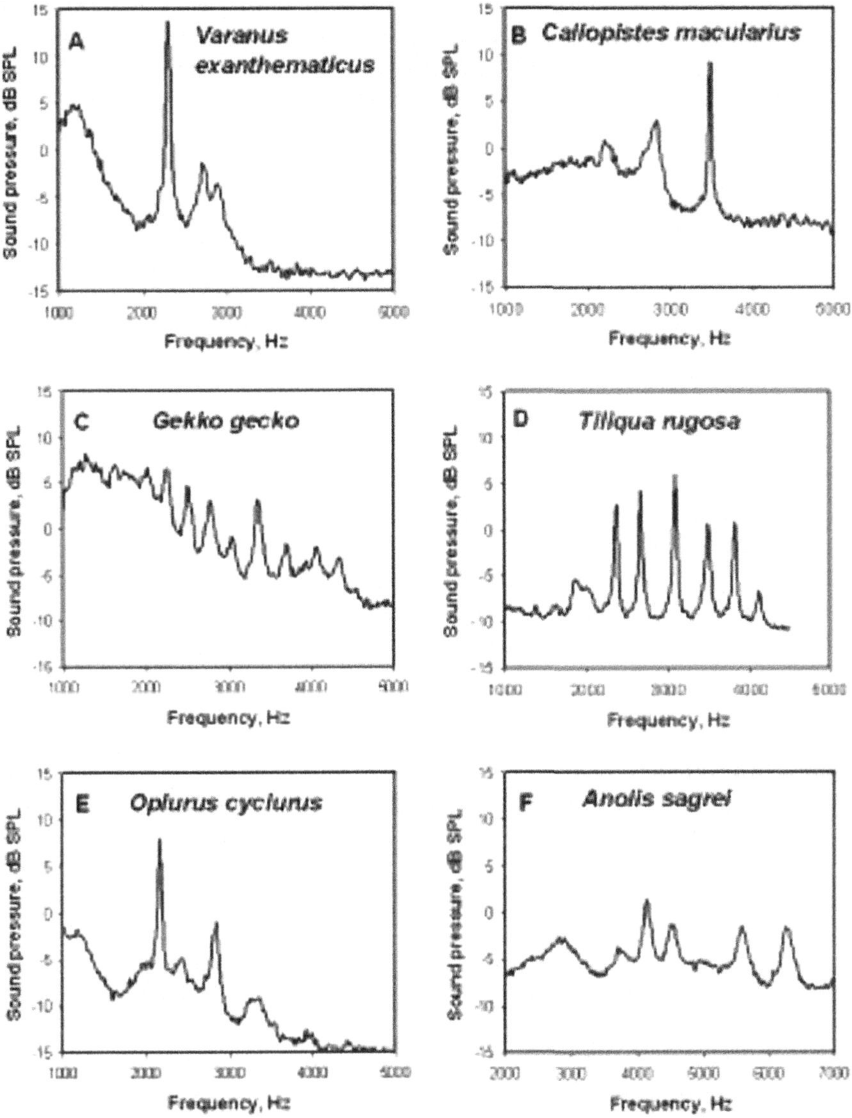

Fig. 5.14 Sample SOAE spectra from different lizard families that have different tectorial structures in the high-frequency papillar areas. (**a**) *Varanus* and (**b**) *Callopistes* have continuous tectorial membranes, (**c**) *Gekko* and (**d**) *Tiliqua* have sallets, (**e**) *Oplurus* and (**f**) *Anolis* lack a tectorium in that area. (From Manley and van Dijk (2008) and used with permission)

2006; Manley et al. 2018). The absolute frequency distance between peaks increased with center frequency, as expected if they each originated from a group of hair cells and their sallet (Köppl and Manley 1993b). Compared to SOAE in humans, the peaks were broader in frequency and generally smaller in amplitude. This frequency instability correlated with large shifts in level and center frequency of SOAE when driven by external tones (Köppl and Manley 1994). A detailed examination of SOAE responses to added tones showed the same frequency tuning as seen in primary nerve fibers. The frequencies and amplitudes of SOAE were modulated by the calcium concentration in a similar fashion to in vitro studies of frog saccular hair cells (Manley et al. 2004). This strongly suggested that groups of hair cells coupled by sallets were responsible for both the tuning of auditory nerve fibers and SOAE suppression tuning, and implicated the hair cells in the generation of the acoustic energy of the SOAE. Thus, evidence accumulated that not only humans, but also non-mammals, exhibit an active process in their hair cells (Manley 2000, 2001; Manley et al. 2001), but this is not based on the prestin motor of mammals (Dallos and Fakler 2002).

The largest temperature dependences of center frequencies of skink SOAE were about 0.04 octaves/°C. In lizard families in which the high-frequency hair cell area lacks a tectorial membrane, SOAE amplitudes and temperature effects are smaller (Manley 1997, 2002, 2006). In families with larger papillae and a continuous tectorial membrane, however, SOAE amplitudes were larger (up to 27 dB SPL, Fig. 5.14a) and the temperature effects were much larger. In monitor lizards (Varanidae, Teiidae), for example, the temperature shift was up to 0.1 octaves/°C (Manley 1997, 2004b; Manley et al. 2018), a doubling of frequency over 10 °C. These differences correlated with the different tectorial coverings (Manley 1997, 2002, 2004b). SOAE from hair cells lacking a tectorial covering were similar to those from hair cells covered by sallets—relatively small in amplitude and relatively numerous—but unstable over time when compared to salletal systems (Manley 2003, 2006; Manley and Köppl 1994; Fig. 5.14). SOAE from papillae with a continuous tectorial covering were both larger and fewer per ear; here, hair cells from a longer stretch of papilla would be coupled (Manley 1997, 2002). This distinction is somewhat blurred, however, since there are species with small papillae and a continuous tectorial membrane (e.g., lacertids). Recent models of lizard SOAE confirmed that coupling of groups of hair cells can generate the observed spectral patterns (Vilfan and Duke 2008; Gelfand et al. 2010; Wit et al. 2012, 2019). A study of SOAE in the iguanid *A. sagrei* (the Bahamian anole) aimed to ascertain the amount of energy emitted by the hair cells and, knowing the number of hair cells and their bundle structure, to calculate the power output of molecules possibly involved in the emission of sound (Manley and Gallo 1997). Individual SOAE peaks were assigned to between three and 38 hair cells, and the calculated power output per hair cell was 141 aW (1 aW = $1.0e^{-18}$ W). The number of bundle myosins putatively involved in the generation of each SOAE was estimated and the force generated by each myosin at 1 kHz was calculated to be approximately 0.1 pN. The data suggested that while myosin indeed could produce sufficient power to be the emission motor, there are other candidate molecules.

The consistent occurrence of lizard SOAE offered a new kind of "window" into the inner ear, since manipulations of hair cells could be non-invasively produced and observed. The two bundle orientations of the high-frequency areas of lizard offered a unique tool for examining the location of the active process. To do this, we stimulated hair cells to emit sound by injecting low-amplitude but high-frequency electrical currents into Scala media. In mammals, this induces hair cells to generate sound at the frequency of the current. In the Bobtail skink, such currents only produced very much smaller sound signals, since the current stimulated hair cells not along their lateral membranes, as in mammals, but in the oppositely oriented bundles. The driven movements of the bundles of roughly equally sized hair cell groups were thus of opposite phase and at least partially canceled out. This was confirmed by using very low-frequency sound in addition to the current and observing changes in the emissions during a full cycle of the sound. The amplitude varied and the phase of the emission generated by the current switched by 180° every half cycle (Manley et al. 2001), strongly suggesting that SOAE energy is generated by hair cell bundles.

Although after 2010 it appeared that research on lizard hearing might stop with the impending closure of the last laboratory working on them (Manley), there was new hope. Chris Bergevin (York University, Canada) started studying lizards using stimulus-frequency emissions (SFOAE) that can be generated at any frequency and their phase measured. Like SOAE, they can be suppressed or entrained by added tones. The technique had been widely used in mammal studies (that usually lack SOAE) and the results from lizards can be compared. Bergevin's data support conclusions reached from neural and SOAE studies in different lizard species that, despite different SOAE mechanisms in lizards and mammals, the energy is bundled into emitted spectral peaks in similar ways (Bergevin et al. 2010, 2015; Bergevin and Shera 2010; Bergevin 2011). Jakob Christensen-Dalsgaard and Catherine Carr also began to study the effects of the open middle-ear connection in lizards on the neural processing of sound (Christensen-Dalsgaard and Carr 2018). Using *G. gecko*, they showed that there is already a strong directionality in neural responses at the level of the auditory nerve, influencing the basis for processing in the brain.

Several newer studies provided data on behavioral responses of lizards to various sounds, a much-neglected subject. They illustrate a rich variety of situations in which different lizard species clearly use sound for orientation, predator avoidance, and food capture. Although there is no space to discuss them at length here, readers are referred to the following publications: Sakaluk and Belwood (1984); Vitousek et al. (2007); Hibbits et al. (2007); Huang et al. (2011); Hoare and Labra (2013); Cantwell and Forrest (2013); Baeckens et al. (2019); Pérez-Cembranos and Pérez-Mellado (2020).

After many decades of work on the inner ears of different vertebrate groups, it has become obvious that the fundamental unit of hearing, the hair cell, has not greatly changed during the evolution of the various lineages. The properties of hair cells dominate physiological responses and determine the great similarities across the groups. The study of lizard hearing has contributed greatly to our understanding of how sensory systems evolved and how even a tiny organ with only about 100 sensory cells can contribute frequency-selective, alerting information to the brain. Lizard studies have provided and will continue to provide very important comparative data that inform on models of how all inner-ear hearing organs actually work.

Compliance with Ethics Requirements Geoffrey A. Manley declares that he has no conflict of interest.

References

Adrian ED, Craik KJW, Sturdy RS (1938) The electrical response of the auditory mechanism in cold-blooded vertebrates. Proc R Soc Lond B Biol Sci 125:435–455

Authier S, Manley GA (1995) A model of frequency tuning in the basilar papilla of the Tokay gecko, *Gekko gecko*. Hear Res 82:1–13

Ayers H (1892) Vertebrate cephalogenesis. II. A contribution to the morphology of the vertebrate ear, with a reconsideration of its functions. J Morphol 6:1–360

Baeckens S, Llusia D, García-Roa R, Martín J (2019) Lizard calls convey honest information on body size and bite performance: a role in predator deterrence? Behav Ecol Sociobiol 73:87

Bagger-Sjoback D (1976) The cellular organization and nervous supply of the basilar papilla in the lizard, Calotes versicolor. Cell Tissue Res 165:141–156

Baird IL (1966) Discussion of the reptilian ear. Am Zool 6:431–436

Baird IL (1970) The anatomy of the reptilian ear. In: Gans C, Parsons TS (eds) Biology of the Reptilia, vol 2. Academic Press, New York, pp 193–275

Bergevin C (2011) Comparison of otoacoustic emissions within gecko subfamilies: morphological implications for auditory function in lizards. J Assoc Res Otolaryngol 12:203–217

Bergevin C, Shera C (2010) Coherent reflection without traveling waves: on the origin of long-latency otoacoustic emissions in lizards. J Acoust Soc Am 127:2398–2409

Bergevin C, Velenovsky DS, Bonine KE (2010) Tectorial membrane morphological variation: effects upon stimulus frequency otoacoustic emissions. Biophys J 99:1064–1072

Bergevin C, Manley GA, Köppl C (2015) Salient features of otoacoustic emissions are common across tetrapod groups and suggest shared properties of generation mechanisms. Proc Natl Acad Sci U S A 112(11):3362–3367

Camparetti A (1789) Observationes anatomincae de aure interna comparata. Patavii Apud S. Bartholomaeum, Typographia jo: Antonii Conzatti

Cantwell LR, Forrest TG (2013) Response of *Anolis sagrei* to acoustic calls from predatory and nonpredatory birds. J Herpetol 47(2):293–298

Chiappe ME, Kozlov AS, Hudspeth AJ (2007) The structural and functional differentiation of hair cells in a lizard's basilar papilla suggests an operational principle of amniote cochleas. J Neurosci 27(44):11978–11985

Christensen-Dalsgaard J, Carr CE (2018) Processing of directional information in the gecko auditory nerve. Acta Acust United Acust 104:848–851

Christensen-Dalsgaard J, Manley GA (2005) Directionality of the lizard ear. J Exp Biol 208:1209–1217

Christensen-Dalsgaard J, Manley GA (2008) Acoustical coupling of lizard eardrums. J Assoc Res Otolaryngol 9:407–416

Clack JA (1997) The evolution of tetrapod ears and the fossil record. Brain Behav Evol 50:198–212

Clason E (1873) Die Morphologie des Gehörorgans der Eidechsen. In: Hasse C (ed) Anatomische Studien. Verlag W. Engelmann, Leipzig, pp 300–376

Corti A (1851) Recherches sur l'organe de l'ouie des mammifères. Première partie: Limacon. Zeits. wiss. Zool 3:109–169

Dallos P, Fakler B (2002) Prestin, a new type of motor protein. Nat Rev Mol Cell Biol 3:104–111

De Burlet HM (1929) Zur vergleichenden Anatomie und Physiologie des perilymphatischen Raumes. Acta Otolaryngol 13(2):153–187

Dierkes K, Lindner B, Jülicher F (2008) Enhancement of sensitivity gain and frequency tuning by coupling of active hair bundles. Proc Natl Acad Sci U S A 105:18669–18674

Eatock RA, Manley GA (1981) Auditory-nerve fibre activity in the Tokay gecko: II. Temperature effect on tuning. J Comp Physiol A 142:219–226
Eatock RA, Manley GA, Pawson L (1981) Auditory-nerve fibre activity in the Tokay gecko: I. Implications for cochlear processing. J Comp Physiol A 142:203–218
Eatock RA, Weiss TF, Otto KL (1991) Dependence of discharge rate on sound pressure level in cochlear nerve fibers of the alligator lizard: implications for cochlear mechanisms. J Neurophysiol 65:1580–1597
Eatock RA, Saeki M, Hutzler MJ (1993) Electrical resonance of isolated hair cells does not account for acoustic tuning in the free-standing region of the alligator lizard's cochlea. J Neurosci 13:1767–1783
Estes R, De Queiroz K, Gauthier J (1988) Phylogenetic relationships within Squamata. In: Estes R, Pregill G (eds) Phylogenetic relationships of the lizard families. Stanford University Press, Stanford, pp 119–281
Evans LT (1936) The development of the cochlea in the gecko, with special reference to the cochlea-lagena ratio and its bearing on vocality and social behaviour. Anat Rec 64:187–201
Fettiplace R (1987) Electrical tuning of hair cells in the inner ear. Trends Neurosci 10:421–425
Frishkopf LS, DeRosier DJ (1983) Mechanical tuning of the free-standing stereociliary bundles and frequency analysis in the alligator lizard cochlea. Hear Res 12:393–404
Gaupp E (1898) Ontogenese und Phylogenese des schall-leitenden Apparates bei den Wirbeltieren. Erg Anat Entwicklungsgesch 8:900–1149
Gelfand M, Piro O, Magnasco MO, Hudspeth AJ (2010) Interactions between hair cells shape spontaneous otoacoustic emissions in a model of the Tokay gecko's cochlea. PLoS One 5(6):e11116
Grey O (1955) A brief survey of the phylogenesis of the labyrinth. J Laryngol Otol 69:151–179
Hamilton DW (1960) Observations on the morphology of the inner ear in certain gekkonoid lizards. Univ Kansas Sci Bull 41:983–1024
Hamilton DW (1964) The inner ear of lizards: gross structure. J Morphol 115:255–272
Hepp-Reymond M-L, Palin J (1968) Patterns in the cochlear potentials of the Tokay gecko (*Gekko gecko*). Acta Otolaryngol 65:270–292
Hibbits TJ, Whiting JM, Stuart-Fox DM (2007) Shouting the odds: vocalization signals status in a lizard. Behav Ecol Sociobiol 61:1169–1176
Hoare M, Labra A (2013) Searching for the audience of the weeping lizard's distress call. Ethology 119:860–868
Holton T, Hudspeth AJ (1983) A micromechanical contribution to cochlear tuning and tonotopic organization. Science 222:508–510
Holton T, Weiss TF (1983a) Receptor potentials of lizard cochlear hair cells with free-standing stereocilia in response to tones. J Physiol 345:205–240
Holton T, Weiss TF (1983b) Frequency selectivity of hair cells and nerve fibers in the alligator lizard cochlea. J Physiol 345:241–260
Huang B, Lubarsky K, Tent T, Blumstein DT (2011) Take only pictures, leave only…fear? The effects of photography on the West Indian anole *Anolis cristatellus*. Curr Zool 57:77–82
Johnstone JR, Johnstone BM (1969) Unit responses from the lizard auditory nerve. Exp Neurol 24:528–537
Kemp TS (2005) The origin and evolution of mammals. Oxford University Press, Oxford
Köppl C (1988) Morphology of the basilar papilla of the bobtail lizard *Tiliqua rugosa*. Hear Res 35:209–228
Köppl C, Authier S (1995) Quantitative anatomical basis for a model of micromechanical frequency tuning in the Tokay gecko, *Gekko gecko*. Hear Res 82:14–25
Köppl C, Manley GA (1990a) Peripheral auditory processing in the bobtail lizard *Tiliqua rugosa*: II. Tonotopic organization and innervation pattern of the basilar papilla. J Comp Physiol A 167:101–112
Köppl C, Manley GA (1990b) Peripheral auditory processing in the bobtail lizard *Tiliqua rugosa*: III. Patterns of spontaneous and tone-evoked nerve-fibre activity. J Comp Physiol A 167:113–127

Köppl C, Manley GA (1992) Functional consequences of morphological trends in the evolution of lizard hearing organs. In: Webster DB, Fay RR, Popper AN (eds) The evolutionary biology of hearing. Springer, New York, pp 489–509

Köppl C, Manley GA (1993a) Acoustic distortion products in the ear canal of the bobtail lizard. II: Suppression tuning characteristics. J Acoust Soc Am 93:2834–2844

Köppl C, Manley GA (1993b) Spontaneous otoacoustic emissions in the bobtail lizard. I: General characteristics. Hear Res 71:157–169

Köppl C, Manley GA (1994) Spontaneous otoacoustic emissions in the bobtail lizard. II: Interactions with external tones. Hear Res 72:159–170

Köppl C, Manley GA, Johnstone BM (1990) Peripheral auditory processing in the bobtail lizard *Tiliqua rugosa*: V. Seasonal effects of anaesthesia. J Comp Physiol A 167:139–144

Lee MSY (1998) Convergent evolution and character correlation in burrowing reptiles: toward a resolution of squamate phylogeny. Biol J Linn Soc 65:369–453

Luo ZX (2007) Transformation and diversification in early mammal evolution. Nature 450:1011–1019

Manley GA (1972a) The middle ear of the Tokay gecko. J Comp Physiol 81:239–250

Manley GA (1972b) Frequency response of the middle ear of geckos. J Comp Physiol 81:251–258

Manley GA (1976) Auditory responses from the medulla of the monitor lizard *Varanus bengalensis*. Brain Res 102:329–334

Manley GA (1977) Response patterns and peripheral origin of auditory nerve fibres in the monitor lizard, Varanus bengalensis. J Comp Physiol 118:249–260

Manley GA (1981) A review of the auditory physiology of the reptiles. Prog Sens Physiol 2:49–134

Manley GA (1990) Peripheral hearing mechanisms in reptiles and birds. Springer, Heidelberg

Manley GA (1997) Diversity in hearing-organ structure and the characteristics of spontaneous otoacoustic emissions in lizards. In: Lewis ER, Long GR, Lyon RF, Narins PM, Steele CR (eds) Diversity in auditory mechanics. World Scientific, Singapore, pp 32–38

Manley GA (2000) Do non-mammals have a cochlear amplifier? In: Wada H, Takasaka T, Ohyama K, Ikeda K, Koike T (eds) Recent developments in auditory mechanics. World Scientific, Singapore, pp 499–505

Manley GA (2001) Evidence for an active process and a cochlear amplifier in non-mammals. J Neurophysiol 86:541–549

Manley GA (2002) Evolution of structure and function of the hearing organ of lizards. J Neurobiol 53:202–211

Manley GA (2003) The tectorial membrane stabilizes spontaneous otoacoustic emissions. In: Gummer AW, Dalhoff E, Nowotny M, Scherer MP (eds) Biophysics of the cochlea: from molecules to models. World Scientific, Singapore, pp 480–487

Manley GA (2004a) The lizard basilar papilla and its evolution. In: Manley GA, Popper AN, Fay RR (eds) Evolution of the vertebrate auditory system. Springer, New York, pp 200–223

Manley GA (2004b) Spontaneous otoacoustic emissions in monitor lizards. Hear Res 189:41–57

Manley GA (2006) Spontaneous otoacoustic emissions from free-standing stereovillar bundles of ten species of lizard with small papillae. Hear Res 212:33–47

Manley GA (2010) An evolutionary perspective on middle ears. Hear Res 263:3–8

Manley GA (2011) Lizard auditory papillae: an evolutionary kaleidoscope. Hear Res 273:59–64

Manley GA (2016) Comparative auditory neuroscience: understanding the evolution and function of ears. J Assoc Res Otolaryngol 18:1–24

Manley GA, Clack JA (2004) An outline of the evolution of vertebrate hearing organs. In: Manley GA, Popper AN, Fay RR (eds) Evolution of the vertebrate auditory system. Springer, New York, pp 1–26

Manley GA, Gallo L (1997) Otoacoustic emissions, hair cells and myosin motors. J Acoust Soc Am 102:1049–1055

Manley GA, Johnstone BM (1974) Middle-ear function in the Guinea pig. J Acoust Soc Am 56:571–576

Manley GA, Köppl C (1994) Spontaneous otoacoustic emissions in the bobtail lizard. III: Temperature effects. Hear Res 72:171–180
Manley GA, Köppl C (2008) What have lizard ears taught us about auditory physiology? Hear Res 238:3–11
Manley GA, Kraus JEM (2010) Exceptional high-frequency hearing and matched vocalizations in Australian pygopod geckos. J Exp Biol 213:1876–1885
Manley GA, Van Dijk P (2008) Otoacoustic emissions in amphibians, lepidosaurs and archosaurs. In: Manley GA, Fay RR, Popper AN (eds) Active processes and otoacoustic emissions in hearing, Springer handbook of auditory research, vol 30. Springer, New York, pp 211–260
Manley GA, Yates G, Köppl C (1988) Auditory peripheral tuning: evidence for a simple resonance phenomenon in the lizard Tiliqua. Hear Res 33:181–190
Manley GA, Köppl C, Johnstone BM (1990a) Peripheral auditory processing in the bobtail lizard *Tiliqua rugosa*: I. Frequency tuning of auditory-nerve fibres. J Comp Physiol A 167:89–99
Manley GA, Yates GK, Köppl C, Johnstone BM (1990b) Peripheral auditory processing in the bobtail lizard *Tiliqua rugosa*: IV. Phase locking of auditory-nerve fibres. J Comp Physiol A 167:129–138
Manley GA, Haeseler C, Brix J (1991) Innervation patterns and spontaneous activity of afferent fibres to the lagenar macula and apical basilar papilla of the chicken's cochlea. Hear Res 56:211–226
Manley GA, Köppl C, Johnstone BM (1993) Acoustic distortion products in the ear canal of the bobtail lizard. I: General characteristics. J Acoust Soc Am 93:2820–2933
Manley GA, Köppl C, Sneary M (1999) Reversed tonotopic map of the basilar papilla in *Gekko gecko*. Hear Res 131:107–116
Manley GA, Kirk D, Köppl C, Yates GK (2001) In-vivo evidence for a cochlear amplifier in the hair cell bundle of lizards. Proc Natl Acad Sci U S A 98:2826–2831
Manley GA, Sienknecht U, Köppl C (2004) Calcium modulates the frequency and amplitude of spontaneous otoacoustic emissions in the bobtail skink. J Neurophysiol 92:2685–2693
Manley GA, Köppl C, Sienknecht UJ (2014) The remarkable ear of geckos and pygopods. In: Köppl C, Manley GA, Popper AN, Fay RR (eds) Insights from comparative hearing research. Springer, New York, pp 111–132
Manley GA, Köppl C, Bergevin C (2015) Common substructure in otoacoustic emission spectra of land vertebrates. In: Karavitaki KD, Corey DP (eds) Mechanics of hearing: protein to perception, American Institute of Physics conference proceedings, vol 1703. AIP, Melville
Manley GA, Wartini A, Schwabedissen G, Siegl E (2018) Spontaneous otoacoustic emissions in teiid lizards. Hear Res 363:98–108
Miller MR (1966) The cochlear duct of lizards. Proc Calif Acad Sci 33:255–359
Miller MR (1973a) Scanning electron microscope studies of some lizard basilar papillae. Am J Anat 138:301–330
Miller MR (1973b) A scanning electron microscope study of the papilla basilaris of *Gekko gecko*. Zeitschr Zellforsch 136:307–328
Miller MR (1974) Scanning electron microscope studies of some skink papillae basilares. Cell Tissue Res 150:125–141
Miller MR (1978a) Scanning electron microscope studies of the papilla basilaris of some turtles and snakes. Am J Anat 151:409–435
Miller MR (1978b) Further scanning electron microscope studies of lizard auditory papillae. J Morphol 156:381–418
Miller MR (1980) The reptilian cochlear duct. In: Popper AN, Fay RR (eds) Comparative studies of hearing in vertebrates. Springer, Berlin, pp 169–204
Miller MR (1981) Scanning electron microscope studies of the auditory papillae of some iguanid lizards. Am J Anat 162:55–72
Miller MR (1985) Quantitative studies of auditory hair cells and nerves in lizards. J Comp Neurol 232:1–24

Miller MR (1992) The evolutionary implications of the structural variations in the auditory papilla of lizards. In: Webster DB, Fay RR, Popper AN (eds) The evolutionary biology of hearing. Springer, New York, pp 463–487
Miller MR, Beck J (1988) Auditory hair cell innervational patterns in lizards. J Comp Neurol 271:604–628
Mulroy MJ (1974) Cochlear anatomy of the alligator lizard. Brain Behav Evol 10:69–87
Mulroy MJ, Williams RS (1987) Auditory stereocilia in the alligator lizard. Hear Res 25:11–21
Mulroy MJ, Altmann DW, Weiss TF, Peake WT (1974) Intracellular electric responses to sound in a vertebrate cochlea. Nature 249:482–485
Nielsen DW, Turner RG (1983) Micromechanics of the reptilian ear. Audiology 22:530–544
Peake WT, Ling A (1980) Basilar-membrane motion in the alligator lizard: its relation to tonotopic organization and frequency selectivity. J Acoust Soc Am 67:1736–1745
Pérez-Cembranos A, Pérez-Mellado V (2020) It sounds like food: phonotaxis of a diurnal lizard. Behav Process 179:104217
Pickles JO, Brix J, Comis O, Gleich O, Köppl C, Manley GA, Osborne MP (1989) The organization of tip links and stereocilia on hair cells of bird and lizard basilar papillae. Hear Res 41:31–42
Reichert C (1837) Ueber die Visceralbogen der Wirbelthiere im allgemeinen und deren Metamorphosen bei den Vögeln und Säugethieren. Müllers Arch Physiol 1837:120–222
Retzius G (1884) Das Gehororgan der Wirbelthiere. II. Das Gehororgan der Reptilien, Vogel und Saugethiere. Commission bei Samson & Wallin, Stockholm
Rosowski JJ, Peake WT, White JR (1984) Cochlear nonlinearities inferred from two-tone distortion products in the ear canal of the alligator lizard. Hear Res 13:141–158
Sakaluk S, Belwood J (1984) Gecko phonotaxis to cricket calling song: a case of satellite predation. Anim Behav 32:659–662
Schmidt RS (1964) Phylogenetic significance of lizard cochlea. Copeia 3:542–549
Shute CCD, d'A Bellairs A (1953) The cochlear apparatus of Geckonidae and Pygopodidae and its bearing on the affinities of these groups of lizards. Proc Zool Soc Lond 123:695–709
Suga N, Campbell HW (1967) Frequency sensitivity of single auditory neurons in the gecko Coleonyx variegatus. Science 157:88–90
Szpir MR, Sento S, Ryugo DK (1990) Central projections of cochlear nerve fibers in the alligator lizard. J Comp Neurol 295:530–547
Turner RG (1987) Neural tuning in the granite spiny lizard. Hear Res 26:287–299
Turner RG, Muraski AA, Nielsen DW (1981) Cilium length: influence on neural tonotopic organization. Science 213:1519–1521
Underwood G (1957) On the lizards of the family Pygopodidae. J Morphol 100(2):207–268
Utz P, Freed P, Aguilar R, Hošek J (eds) (2022) The Reptile Database. http://www.reptile-database.org. Accessed 20 May 2022
Van Dijk P, Manley GA (2013) The effects of air pressure on spontaneous otoacoustic emissions of lizards. J Assoc Res Otolaryngol 14:309–319
Versluys J (1898) Die mittlere und äussere Ohrsphäre der Lacertilia und Rhynchocephalia. Thesis, Gustav Fischer Verlag, Jena
Vidal N, Hedges SB (2009) The molecular evolutionary tree of lizards, snakes, and amphisbaenians. C R Biol 332:129–139
Vilfan A, Duke T (2008) Frequency clustering in spontaneous otoacoustic emissions from a lizard's ear. Biophys J 95:4622–4630
Vitousek M, Adelman JS, Gregory NC, St. Clair JJH (2007) Heterospecific alarm call recognition in a non-vocal reptile. Biol Lett 3:632–634
Weiss TF, Leong R (1985) A model for signal transmission in an ear having hair cells with free-standing stereocilia. III. Micromechanical stage. Hear Res 20:157–174
Weiss TF, Mulroy MJ, Turner RG, Pike CL (1976) Tuning of single fibres in the cochlear nerve of the alligator lizard: relation to receptor morphology. Brain Res 115:71–90

Werner YL (1972) Temperature effects on inner-ear sensitivity in six species of iguanid lizards. J Herpetol 6:147–177

Werner YL, Wever EG (1972) The function of the middle ear in lizards: Gekko gecko and Eublepharis macularius (Gekkonidae). J Exp Zool 179:1–16

Werner YL, Igić PG, Seifan M, Saunders JC (2002) Effects of age and size in the ears of gekkonomorph lizards: middle ear sensitivity. J Exp Biol 205:3215–3223

Wever EG (1970) The lizard ear: Scincidae. J Morphol 132:277–292

Wever EG (1974) The evolution of vertebrate hearing, Ch. 13. In: Autrum H, Jung R, Loewenstein WR, Mackay DM, Teuber HL (eds) Handbook of sensory physiology, vol V/II. Springer, Berlin, Heidelberg, New York

Wever EG (1978) The reptile ear. Princeton University Press, Princeton

Wever EG, Bray CW (1930) Action currents in the auditory nerve in response to acoustical stimulation. Proc Natl Acad Sci U S A 16(5):344–350

Wever EG, Werner YL (1970) The function of the middle ear in lizards: Crotaphytus collaris (Iguanidae). J Exp Zool I75:327–342

Wever EG, Vernon JA, Peterson EA, Crowley DE (1963) Auditory responses in the Tokay gecko. Proc Natl Acad Sci U S A 50:806–811

Wever EG, Vernon JA, Crowley DE, Peterson EA (1964) Electrical output of lizard ear: relation to hair cell population. Science 150:1172–1174

Wit HP, van Dijk P, Manley GA (2012) A model for the relation between stimulus frequency and spontaneous otoacoustic emissions in lizard papillae. J Acoust Soc Am 132:3273–3279

Wit H, Manley GA, van Dijk P (2019) Modeling the characteristics of spontaneous otoacoustic emissions in lizards. Hear Res 385:107840

Chapter 6
Birds as a Model in Hearing Research

Robert J. Dooling and Georg M. Klump

6.1 Introduction

Speculations about the hearing capabilities of birds undoubtedly arose from early fascination with the prominence and complexity of their vocalizations. Except for assumptions that birds could probably hear the sounds that they produce, there are no records of any systematic studies of hearing in birds prior to about the 1930–1940s. Modern methods of studying hearing reveal that there are many different aspects of hearing and auditory perception and many different methodological approaches to determining what birds can hear. For the most part, the earliest attempts at studying hearing in birds focused on simple measures of hearing, and we begin our review here.

Authors Robert J. Dooling and Georg M. Klump are shared first authors

R. J. Dooling (✉)
Department of Psychology and Program in Neuroscience and Cognitive Science, University of Maryland, College Park, MD, USA
e-mail: rdooling@umd.edu

G. M. Klump
Carl von Ossietzky University Oldenburg, Division Animal Physiology and Behavior, Department for Neuroscience, School of Medicine and Health Sciences and Cluster of Excellence Hearing4all, Oldenburg, Germany
e-mail: georg.klump@uol.de

D. R. Ketten et al. (eds.), *A History of Discoveries on Hearing*, Springer Handbook of Auditory Research 77,
https://doi.org/10.1007/978-3-031-41320-9_6

6.2 Hearing in Birds: Detection, Discrimination, and Categorization

Information about hearing in birds can be gleaned from casual observation going back into antiquity as far as Aristotle (c.f., Peck 1984). These observations suggested that birds can hear the sounds that they produce, that the "pitch" or spectral range of bird songs falls well within the range of human hearing, and that many birds learned their vocalizations by reference to auditory information. The earliest objective descriptions of birdsong arose in the latter part of the nineteenth century and involved using musical notation. This approach was only moderately successful because the temporal features of the songs of many species were too fast to be adequately captured by such a technique (e.g., Barrington 1773). This approach, moreover, undoubtedly also fueled informal speculation about the resolving power of the avian ear.

The latter part of the nineteenth century also saw the first anatomical description of the bird's inner ear by Gustaf Retzius (1884). But it was not until the middle of the twentieth century that speculation about what birds could hear and discriminate was linked to the anatomy of the ear and the sensory epithelium of the bird's inner ear. This includes a single bone middle ear, or columella, and a relatively short basilar papilla, a relatively thick basilar papilla and tectorial membrane (supporting excellent temporal resolution), and a relatively large number of hair cells across the width of the papilla, thereby allowing better frequency and/or intensity discrimination (e.g., Johann Schwartzkopff 1949; Pumphrey 1961; Gleich and Manley 2000). At the same time, there is also compelling evidence that the single bone columella in birds is not very efficient above about 10 kHz (Saunders 1985) and that the middle ears in birds are connected by an interaural pathway (e.g., Yoshitsune Wada 1923; Rosowski and Saunders 1980; Köppl 2019; see also review by Starck 1995), which affects various aspects of hearing.

6.2.1 Detection

The first serious experimental attempts to measure pure-tone thresholds in birds (i.e., the audibility curve) had to await sufficient technological development. The decades of the 1930s–1950s represented a watershed period in the history of the comparative hearing of birds. For the first time, investigators made serious efforts to control the precision of both the frequency and intensity of the stimulus. They also strove to control for switching and acoustic artifacts, conducted tests in sound-treated testing chambers, tested different species using the exact same methods, and sometimes even compared their results with humans tested with the same equipment (e.g., Brand and Kellogg 1939a, b; Edwards 1943; Trainer 1946). These early studies all involved some version of instrumental avoidance conditioning with shock, for example, when shock was paired with a tone and delivered to a feeding tray. Soon birds learned to flee the feeding tray upon hearing the tone. Variations of this avoidance conditioning technique used by these early investigators were used with a variety of species including the European starling (*Sturnus vulgaris*), canary

(*Serinus canaria*), canvas-back (*Nyroca valisineria*), great horned owl (*Bubo virginianus*), prairie horned larks (*Otocoris alpestris praticola*), and snow bunting (*Plectrophenax nivalis*) (Trainer 1946).

These types of studies were continued with other species by other investigators of the time including Sigrid Knecht (1939), Johann Schwartzkopff (1949), and John Ezra Trainer (1946), a student of Brand and Kellog. There was considerable variation across these studies, in large part due to methodological limitations. However, the general picture of avian hearing that emerged from these studies was of a narrow range of best hearing between 1 and 5 kHz with sensitivity approaching that of humans and with sensitivity declining rapidly about 5 kHz and somewhat less rapidly below 1 kHz (e.g. Heise, 1953). Taken together, the understanding of bird hearing presented by these earliest studies has stood the test of time quite well.

There are now audibility curves for over 50 species of birds, going back to this period of the 1930s–1950s. Citations to most of these audiograms are available by common name, genus, and species in various places (e.g., Dooling et al. 2000). Because a considerable number of avian species have been tested compared with other vertebrate groups, some relationships have emerged between hearing and other biological variables.

Remarkably, a tight relationship in birds has been discovered between the range of hearing, body mass, and the length of the basilar papilla (Gleich et al. 2005; Gleich and Langemann 2011). In general, small birds hear better at high frequencies, large birds hear better at low frequencies, and nocturnal predators hear much better at midrange frequencies than these other birds. For comparison, Fig. 6.1

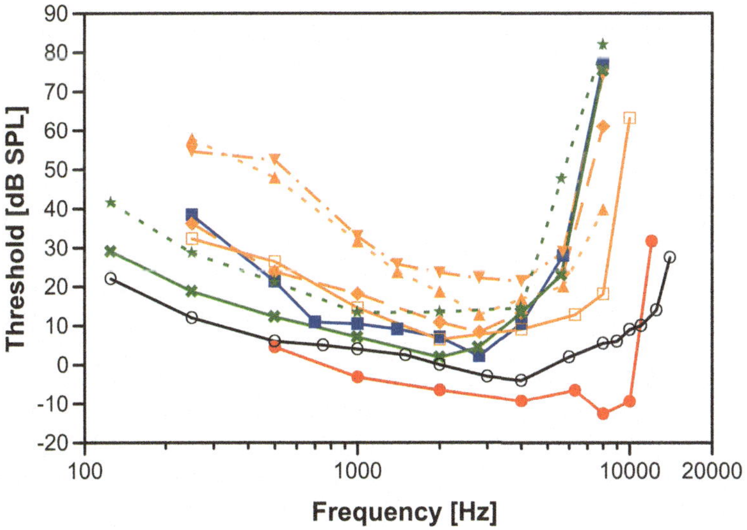

Fig. 6.1 Audiogram in quiet of different bird species, such as ● barn owl (Krumm et al. 2017); ✖ chicken, ★ pigeon (Hill et al. 2014); □ great tit (Langemann et al. 1998); ■ budgerigar, ▲ canary, ▼ zebra finch, ◆ European starling (Okanoya and Dooling 1987), and of humans ○ (Fastl and Zwicker 2007)

shows the behavioral audiogram for exemplary bird species and humans (for a general overview, see Dooling et al. 2000). There are only very limited data on the hearing of very small birds such as hummingbirds, but the evidence suggests that they have relatively good hearing at high frequencies (Lohr and Dooling 2004). Beyond the audibility curve, the complexity of avian vocalizations and their use in territorial defense and mate attraction have often provoked informal speculation among ecologists and field biologists about the ability of birds to discriminate among sounds.

6.2.2 Frequency and Intensity Discrimination

The observation that birds can accurately reproduce small frequency intervals in their songs and can accurately mimic other vocalizations, and even human melodies, prompted curiosity about birds' sense of hearing and musicality. Although Knecht (1939) used rudimentary mechanical sound devices such as Edelmann whistles or tuning forks, she achieved good stimulus control and was able to produce small frequency intervals. While this was before the time Skinner box-like experimental setups provided a standard, Knecht (1939) carefully designed a setup that prevented the birds from using unwanted cues (Fig. 6.2). In the experiment, the experimenter was hidden behind a screen while observing bird subjects through a periscope-like mirror system. This avoided the problem of the bird observing Knecht's movements during testing that might cue the bird during testing and she remotely provided access to the feeder providing the reward.

By presenting rewarded and unrewarded tones differing in frequency and observing the bird's approach to the feeder, Knecht determined the minimum distinguishable frequency interval in several small bird species (among these were budgerigars (*Melopsittacus undulatus*) and canaries, two species used frequently). Knecht (1939) observed considerable variation in the performance of different species. Budgerigars performed very well and could clearly discriminate tones with a frequency interval of 5 Hz from a reference of 659 Hz (i.e., a threshold at a Weber fraction of 0.008). The smallest Weber fraction for successfully discriminating frequencies that Knecht observed was 0.003. Knecht (1939) also concluded that the birds based their decision on absolute pitch, a finding that was later confirmed by Hulse and colleagues (e.g., Hulse and Cynx 1985). In the 1970s and 1980s, several studies measured frequency difference limens in birds using techniques that generally met the requirements of modern psychophysical methods and stimulus control (summarized in Fay 1988; Langemann and Klump 1992; Klump et al. 1995). These included frequency difference limens in budgerigars, European starlings, redwing blackbirds (*Agelaius phoenicus*), cowbirds (*Molothrus ater*), pigeons (*Columba livia domestica*), and young chickens (*Gallus gallus domesticus*). Generally, in the region of best hearing of birds between 1 and 5 kHz, frequency difference limens approached those of humans with the best difference limens below 10 Hz (i.e.,

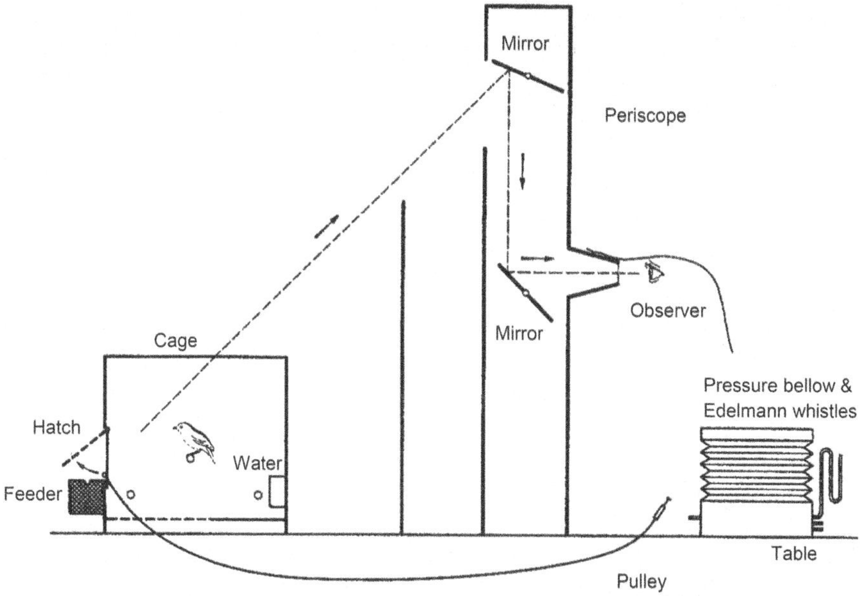

Fig. 6.2 Experimental setup for investigating frequency discrimination in small birds. (After Knecht (1939)). The responses of the bird perched in a cage are observed via a periscope by the investigator hidden behind a screen. The observer rewards the bird by remotely releasing a hatch that then allows access to the feeder. Tones of different frequencies are produced by a sophisticated air-pressure control system driving precision Edelmann whistles producing tones with a specific frequency and sound-pressure level. (Figure reprinted with permission of Springer Nature Society)

Weber fractions below 0.01; Dooling et al. 2000). These results are generally in line with the earlier work on several species by Knecht (1939).

Intensity discrimination has not historically been an interest in hearing studies in birds though it was suggested early on that the large number of hair cells across the width of the basilar papilla in birds could support excellent intensity discrimination (Pumphrey 1961). Pure-tone intensity difference limens have now been measured in several species including canaries, budgerigars, starlings, redwing blackbirds, cowbirds, and pigeons (summarized in Fay 1988; Dooling et al. 2000). These results generally show intensity difference limens for pure tones in the range of 2–3 dB in birds compared with about 1–2 dB in humans with even larger intensity difference limens for birds at lower intensity levels.

6.2.3 Masking

Masking is important because almost everything we hear, unless in the quiet of an auditory testing booth, is to some degree masked by environmental noise. But the earliest comparative studies of masking in birds, and those of other animals, were motivated to a large degree by the classic understanding of the place theory of hearing and the related predictions of a correlation between critical bands, critical ratios, and frequency difference limens. To this end, broadband-noise masking studies in over a dozen species of birds show that the threshold for a pure tone increases by 1 dB for every 1 dB increase in noise level and that this signal-to-noise (S/N) ratio generally increases systematically with increasing frequency (Okanoya and Dooling 1987; Fay 1988; Dooling et al. 2000). Beyond this, there are some interesting species differences in birds on how masking varies across frequency (Fig. 6.3). Critical ratios and critical bands in budgerigars do not increase in size with frequency as would be expected by the place theory but instead show much smaller S/N ratios (i.e., narrower bandwidths) in the region of 3 kHz—the region of best hearing and where the peak in the power spectrum of vocalizations occurs (Dooling and Saunders 1975). The great tit (*Parus major*) also shows an unusual pattern of masked thresholds with a fairly constant critical ratio over a broad frequency range rather than increasing at the rate of 3 dB/octave, suggesting a specialization for hearing high-frequency signals in noise (Langemann et al. 1998).

The barn owl (*Tyto alba*) also shows much smaller critical ratios compared with other birds (Dyson et al. 1998) more in line with humans. Tone-on-tone masking studies in the budgerigar (Saunders et al. 1979; Dooling et al. 2000) show psychophysical tuning curves that are narrower than those of humans in the region of best hearing and also tend not to broaden with increasing level as they do in humans and

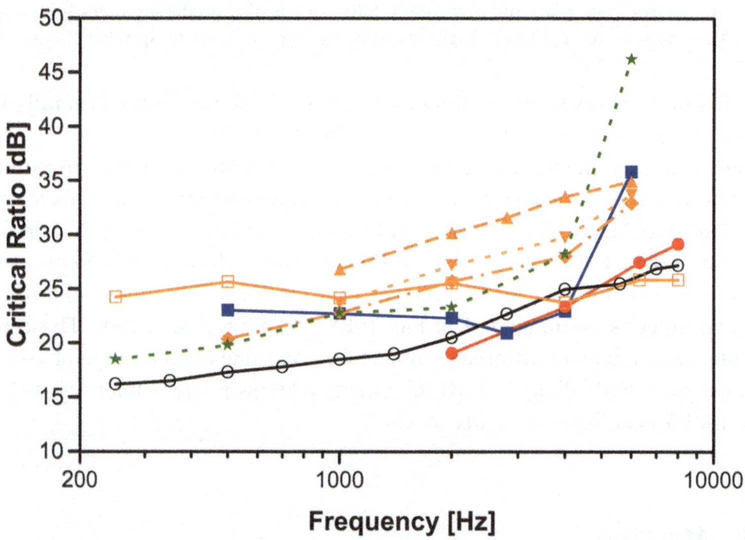

Fig. 6.3 Critical masking ratio of birds, such as ● barn owl (Dyson et al. 1998); □ great tit (Langemann et al. 1998); ★ pigeon (Hienz and Sachs 1987); and ■ budgerigar, ▲ canary, ▼ zebra finch, ◆ European starling (Okanoya and Dooling 1987), and of humans ○ (Hawkins and Stevens 1950)

other mammals. Taken together, these studies for several species suggest a higher degree of spectral resolving power in certain frequency regions that are important for vocal communication or prey and predator detection.

In an extension of the above studies using pure tones, psychophysical masking studies using vocalizations as stimuli show that for several species of birds the amount of masking of vocalizations can be reasonably predicted from the peak in the power spectrum of the vocalization and the detection threshold for a pure tone at that same frequency in broadband noise (i.e., the critical ratio). Larger signal-to-noise ratios are required for discrimination and recognition (Klump 1996; Dooling and Blumenrath 2013; Dooling and Leek 2018). Precise masking studies such as these using species' vocalizations provide a way of moving from the laboratory to the field and enable investigators to estimate communication distance in noisy, natural habitats (Klump 1996).

6.2.4 Temporal Processing

There are numerous ways to measure how the auditory system handles the temporal aspects of an acoustic signal. Again, the long-standing fascination with bird song and the tractability of birds as psychophysical subjects have led to more studies on temporal auditory processing in birds than in any other vertebrate group. In terms of traditional measures of gap detection (i.e., minimum temporal integration time), birds, including barn owls, budgerigars, European starlings, and zebra finches (*Taeniopygia guttata*), compare favorably with humans in being able to detect gaps in noise that are larger than about 2 ms (Dooling et al. 2000). In studies of the improvement of threshold with increasing duration (i.e., maximum temporal integration time), there is a remarkable consistency among birds with all showing a time constant of about 200 ms, much like humans and most other vertebrates that have been tested (Dooling et al. 2000).

On another traditional measure of auditory temporal resolution, duration discrimination, budgerigars (Dooling and Haskell 1978) and European starlings (Maier and Klump 1990) can discriminate about a 10% change in the duration of a pure tone roughly on par with humans. And finally, another traditional measure of temporal resolving power, the temporal modulation transfer function, measures the threshold for detecting the modulation of white noise at different modulation frequencies. Birds and mammals are remarkably similar, with the barn owl being as sensitive as humans at low modulation frequencies (Dent et al. 2002). It is likely that there is a relationship between the temporal modulation transfer function and birds' ability for intensity discrimination (Dooling and Searcy 1981).

By the conventional measures of temporal processing described above, birds have not been shown to be dramatically different from humans. On another measure of temporal resolution, however, birds may be quite superior to humans. Lohr and Dooling (1998) discovered that birds, and especially zebra finches, were ten times more sensitive than humans to the mistuning of a single tone in a harmonic complex. Originally considered a kind of enhanced spectral sensitivity, it quickly became apparent that a more accurate description would be an unusual sensitivity to changes in the temporal fine structure of complex harmonic sounds. This enhanced temporal discrimination has been most clearly demonstrated in experiments measuring the discrimination of forward and reversed Schroeder wave harmonic complexes (Dooling et al. 2002). These waveforms are constructed by adjusting the phases of individual components in a harmonic complex, so that the resulting waveform has a flat envelope. Forward and reversed Schroeder waveforms, which sweep up or down in frequency with the period of the fundamental, have the same overall level and spectrum that removes pitch as a cue.

Figure 6.4 shows an example of a positive and negative Schroeder waveform complex and the results for humans and three species of birds tested on Schroeder complexes of different fundamental frequencies. Humans lose the ability to discriminate between these complexes at fundamental frequencies above about 200 Hz, while birds do much better with the zebra finch discriminating between positive and negative complexes at fundamentals of 1000 Hz (Dooling et al. 2002). Subsequent experiments with zebra finches show, not surprisingly, that this enhanced ability to discriminate temporal fine structure extends to the discrimination of the temporal fine structure in the harmonic vocalizations of this species with even the slightest change in syllable structure (i.e., from rendition to rendition of the same syllable) being highly discriminable and salient (Prior et al. 2018; Fishbein et al. 2020, 2021). This extreme sensitivity suggests that birds may use acoustic fine structure to communicate important biological information (Prior et al. 2018).

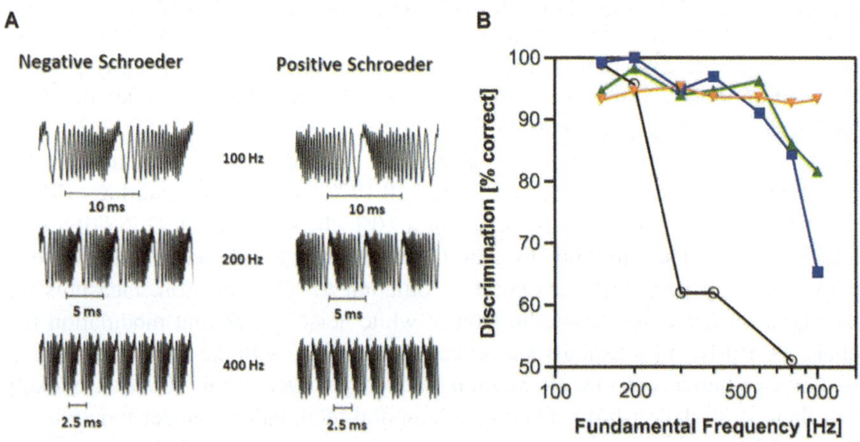

Fig. 6.4 Perception of temporal fine structure of Schroeder-phase stimuli. (**a**) Examples of positive and negative Schroeder waveform complexes with fundamental frequencies of 100, 200, and 300 Hz. (**b**) Discrimination performance of three species of birds (□ canary, □ zebra finch, ■ budgerigar) and human subjects (•). (After Dooling et al. (2002) and Dooling and Prior (2017)). In the discrimination procedure, 50% correct is a chance performance

6.2.5 Song Learning and Perceptual Categories

Birds are known for their complex vocalizations and widespread vocal learning, including the mimicry of human speech sounds. It is not surprising, therefore, that this group of vertebrates has enjoyed special attention when considering the perception and categorization of complex sounds. The phenomenon of song learning in birds, nearly unique in the animal world, provided some of the first hints of what the auditory world of birds was like. Aristotle (c.f., Peck 1984) noted that birds are the most vocal of animals, that they have the most vocal variety of any animal, and that young birds can be observed learning vocalizations from their parents. Charles Darwin (1859) also recognized that some songbirds, like children, acquire sophisticated vocalizations in part by listening to adult "tutors" and in part from prior predispositions, suggesting that the sounds uttered by birds may be the closest animal parallels to human speech and language.

The focus on species differences in birdsong in the early to mid-1900s led to increasing interest in how, what, and when birds learn their song and revealed an incredible diversity (e.g., Kroodsma and Miller 1996). Attention naturally turned to the role that hearing plays in song learning in a series of comparative studies of song development by Peter Marler and his colleagues (e.g., Marler and Peters 1989). These investigators pioneered the use of techniques restricting the hearing of adult song by young birds such as isolation, hand-rearing, selective tutoring during the sensitive period, and deafening techniques to delineate the role hearing in song development with demonstrations of early perceptual predispositions for attending to, and learning, species-specific song elements (e.g., Marler and Peters 1977; Marler, 1970). This was confirmed by showing differences in the cardiac orienting response between conspecific and other species song syllables in young nestling sparrows (Dooling and Searcy 1980).

This early vocal learning work in birds made it clear that the basic hearing capabilities described earlier do not capture how birds use their hearing in the real world any more than a human audiogram can capture the complexities of human speech perception. Hearing is obviously much more than simply detecting faint sounds or discriminating subtle differences between sounds. The auditory system also parses the acoustic world into perceptual groups or categories, recognizing important acoustic similarities among sounds belonging to one category, while also attending to the differences between sounds that define different acoustic perceptual categories.

The classic example is the development and maintenance of robust perceptual categories of human speech sound across talkers and contexts (e.g., Kuhl et al. 2008). In birds, as in humans, hearing functions to maintain vocal precision illustrated by the fact that many birds deafened as adults show both immediate and more gradual deterioration in their vocalizations (e.g., Konishi 1965; Dooling et al. 1997). By the 1960s, the tractability of birds for operant conditioning, combined with modern advances in animal psychophysical methods and the application of signal-detection theory, made it practical to begin asking more refined questions about the perception of complex sounds in birds rivaling those that were used to test speech perception in humans (Dooling and Okanoya 1995). Speech perception experiments in birds over the following several decades show that birds also perceive human speech vowels and consonants categorically, much as humans do—not a surprising finding given that some birds are excellent speech mimics (e.g., Dooling and Okanoya 1995; Dooling and Brown, 1990; Dooling, 1992).

From a completely different perspective, the complex, sequential nature of most bird song has also long suggested parallels with human language in the coding of information in the sequence of notes, syllables, and phrases that make up song. However, evidence for a strong perceptual salience of sound sequences in birds is surprisingly weak both from field studies and from song tutoring experiments (e.g., Emlen 1972; Bredman 1976; Marler and Peters 1977). Nevertheless, there is persistent speculation about such salience and evidence in starlings shows that the learning of particular sequences of sounds can indeed influence how song elements within these sequences are subsequently categorized (e.g., Comins and Gentner 2014, 2015).

The question of the perceptual salience of sequential information in song has been directly addressed in the zebra finch, which in the last 25 years has become the premier model for addressing both behavioral and brain mechanisms underlying vocal learning and song perception. Male zebra finches learn one song, consisting of a few introductory notes followed by a motif of 4–6 syllables, from their father within the first few months of life. The bird learns both the syllable sequence and syllable structure of the motif with a high degree of precision and sings this motif, unaltered, for the rest of his life (e.g., Zann 1996). Though there is evidence that these birds can discriminate rhythmic patterns (Rouse et al. 2021), the evidence that they attend to the sequence of syllables in their songs is much weaker. In budgerigars, on the other hand, there is compelling evidence of sequential dependencies both in the production and perception of the long, rambling male warble song (Tu and Dooling 2012).

Using zebra finches, budgerigars, Bengalese finches, and canaries, Fishbein et al. (2020) measured the relative discriminability of syllable structural changes versus syllable sequence changes in natural zebra finch song motifs. As expected, all four species were very good at discriminating syllable reversals (i.e., largely a temporal fine structure change due to reversing the internal structure of the syllable), while only budgerigars were able to easily discriminate between a normal motif and one in which the order of syllables was randomized (Fig. 6.5).

The excellent performance on syllable reversals is perfectly in line with the established ability of birds in psychophysical tests to discriminate changes in the temporal fine structure of complex sounds like Schroeder complexes—a performance that is beyond human capability (Dooling et al. 2002). The poor performance of songbirds on sequence changes, especially zebra finches, remains surprising. This is so given evidence from other studies of some degree of sensitivity to syllable sequence (e.g., Okanoya 2004; Berwick et al. 2011). Male zebra finches learn their song syllable sequence early in life and sing this sequence with a high degree of precision throughout adulthood. Whether these species differences in sequence

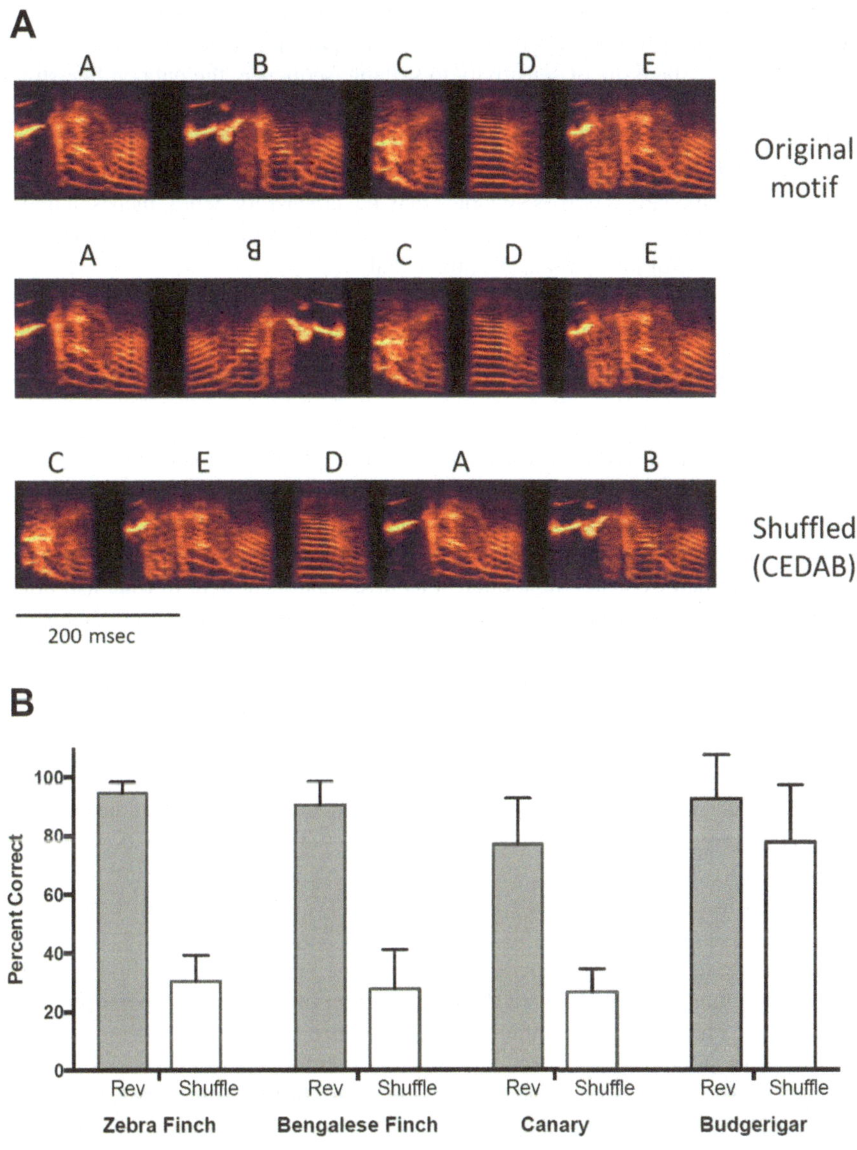

Fig. 6.5 Discrimination of zebra finch song syllables by four different species of birds: zebra finches, budgerigars, Bengalese finches, and canaries. (**a**) Top row shows a normal zebra finch song motif with 5 syllables labeled A–E. The middle row shows the same motif with the syllable "B" reversed. The third row shows the song motif with the order of the syllables shuffled. (**b**) Species differences in the ability to discriminate changes in syllable structure and syllable sequence. (Modified after Fishbein et al. (2020)). (Figure reprinted with permission of The Royal Society)

perception reflect a difference between songbirds and parrots or a difference between species with rigid versus flexible vocal repertoires is not yet clear.

Putting aside the issue of sensitivity to syllable sequence, the data on fine structure discrimination strongly suggest that birds can communicate information across song renditions of the same notes and syllables very well using extremely subtle acoustic features in the syllable that are beyond the range of human hearing. These subtle structural changes across syllables may be, at least for zebra finches, a major avenue of information transfer—far more important than the sequence of elements as was suspected for many years (Fishbein et al. 2020).

6.3 Recovery of Function Following Hair Cell Regeneration

Early developmental studies of vocal learning in songbirds demonstrated the crucial role that hearing plays in song learning and development (e.g., Marler and Peters 1987; Marler 1997). In the 1970s–1980s, comparative studies of temporary and permanent threshold shifts from acoustic overexposure in humans and other mammals was a topic of considerable interest as investigators sought to establish noise exposure limits for human health (e.g., Miller 1974). These early mammalian studies were able to clarify the role of noise level, spectrum, duration, and temporal pattern of the exposure in causing both temporary and permanent threshold shifts and, in many cases, to also identify the anatomical correlates of hearing loss and recovery.

Solely out of comparative curiosity, several studies at that time also examined threshold changes in birds following acoustic overexposure. These studies showed that birds were more resistant than mammals to both temporary and permanent threshold shifts and that hearing loss was symmetrical about the exposure frequency as opposed to a half-octave shift in hearing loss typically observed in mammals (e.g., Saunders and Dooling 1974; Dooling 1980). The reason for this resistance to permanent hearing loss remained unexplained for another decade. In 1988, two studies simultaneously discovered the phenomenon of hair cell regeneration in the inner ear of chickens and common quail (*Coturnix coturnix*) following acoustic overexposure (Corwin and Cotanche 1988; Ryals and Rubel 1988).

The interest in hair cell regeneration and recovery of function in birds received a great deal of attention in the several decades following this discovery. Other methods of inducing hair cell loss were developed such as high doses of aminoglycoside antibiotics (e.g., Marean et al. 1998). While high levels of acoustic overexposure can likely damage other structures in the inner ear, high doses of aminoglycosides only damage hair cells enabling investigators to isolate the functional effects of hair cell damage and recovery alone. Nevertheless, the picture remains complicated with considerable recovery from noise-induced threshold shift occurring before hair cell regeneration occurs (Saunders 2010; Saunders and Dooling, 2018). Hair cell loss from acoustic overexposure is related to the spectral characteristics of the exposing stimulus, whereas hair cell loss from aminoglycoside administration affects the high-frequency region of the avian papilla first, gradually extending toward the lower frequency region of the papilla with higher doses (e.g., Ryals et al. 2013). In both cases, unless there is collateral damage to other structures in the inner ear, the full complement of hair cells is generally replaced within 4–6 weeks. However, regardless of the method of causing hair cell loss, in both cases the orientation pattern of the stereovilli bundles in regenerated hair cells is radically and permanently disturbed (Salvi et al. 2008).

The discovery of hair cell regeneration in birds, with the potential for hearing recovery extending to other species, led to numerous physiological and psychophysical studies over the following decades. These studies documented the extent of behavioral recovery in a suite of psychophysical measures including audibility curves, frequency and intensity difference limens, critical bands, critical ratios, psychophysical tuning curves, minimum and maximum temporal integration, and modulation transfer functions (summarized in Salvi et al. 2008). It is fair to say that hair cell regeneration generally leads to a nearly complete recovery of pure-tone thresholds and a nearly complete recovery of hearing on other psychophysical measures such as frequency and intensity difference limens, measures of filter bandwidths, and the ability to categorize species-specific vocalizations (Dooling et al. 1997; Ryals et al. 2013). This occurs despite the universal finding of disordered orientation of stereovilli on the hair cells (Salvi et al. 2008).

Comparative studies also reveal that acoustic overexposure in birds, even when exposure and test conditions are identical, still results in considerable species differences in the amount of damage and the time course of loss and recovery from acoustic trauma (Ryals et al. 1999; Dooling et al. 2008). Some of these species differences in response to long-term noise exposures may be due to the ability of some birds to regulate the air pressure in their interaural canal by closing the Eustachian tube, thereby reducing the efficiency of the tympanic membrane. Experiments show that when this interaural space is vented with a canula, birds experience much greater threshold shift and hair cell damage from long-term noise exposure (Larsen et al. 1996).

The fact that many birds learn and maintain their vocalizations by reference to auditory information also presented a unique opportunity to test the effect of profound hearing loss, followed by inner ear repair and threshold recovery, on the production of learned vocalizations. In other words, do birds experience something like "deaf" speech, and do these vocalizations recover with inner ear repair and hair cell regeneration? Behavioral tests show that after an 8-day course of aminoglycoside antibiotics, budgerigars experienced not only hair cell loss and hearing loss but also lost the ability to produce a call that precisely matched their previous vocalization (Dooling et al. 1997). Interestingly, this precision of vocal production returned several weeks before hearing fully recovered, illustrating that mild hearing loss in birds, as in humans, is not sufficient to disrupt vocal precision. So, even in the absence of veridical auditory feedback, budgerigars, like humans, can also rely on long-term memory combined perhaps with feedback from other sensory modalities, to guide vocal production.

These studies of changes in hearing and vocal production with hair cell loss also raise the question of whether the world sounds different with a new set of hair cells. This question was answered in budgerigars, showing that while the ability to discriminate among acoustically distinct vocalizations was not impaired following hair cell regeneration, the ability to make subtle, fine-grain discriminations among acoustically similar vocalizations was affected, even weeks after the basilar papilla had been repopulated with new hair cells. Perhaps more interesting, the birds' ability to recognize previously familiar vocalizations was initially impaired by new hair cells (Dooling et al. 2006). Eventually, with experience, the ability for vocal discrimination and perceptual recognition of vocal signals returns to original levels. Thus, at least in budgerigars, there appear to be little or no long-term effects of temporary hearing loss on auditory perception, on the recognition of species-specific vocalizations, or on other aspects of acoustic communication. And perhaps related to the ability to regenerate auditory hair cells, there also appears to be no age-related hearing loss that is so prevalent in mammals (Langemann et al. 1999; Krumm et al. 2017).

Birds continue to provide an interesting preparation to examine recovery from hearing loss by simulating hair cell regeneration. For instance, it has been known for some time that Belgian Waterslager canaries have an inherited hearing loss due to missing and damaged hair cells despite ongoing hair cell regeneration (Okanoya and Dooling 1987; Gleich et al. 1994, 1996). Subsequent work has shown that this impairment is due to a sex-linked factor on the Z chromosome (Wright et al. 2004) involving hair cell genes common to hearing-impaired mammals and thus may offer new possibilities for exploring potential cures for human deafness. The fact that Belgian Waterslager canaries, when treated with aminoglycosides, experienced an additional threshold shift at high frequencies but then, upon recovery, showed 5–10 dB lower thresholds at high frequencies than before treatment is also quite intriguing (Ryals et al. 2013). Whether this is due to an increased number of functional hair cells in recovered birds or to some other process is presently unknown but worthy of future investigation. Current work on hair cell regeneration in birds is aimed at understanding the regenerative capacity of avian support cells and whether events at the molecular level in these support cells provide clues that could activate a similar response in mammalian support cells (Janesick and Heller 2018; Roccio et al. 2020).

6.4 Binaural Hearing and Sound Localization

The marriage of operant techniques and modern psychophysical procedures resulted in an explosion of studies on basic hearing mechanisms in birds in the decades after the 1940s. It was not long before investigators began to turn their attention to more complicated aspects of hearing in the real world such as sound localization. The importance of acoustic communication for birds coupled with their small heads and closely spaced ears have also made them excellent subjects for understanding binaural hearing and sound localization. Bird studies have contributed prominently to unraveling the neural mechanisms of sound localization. Beginning with the work of Masakazu Konishi and his group (Konishi 1973, 2003), the barn owl quickly became the primary model organism for investigating the neural mechanisms of sound localization as proposed by Lloyd Jeffress (1948) for the underlying neuronal machinery. The barn owl (Payne 1971), other owls, some birds of prey (e.g., harriers, Rice 1982), even chickens (Krumm et al. 2022), and small songbirds (e.g., Klump et al. 1986) are capable of accurate sound localization despite no prominent external ears and a lack of pinnae. Waldemar Engelmann (1928) reported a localization acuity in chicks of about 4° when responding to a clucking hen. Although it was difficult to control the stimulus parameters in these early observations and repeated stimuli were used, the data suggest a fairly good localization acuity for chickens. Using tones produced by Edelmann whistles at different locations, Granit (1941) observed that pine grosbeak (*Pinicola enucleator*) may discriminate angles of sound incidence of 20°–23°. Beginning around the 1950s, studies in a number of species using electro-acoustically generated stimuli have established beyond doubt that birds have an excellent ability to localize sounds.

6.4.1 Exploiting Natural Orienting Responses for Investigating Sound Localization

In contrast to early psychophysical studies of hearing in birds, the study of sound localization required considerable creativity and an adjustment in techniques and understanding of a bird's natural environment as exemplified by early studies on the barn owl. Payne and Drury (1958) provided the first reports that barn owl can hunt in total darkness guided solely by their auditory system. Payne (1971) subsequently extended these experiments and estimated the binaural cues that allowed the owl to accurately strike a target.

The work by Konishi and his group further developed the barn owl as an excellent model for revealing the mechanisms of sound localization (for a review, see Konishi 2003). Refining the earlier method used by Payne (1971), Konishi (1973) was able to observe a barn owl's attack on a small loudspeaker broadcasting rustling sounds under infrared illumination and with no visible light. This enabled him to determine with a high degree of precision the accuracy of a barn owl flying to and then striking a sound source (Fig. 6.6a–c). When flying toward the target loudspeaker from more than 3 m, the owl's minimum localization error for tones was as small as 10 cm, which, coincidently, corresponds nicely to the range covered by the talons when they are spread out to grab the target (Fig. 6.6c).

Konishi and colleagues extended the early barn owl studies by exploiting a reflexive saccadic head turn by a perched barn owl toward the location of a sound source to measure the localization accuracy (Knudsen et al. 1979; Knudsen and Konishi 1979). Unlike in the paradigm with the owl flying at the target (which

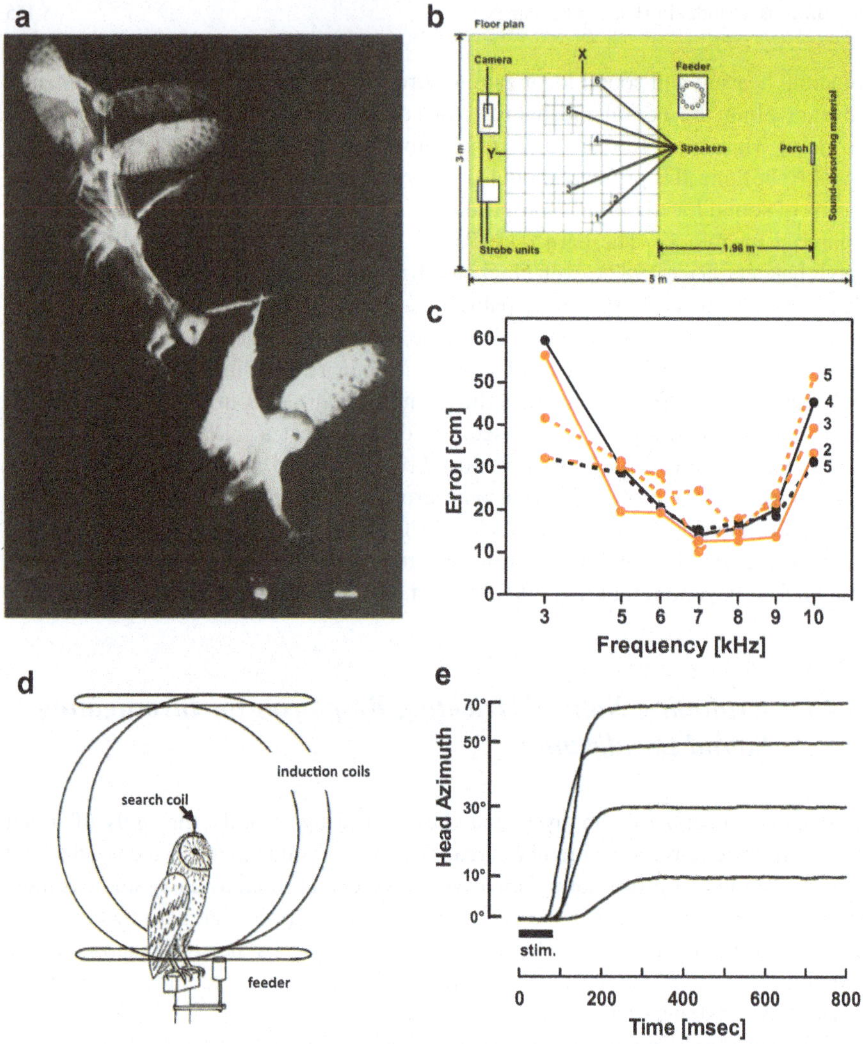

Fig. 6.6 Behavioral studies of barn owl sound localization. (**a**) Multiexposure IR picture of an owl striking a paper target that is tied to the tail of a mouse. The owl is homing in on the paper that produces rustling noise when the mouse moves indicating a sound-guided hunting behavior in total darkness. (**b**) Schematic of the experimental setup employed for determining the accuracy of the owl's strike in the sound-guided target approach. Here, the target is a loudspeaker hidden in the floor broadcasting the sound. The owl waits on the perch, and it flies down and strikes the target speaker once it starts playing. The owl is rewarded with a feeder for approaching the correct loudspeaker. (**c**) Striking error in relation to the frequency of a pure tone that generally lasted until the owl's landing. Numbers designate the target speaker positions in the setup shown in subpanel b. The dashed colored line shows data for a short tone beep presented at position 5. (**d**) Experimental setup for measuring the owl's head orientation during the reflexive orienting response elicited by a sound. The alternating magnetic field produced by the large coils induces a voltage in the search coil mounted on the owl's head that reflects the head orientation. (**e**) Change of head azimuth angle over time. The stimulus is so brief that the owl must rely on the cues analyzed before the saccadic head motion is initiated. (Subpanels **a–c** from Konishi (1973); subpanels **d**, **e** from Knudsen and Konishi (1979) and Knudsen et al. (1979), respectively. Subpanels **a–c** reprinted with permission of *American Scientist*, magazine of Sigma Xi, The Scientific Research Society. Subpanels **d**, **e** reprinted with permission of Springer Nature)

resulted in changing acoustic cues during the approach), Konishi's head-turn paradigm, with a perched owl, allowed a much better control of binaural cues related to both azimuth and elevation of the sound source. In measuring head orientation in both azimuth and elevation with the search-coil technique (Fig. 6.6d), the barn owl showed very precise sound localization ability even under open-loop stimulation, that is, when the sound was switched off before the head started turning (Fig. 6.6e). This precluded the bird homing in on the sound source position. In this paradigm, the observed accuracy in azimuth evident in the error of the head-turning response was 3.6° for broadband noise, 7.9° for 4-kHz pure tone, and it deteriorated above and below that frequency. These creative techniques established the barn owl's exceptional ability in localizing sounds presented from a single source.

6.4.2 *The Birds' Ability for Discriminating Sounds from Two Directions*

The methodological advances used in absolute localization studies above needed to be modified again in order to study the ability to discriminate between sounds coming from different directions—a measure called the minimum resolvable angle (MRA). Here a critical issue is sound reflection. Schwartzkopff (1950) determined the MRA (Feinkohl et al. 2016) for locating the sources of two sounds (loudspeakers) differing in azimuth in two Eurasian bullfinches (*Pyrrhula pyrrhula*) with the test cage outdoors. Since only a single sound source was used in this study design, the MRA was considered to reflect absolute localization accuracy. The outdoor placement of the setup on a patio made sure that reflections from surfaces (i.e., echoes) could not affect localization performance. By comparing the localization performance for a series of 300-ms tone pulses with transient onsets with results from long-duration persisting tones with a gradual onset, Schwartzkopff was able to differentiate between localization mechanisms based on the transient interaural time differences (ITD), interaural phase differences (IPD), or interaural level differences (ILD).

Results showed that an angle of more than 25° was required for the birds to identify which of the two loudspeakers had played the 1500-Hz or 3000-Hz tone. For the series of 850-Hz tone pulses, however, an angle of 30° was not sufficient to identify which speaker produced the tone. Finding no difference in the discrimination ability between the pulsed and persistent 1500-Hz or 3000-Hz tones led Schwartzkopff (1950) to conclude that ITDs are not important. Since he found it difficult to train the birds to localize 850-Hz tones, Schwarztkopff also argued that was unlikely that the birds were exploiting IPDs. From failing to train a bullfinch that had one ear deafened, he concluded that it must be a binaural comparison enabling the localization considering ILDs as the most likely cue.

More recent experiments (Klump et al. 1986; Park and Dooling 1991; Feinkohl et al. 2016) were able to estimate the MRA in other species of small birds using

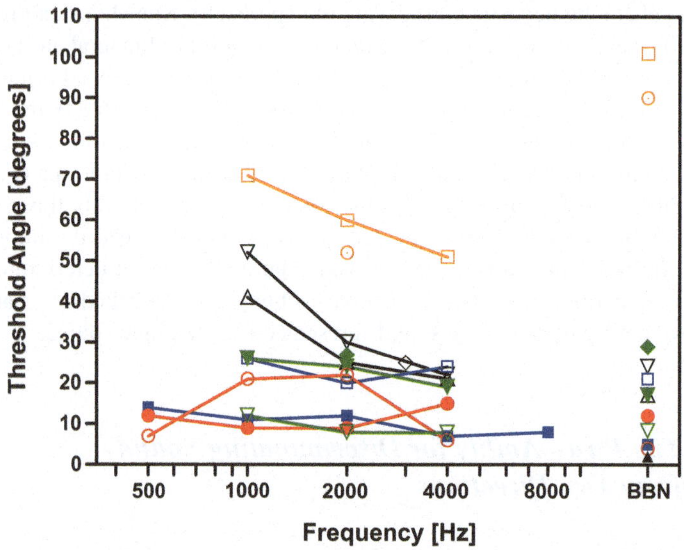

Fig. 6.7 Sound localization ability of birds for tones of different frequencies and broadband noise (BBN). Minimum audible angle (MAA) and minimum resolvable angle (MRA) reflecting localization acuity and accuracy, respectively. Filled symbols indicate the MAA and open symbols the MRA. Species are indicated by the symbols: ■ barn owl (Krumm et al. 2019); ● chicken (Krumm et al. 2022); ▼ European starling (closed loop) (Feinkohl and Klump 2013; Feinkohl et al. 2016); ◆ European starling (open loop); ▽ European starling (open loop), ▲ marsh hawk (*Falco uliginosus*) (Rice 1982); ○ bobwhite quail (*Colinus virginianus*) (Gatehouse and Shelton 1978); □ zebra finch, ▽ budgerigar, △ canary (Park and Dooling 1991); □ great tit (Klump et al. 1986); ○ pigeon (Lewald 1987)

pure-tone frequencies between 750 Hz and 8 kHz, broadband noises, and digital renderings of bird vocalization (see Fig. 6.7). These studies confirmed that small songbirds in an acoustic environment with little sound reflections from walls or other surfaces are capable of accurate sound localization with the best localization performance observed for broadband signals. The smallest ITD cues corresponding to the best localization accuracy in small birds are in the range of 20–30 μs and the ILD cues are in the range of 0.5–2 dB. This ability of small birds for processing differences between the sounds impinging onto the two ears being separated by 15 mm or less corresponds well to the ability of human subjects for analyzing interaural time and intensity-difference cues. The unexpected good sensitivity of birds in this task may partially be explained by the enhancement of these differences by the interaural canal acoustically connecting the birds' middle ears (e.g., Köppl 2019).

Yet another experimental paradigm to understand how hearing works in a natural environment measures the minimum audible angle or MAA (Mills 1958). This procedure asks whether the bird can detect a change in sound source position. This is also referred to as a relative localization task. The MAA was determined in the European starling (Feinkohl and Klump 2013), the chicken (Krumm et al. 2022), and the barn owl (Bala et al. 2003; Krumm et al. 2019). The starling's MAA (Fig. 6.7) was like its MRA. The smallest MAA has been observed in the barn owl. Using a pupillary-dilation reflex response and a 100-ms pulse of broadband noise as the stimulus, Bala et al. (2003) determined an MAA of only 3°. Later, Krumm et al. (2019) determined the MAA of the barn owl for narrowband and broadband noise signals. The MAA for broadband noise presented from the front was 3.4° confirming the results obtained by Bala et al. (2003). Thus, the barn owl has an MAA that is in the range of the human MAA despite having a much smaller interaural distance confirming a highly sensitive mechanism for neural computation of ITDs in this species. Surprisingly, the chicken MAA is also quite small and rivals the barn owl MAA at frequencies up to 2 kHz (Krumm et al. 2022).

6.4.3 Comparison of Localization Cues Across Species

Schwartzkopff (1950, 1952) was the first to speculate that the open connection between the two middle-ear cavities provided by the interaural canal (described by Wada 1923) must affect the binaural cues for localizing the sound. Conducting cochlear microphonic (CM) measurements in a few small songbird species, he found that there was little difference between the amplitudes of the CM with the ear canal contralateral to the recorded ear being open or acoustically blocked leading him to the conclusion that in the Eurasian bullfinch the interaural canal does not enhance ILDs. The large change in amplitude of the CM with the angle of sound incidence resulted in ILDs that could explain the localization performance of the bullfinch (Fig. 6.8a). Then, using an apparatus that allowed the independent stimulation of the two ears of the bullfinch with sounds of different intensity, he demonstrated that an ILD of 1.4 dB corresponding to the MRA could be perceived by the bullfinch confirming that the azimuth sound localization threshold of the bullfinch could be explained by the acoustic ILDs created by the head shadow.

Later, miniature transducers positioned at the entrance of the budgerigar's ear canal (Welch and Dent 2011) or in the ear canal of the barn owl (e.g., Moiseff and Konishi 1981) were used to allow for dichotic headphone stimulation. In the budgerigar, the best mean ITD thresholds for broadband noise bursts and 2 kHz tones were 15 and 17 μs, respectively. The budgerigar's mean best ILD thresholds were 2.3–3.4 dB for tones and 1.5 dB for broadband noise. In the barn owl, ITDs of 10 μs were sufficient for reliably eliciting a head turn toward the expected azimuthal direction (Moiseff and Konishi 1981). ILDs are perceived by the barn owl as cues representing the elevation of the sound source reflecting the effect of the facial ruff and asymmetries in the morphology of the left and right ear (Moiseff 1989).

The role of the interaural canal has continued to be a hot topic in the debate on bird sound localization mechanisms. One suggestion is that the interaural canal serves to establish a pressure-gradient system at each ear enhancing the ITD and ILD. CM measurements for revealing directional effects created by a functional pressure-gradient system have been conducted in a number of bird species (e.g.,

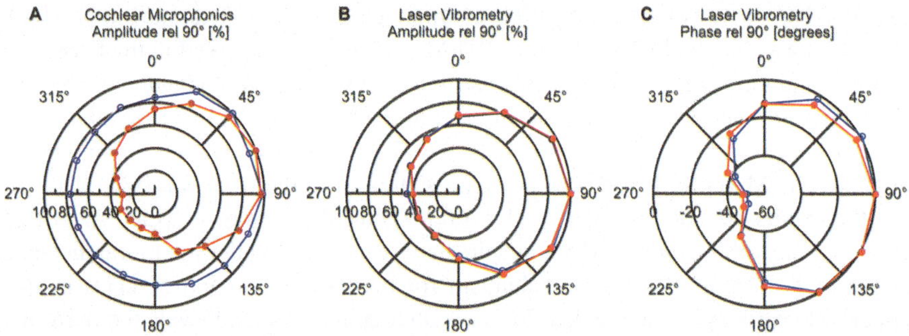

Fig. 6.8 Directional patterns of sound-localization cues of birds: (**a**) normalized cochlear microphonic amplitude (in percent relative to value for 90° angle of sound incidence, i.e., sound from the right side) of the Eurasian bullfinch in relation to the angle of sound incidence. (After Schwartzkopff (1952)). Blue and red lines show data for 800 Hz and 3200 Hz, respectively. (**b**) Normalized amplitude (in percent relative to value for 90° angle of sound incidence) of tympanum vibrations in relation to the angle of sound incidence in the budgerigar. (After Larsen et al. (2006)). Red line and symbols show measured data for 2-kHz tone, blue line and symbols show model results taking into account the sound transmission through the interaural canal. (**c**) Relative phase (in relation to value for 90° angle of sound incidence) of tympanum vibrations in relation to the angle of sound incidence in the budgerigar. (After Larsen et al. (2006)). Red line and symbols show measured data for 2-kHz tone, blue line and symbols show model results

Klump and Larsen 1992; Larsen et al. 2006; Kettler et al. 2016; Köppl 2019). The common finding from these studies is that the interaural canal is mostly effective at low frequencies (i.e., frequencies below 4 kHz). Kettler et al. (2016) used laser vibrometry for measuring the barn owl's tympanum vibration amplitude in relation to the angle of sound incidence. They demonstrated a high directionality of the tympanic response for frequencies between 2 and 3 kHz that was abolished if the ear contralateral to the ear at which the tympanum vibrations were measured was blocked. The smallest observed transmission loss through the interaural canal was 6 dB at 2 kHz, which is sufficiently small to produce a pressure-gradient receiver. Also, a sufficiently low transmission loss for establishing a pressure-gradient receiver enhancing ILDs was observed in other bird species (budgerigar, European starling, and chicken). At the same time, this transmission loss has been found to increase the physiological ITDs enhancing the cues available for discrimination. However, anesthesia remains a possible confound in measuring an enhancement of interaural cues by the interaural canal (Larsen et al. 2016).

6.4.4 The Barn Owl as a Model for Neural Processing of Auditory Spatial Information

The impressive sound localization ability that enables the barn owl to capture a mouse even in total darkness (Konishi 1973) has long attracted interest in the neurophysiological mechanisms underlying sound localization. Pioneering studies by

Konishi using the barn owl made this species a model for the study of neuronal sound localization mechanisms (e.g., see Konishi 2003 for a review). We now know that the direction of sound sources is computed based on ITDs and ILDs by neurons in the owl's midbrain that has a map of auditory space with neurons tuned to a specific direction (Knudsen and Konishi 1978). In the barn owl auditory space, ITD represents the azimuth angle of sound incidence and ILD the elevation (Moiseff 1989). The ratio of direct to reverberant sound provides the barn owl with a monaural cue representing the distance of a sound source (Kim et al. 2008). Up to the level of the midbrain, ITDs and ILDs are processed in parallel pathways with neurons specialized in processing temporal or spectral features of the sound, respectively (Takahashi et al. 1984; Carr 1992). ITDs are already evaluated at the level of the auditory brainstem in the nucleus laminaris that has neurons receiving binaural input functioning as coincidence detectors and as a population forming a map of ITDs (Carr et al. 2015). These neuronal coincidence detectors in combination with axonal delay lines that compensate for the acoustic time differences of sounds reaching the two ears form a structure that has been suggested by Jeffress (1948) as forming the basis of a mechanism for azimuthal sound localization (Ashida and Carr 2011).

Based on the above correspondence, the barn owl's mechanisms for sound localization were viewed, for a long time, as a model for the mechanisms underlying human sound localization. Studies in mammals suggest, however, that mammalian sound localization based on ITDs involves a bilateral comparison of the excitatory responses of neurons in the medial superior olive, the mammalian homolog of nucleus laminaris, in both sides of the brainstem providing input to the auditory midbrain (the so-called two-channel hemispheric model, see review by McAlpine and Grothe 2003). Downstream from the barn owl's midbrain, however, the neural patterns of representing ITDs become more "mammal-like" reconciling the coding schemes in the barn owl and mammalian brain (Peña et al. 2019).

6.5 Bird Hearing in Complex Acoustic Environments

Not surprisingly, birds have proven to be an extremely useful model of how the vertebrate auditory system functions in the real world. It is known from human studies that auditory perception in complex acoustic environments constitutes an enormously complicated problem. Early avian laboratory studies have mostly focused on fundamental auditory processing mechanisms often with sounds being presented in isolation from a single source. But since the 1980s, studies have pioneered investigations in more complex acoustic settings that reflect hearing in real-world situations.

In the real world, signals are typically distorted when being transmitted through the natural environment (e.g., by reflections from objects or by the effect of wind). However, this is not always to the disadvantage of a receiver, since birds have been shown to estimate the distance to another signaling individual by evaluating the distortion (Naguib and Wiley 2001).

In real-world situations, sounds from multiple sources are typically present at the same time, and the bird auditory system must segregate the signals of interest from the acoustic background composed of the sounds from other sources (for a review, see Hulse 2002). When listening to a conspecific singer in a dawn chorus, for example, the listener needs to segregate the songs of the conspecific from those of the other species. In addition, the listener needs to integrate the components and the sequence of conspecific vocalizations to evaluate the whole series of elements presented by the singer. These tasks have been subsumed under the name "auditory streaming" (Bregman 1991). Not only can masking of the signals of interest be due to other biological sound sources, but it can also result from interference with non-biological sources of noise: noise; matters enormously in the real world (Brumm and Slabbekoorn 2005; Wiley 2015).

6.5.1 Masking and the Cocktail-Party Effect

Stationary masking noise, as it is used in traditional critical masking ratio experiments in the laboratory, is rarely found in nature. In the natural environment, masking background noise fluctuates in amplitude. These fluctuations may result from the intermittent activity or varying levels of the sounds produced by the sources (e.g., in a dawn chorus, an individual bird sings for a few seconds and then is silent again, and song elements frequently have inherent level fluctuations). Furthermore, fluctuations in the level of the background noise may be imposed on the sounds when traveling through the environment (e.g., by wind-related air turbulence). The European starling was the first animal model in which masking by fluctuating background noise was studied (e.g., Klump and Langemann 1995). In starlings, like humans at the cocktail party, the fluctuating level of background noise can be exploited to substantially improve sensitivity in signal detection (Fig. 6.9).

Bee and Micheyl (2008) suggest that the improvement in sensitivity due to this cocktail-party effect is a widespread important feature in auditory processing having evolved to enable animals communicating in noisy natural environments. Thresholds in noise can be lowered by up to 20 dB if the level fluctuations in the noise are modulated at rates of 10 Hz or below. Two mechanisms are hypothesized to explain this gain in sensitivity: temporal processing of signals within an auditory frequency filter (i.e., within-channel cues) or a comparison of the signal envelopes across different auditory frequency filters (i.e., across-channel cues). So far, the evidence suggests that in the European starling, as in human subjects, both within- and across-channel mechanisms may contribute to this effect (Klump 2016).

Our knowledge of the frequency analysis in the birds' ear, and the birds' exploitation of the level fluctuations in the background noise, can be used to predict the birds' ability to communicate over distance in a natural environment (e.g., Klump 1996; Dooling and Leek 2018). Detection threshold for a signal in steady-state

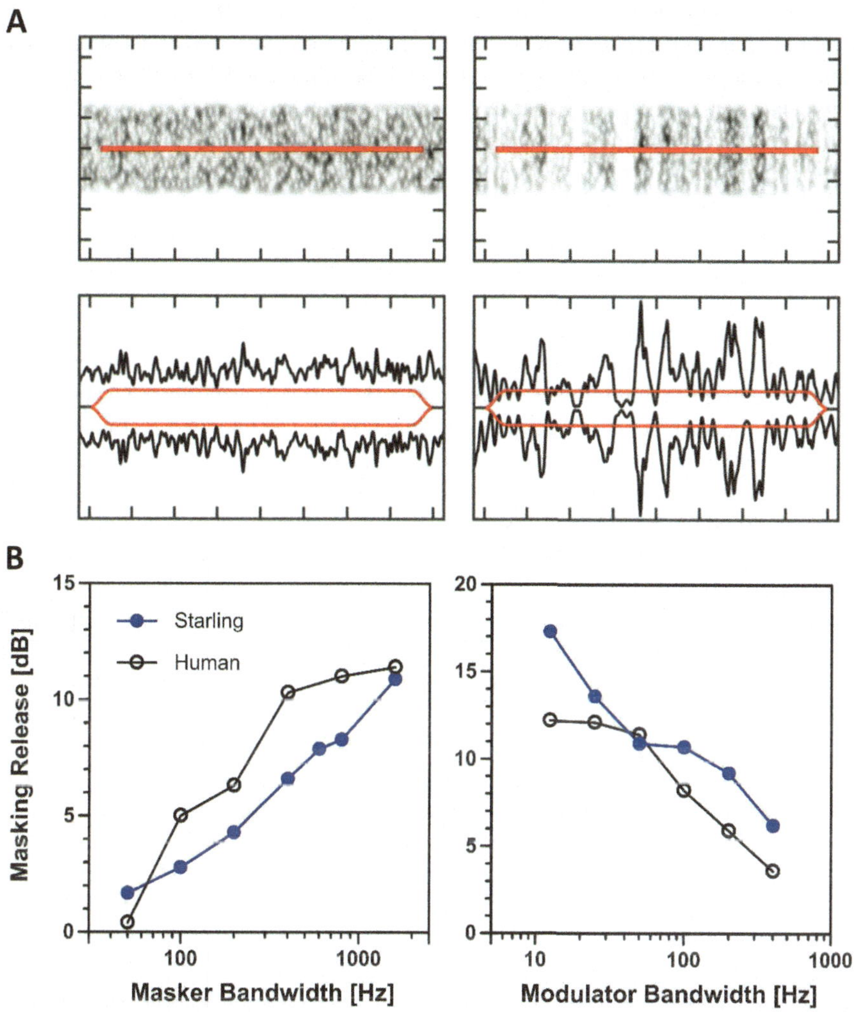

Fig. 6.9 Threshold improvement for the detection of a tone in noise by exploiting temporally coherent amplitude fluctuations in different frequency bands (also termed comodulation masking release, CMR). (**a**) Upper panels show spectrogram of a tone (red line) embedded in a broadband masking noise with (left panel) or without (right panel) comodulation of the masker. Lower panels show envelopes of the waveforms of the tone (red lines) and maskers (black lines) with (left panel) or without (right panel) comodulation of the masker. (**b**) Masking release observed in the European starling and humans resulting from comodulation of the masker. (Data from Klump and Langemann (1995), Schooneveldt and Moore (1989)). Left panel shows the CMR in relation to the masker bandwidth. Right panel shows the CMR in relation to the fluctuation speed of the envelope that was determined by the modulator bandwidth (small bandwidth indicates slow envelope fluctuations and large bandwidth indicates fast envelope fluctuation)

wideband background noise is typically between 20 and 27 dB above the spectrum level of the noise (SNR as predicted by the critical ratio, see above). Exploiting the natural level fluctuations, this limit for signal detection can be lowered to an SNR between 0 and 10 dB. Given that the spectrum of natural, wind-generated noise is dominated by low frequencies, the low SNR achieved by exploiting the level fluctuations will allow the birds often to be as sensitive in detecting signals as their absolute threshold permits.

Natural background noise produced by the other singing birds in a dawn chorus will be more of a challenge for song signal detection, especially since the frequency spectrum of the dawn chorus overlaps with the spectrum of the songs of species that contribute to that cacophony. Even in this situation, however, the temporal structure of the dawn chorus permits using short episodes with low noise levels to improve signal detection. In natural background noise, the active space of bird vocal signals appears to be sufficiently large to allow detection by the intended recipients (e.g., the neighbors of a territory holder or potential mates).

In the last 50 years, however, background noise from anthropogenic sources, especially road-traffic noise, has become an interesting topic in bird communication (e.g., Slabbekoorn and den Boer-Visser 2006; but see Nemeth and Brumm 2010; Ware et al. 2015). It has been shown in multiple species that birds can adaptively modify the amplitude of their song (the so-called Lombard effect, see Brumm and Todt 2002; Osmanski and Dooling 2009) and the frequency spectrum of the song elements to enlarge the communication distance (e.g., Slabbekoorn and Peet 2003; Pohl et al. 2009). Though it is clear that traffic noise can reduce the quality of the birds' natural habitat, it is less clear whether birds are consistently able to adapt.

While most studies on bird hearing in noisy environments have focused on signal detection, we know that discrimination and recognition of vocal signals are also necessary for successful communication. As an example, it is one thing to detect speech and quite another to be able to understand what is being said. In general, the SNR for discriminating a signal in background noise appears to be about 2–3 dB larger than for mere signal detection, and recognition requires another 2–3 dB increase in SNR (e.g., Dooling and Leek 2018). In most environments with natural background noise, this may not severely constrain communication. However, in environments that are dominated by high levels of anthropogenic noise, this could easily have a dramatic effect on acoustic communication and the detection of other biological behaviors such as the detection of predators.

6.5.2 Auditory Scene Analysis

The dawn chorus of birds results from multiple singers being active at the same time. Listeners in this dawn chorus are confronted with the task of identifying the sounds produced by individual singers in such a complex acoustic scene. This requires that sound signals from specific sources are separated from those of other sources, and at the same time, the sequence of sounds and their components originating from one source are integrated into a complex percept. Such a percept of a song may provide information on the species of the singer, its individual identity, and attributes reflecting its quality being important for choosing a mating partner. In his seminal review on auditory scene analysis by human subjects, Bregman (1991) referred to the sounds originating from a specific source as an auditory stream and the associated percepts as auditory objects. Since the ability of birds to communicate in a complex acoustic scene is of utmost importance, they have provided an excellent animal model for studying the mechanisms underlying auditory scene analysis (Hulse 2002).

In several pioneering studies, Hulse and colleagues investigated the natural propensity of European starlings to segregate auditory objects in complex acoustic scenes (Hulse et al. 1997; Wisniewski and Hulse 1997). Starlings were trained to discriminate between two categories of bird song snippets, one category containing a starling song together with the song of one other bird species and the other category being composed of the songs of other species excluding starlings. Results showed that the birds were able to form two categories (with/without starling song) despite a strongly varying composition of the sound snippets. The starlings were able to transfer this ability to discriminate immediately to novel exemplars of the two stimulus types. Furthermore, when being tested in unrewarded probe trials with novel single-species song snippets, they still were able to successfully categorize novel songs (starling song/song of other species). The starlings would even maintain the discrimination of the two-species song mixtures when recordings of a dawn chorus were added as distractors. This study convincingly demonstrates that European starlings can form auditory objects from starling song allowing them to single out songs from an auditory mixture. The starlings were also able to discriminate song elements from two different individual starlings (Wisniewski and Hulse 1997). The discrimination performance was reduced, but still significantly above chance when song snippets of up to four other starlings were added as distractors providing no information for the discrimination. This task complexity surely rivals the ability of human subjects to discriminate familiar voices in the cacophony of talkers at a cocktail party (Hulse 2002), although this ability is not unexpected given that starlings possibly learn songs from each other when assembling at the roost (Hausberger et al. 2008).

In an extension of these experiments, starlings were tested with pure tones in an auditory streaming paradigm that had been developed for human subjects by van Noorden (1975). In this so-called ABA-paradigm, triplets of A and B tones are followed by a short silent interval with a duration that was identical to the time between the onsets of two tones in the triplet (Fig. 6.10a). If the A and B tones have similar frequencies, human subjects perceive one stream of sounds with a galloping rhythm. If the A and B tones have a large frequency separation, human subjects perceive two separate isochronous streams of A and B tones, respectively, with A tones being perceived as having two times the presentation rate of the B tones. This subjective percept of the human subjects occurs without any prior experience with the signals. In a parallel experiment, European starlings were trained to discriminate sequences of tones of the same frequency that were presented with a galloping rhythm or with

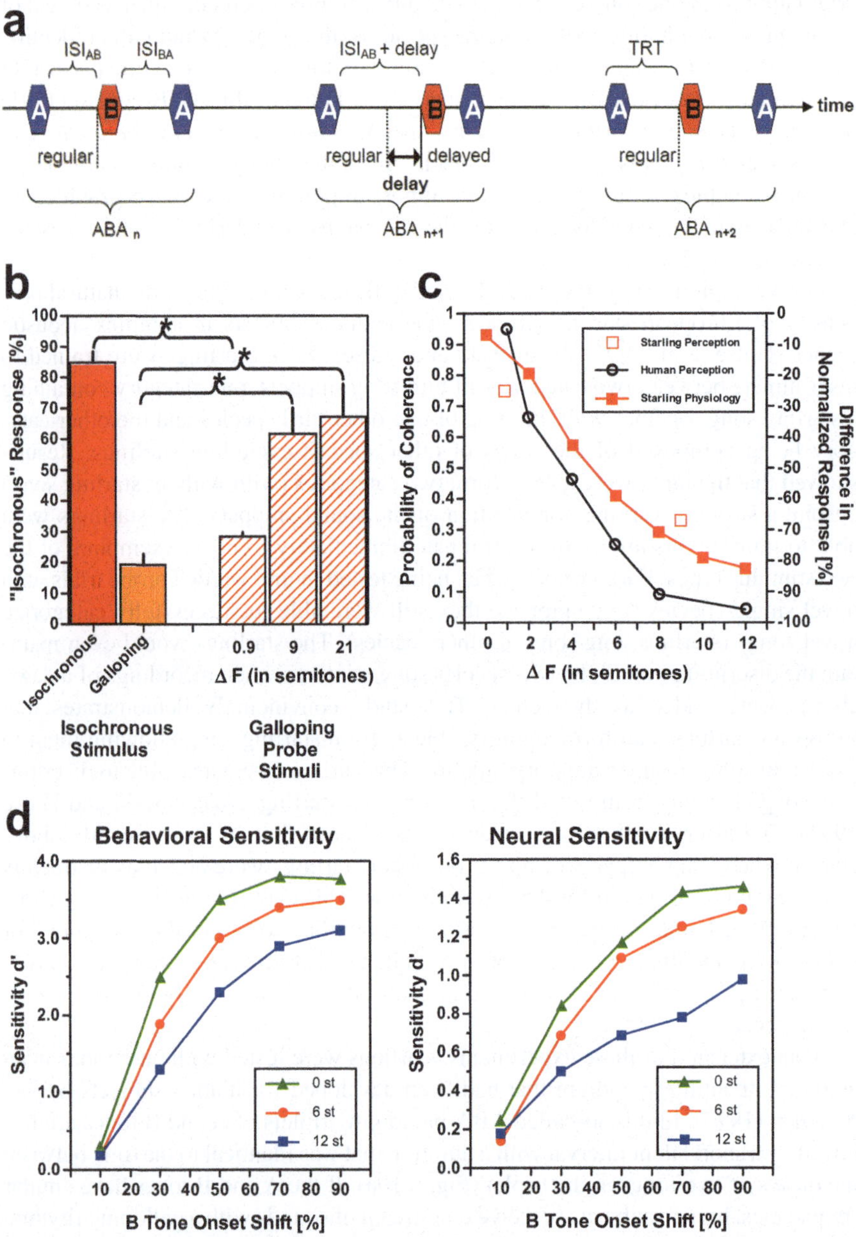

Fig. 6.10 Perception of auditory streams by European starlings and its neural correlates. (**a**) The ABA_ stimulus paradigm that was developed by van Noorden (1975) presents triplets of two different signals (A and B) that are perceived with a galloping rhythm if A and B signals are processed in one common stream and with two isochronous rhythms if A and B signals are processed in **Fig.**

a slow or fast isochronous rhythm (Fig. 6.10b). In this initial training, the identical sequential tones had frequencies of 1000 Hz, 1050 Hz, 1710 Hz, or 4538 Hz. Once they achieved this discrimination of rhythms being 85% correct, unrewarded probe stimuli were introduced. These probe stimuli consisted of triplets of 1000-Hz A tones and B tones differing in frequency by 0.5, 9, or 21 semitones. Even the lowest frequency difference of 0.5 semitones should lie above the birds' frequency difference limen (Langemann and Klump 1992). If the starlings perceived the A and B tones in the probe stimuli as one stream, their response should indicate perceiving a galloping rhythm. If they perceived separate low- and high-frequency streams, their response should indicate perceiving an isochronous rhythm. The results of this test on stream segregation indicate that the birds perceived A and B tones with a 0.5 semitone frequency difference as belonging to one stream and the A and B tones separated by 9 or 21 semitones as belonging to two streams. Thus, starlings appear to have a similar percept of auditory streams as human subjects.

Van Noorden's (1975) ABA-paradigm has also proven suitable for investigating the neural mechanisms underlying auditory stream segregation in the European starling. Using stimuli that allowed testing a larger range of frequency differences, with time periods ranging from 100 to 800 ms and duty cycles between 12.5% and 100%, Bee and Klump (2004) reported response patterns in starling forebrain neurons that reflected the starlings' percept of auditory streams (Fig. 6.10c). The dependence of the neurons' rate responses on the stimulus parameters suggested that suppression provided by the A tones presented at the neurons' best stimulating frequency on the B tone responses resulted in a separated representation of A and B signals by neurons in the starling's auditory forebrain. Adaptation of the tone-evoked neuronal responses could also explain the gradual buildup of stream segregation over time being observed in human subjects (Bee et al. 2010). A separate representation of auditory streams by starling forebrain neurons has also been observed for A and B stimuli differing in other stimulus features such as amplitude

6.10 (continued) separate streams. The likelihood of a percept of galloping or isochronous rhythms is viewed as a subjective measure of stream segregation. The sensitivity for the detection of a time shift of the B signal relative to the A signals is viewed as an objective measure of stream segregation with a high sensitivity indicating preferential processing of both signal types in one stream. (**b**) Probability of an isochronous response when starlings were presented with an ABA-triplet series in relation to the frequency difference between A and B tones (hatched bars, responses in unrewarded probe trials) after the birds had been trained to discriminate signals with a galloping or isochronous rhythm. (After MacDougall-Shackleton et al. (1998)). (**c**) The probability of perceiving one coherent stream in starlings and humans in relation to the frequency difference between A and B signals (right axis) is compared with starling forebrain neuron's difference between the response to A and B signals. These signals in the triplets are represented by separate populations of neurons if perceived as being in separate streams. (After Bee and Klump (2004)). (**d**) Behavioral and neuronal sensitivity for responses to a time shift of the B signal within an ABA-triplet in relation to the size of the time shift and the frequency separation between the A and the B tones. (After Itatani and Klump (2014)) modulation rates (Itatani and Klump 2009) or phase relationships between components in harmonic complex tones (Itatani and Klump 2011).

In humans, an alternative approach to studying auditory stream segregation measured the subject's ability to discriminate a stimulus feature that relied on within-stream comparisons. For example, human subjects are more sensitive in detecting a time shift between the onsets of A and B tones of the triplets if these are processed in the same stream. In the case of a time shift, the galloping rhythm is perceived as being irregular (Micheyl and Oxenham 2010). Itatani and Klump (2014) tested starlings with this paradigm and showed that the birds were more sensitive in detecting the time shift of the B tones if being processed in the same stream as the A tones. Thus, the sensitivity can be used as an objective indicator for auditory stream segregation. The starling's sensitivity for detecting the shift was reduced with the increasing frequency difference between the tones in the ABA-triplet (Fig. 6.10d). The sensitivity of starling forebrain neurons recorded while the birds made their behavioral decision varied with the frequency difference between A and B tones in a similar way as did the behavioral sensitivity.

These studies in birds regarding the processing of sounds in complex natural acoustic scenes with a mixture of sounds from different sources have pioneered our understanding of processing mechanisms relevant to real-world hearing. The European starling has served as a prime model for investigating this topic using both psychophysical and neurophysiological methods. Results from this species show many similarities to human auditory scene analysis (Klump 2016; Hulse 2002) and thus have considerably advanced our general understanding of hearing in natural settings.

6.6 Conclusion

In summary, the vocal virtuosity of birds and their propensity for vocal learning have long fueled intense interest in their hearing capabilities and how birds compare with other vertebrate groups, especially humans. Despite dramatically different peripheral and central auditory systems, birds as a group generally approach the hearing capabilities of humans in a narrow frequency range of 1–5 kHz but outperform humans on some measures of temporal processing. The large number of species tested has revealed systematic relationships between vocalizations, hearing capabilities, and anatomical features of the cochlea in birds. Birds stand out among other vertebrate groups in the sheer number of species tested because of their tractability as laboratory subjects. Compared with other animal groups, this has facilitated many investigations into more sophisticated aspects of hearing such as masking and the perception of complex sounds providing a bridge between laboratory studies of hearing and communication in natural environments. Beyond conventional auditory tests, birds have proven excellent subjects for examining hearing in the real world. Birds exhibit perceptual categories for many complex sounds including their own and other species' vocalizations and the consonants and vowels of human speech. Studies of masking, the cocktail-party effect, and auditory scene analysis reveal an extreme sophistication in dealing with noise and untangling ambiguities when hearing in the real world that is noisy and variable. Birds, especially the barn owl, have provided enormous insights into the mechanisms of sound localization. Adaptations, such as an interaural pathway, compensate for small interaural distances enabling birds to attain localization precision greater than predicted. Finally, these aspects of complex auditory perception, coupled with the phenomenon of hair cell regeneration in the avian ear following damage, offer a promising platform for testing and designing new methods for aiding recovery in human hearing.

References

Aristotle (1984) The history of animals (trans: Peck AL). Heinemann, London

Ashida G, Carr CE (2011) Sound localization: Jeffress and beyond. Curr Opin Neurobiol 21(5):745–751. https://doi.org/10.1016/j.conb.2011.05.008

Bala ADS, Spitzer MW, Takahashi TT (2003) Prediction of auditory spatial acuity from neural images on the owl's auditory space map. Nature 424(6950):771–774. https://doi.org/10.1038/nature01835

Barrington D (1773) XXXI. Experiments and observations on the singing of birds, by the Hon. Daines Barrington, Vice Pres. RS in a letter to Mathew Maty, MD Sec. RS. Philos Trans R Soc 63:249–291

Bee MA, Klump GM (2004) Primitive auditory stream segregation: a neurophysiological study in the songbird forebrain. J Neurophysiol 92(2):1088–1104. https://doi.org/10.1152/jn.00884.2003

Bee MA, Micheyl C (2008) The cocktail party problem: what is it? How can it be solved? And why should animal behaviorists study it? J Comp Psychol 122(3):235–251. https://doi.org/10.1037/0735-7036.122.3.235

Bee MA, Micheyl C, Oxenham AJ, Klump GM (2010) Neural adaptation to tone sequences in the songbird forebrain: patterns, determinants, and relation to the build-up of auditory streaming. J Comp Physiol A 196(8):543–557. https://doi.org/10.1007/s00359-010-0542-4

Berwick RC, Okanoya K, Beckers GJL, Bolhuis JJ (2011) Songs to syntax: the linguistics of birdsong. Trends Cogn Sci 15(3):113–121

Brand AR, Kellogg PP (1939a) Auditory responses of starlings, English sparrows, and domestic pigeons. Wilson Bull 51:38–41

Brand AR, Kellogg PP (1939b) The range of hearing of canaries. Science 90:354

Bregman AS (1991) Auditory scene analysis: the perceptual organization of sound. MIT Press, Cambridge

Bremond J-C (1976) Specific recognition in the song of Bonelli's warbler (*Phylloscopus bonelli*). Behaviour 58:99–116

Brumm H, Slabbekoorn H (2005) Acoustic communication in noise. Adv Study Behav 35:151–209

Brumm H, Todt D (2002) Noise-dependent song amplitude regulation in a territorial songbird. Anim Behav 63:891–897. https://doi.org/10.1006/anbe.2001.1968

Carr CE (1992) Evolution of the central auditory system in reptiles and birds. In: The evolutionary biology of hearing. Springer, New York, pp 511–543

Carr CE, Shah S, McColgan T, Ashida G, Kuokkanen PT, Brill S, Kempter R, Wagner H (2015) Maps of interaural delay in the owl's nucleus laminaris. J Neurophysiol 114(3):1862–1873. https://doi.org/10.1152/jn.00644.2015

Comins JA, Gentner TQ (2014) Auditory temporal pattern learning by songbirds using maximal stimulus diversity and minimal repetition. Anim Cogn 17(5):1023–1030

Comins JA, Gentner TQ (2015) Pattern-induced covert category learning in songbirds. Curr Biol 25(14):1873–1877
Corwin JT, Cotanche DA (1988) Regeneration of sensory hair cells after acoustic trauma. Science 240(4860):1772–1774
Darwin C (1859) On the origin of species by means of natural selection, or the preservation of favoured races in the struggle for life. John Murray, London
Dent ML, Klump GM, Schwenzfeier C (2002) Temporal modulation transfer functions in the barn owl (*Tyto alba*). J Comp Physiol A 187(12):937–943. https://doi.org/10.1007/s00359-001-0259-5
Dooling RJ (1980) Behavior and psychophysics of hearing in birds. In: Comparative studies of hearing in vertebrates. Springer, New York, pp 261–288
Dooling RJ (1992) Perception of speech sounds by birds. In: Cazals Y, Horner K, Demany L (eds) Auditory physiology and perception. Pergamon, pp 407–413. https://doi.org/10.1016/B978-0-08-041847-6.50052-6
Dooling RJ, Blumenrath SH (2013) Avian sound perception in noise. In: Animal communication and noise. Springer, Heidelberg, pp 229–250
Dooling RJ, Brown SD (1990) Speech perception by budgerigars (Melopsittacus undulatus): spoken vowels. Percept Psychophys 47(6):568–574
Dooling RJ, Leek MR (2018) Communication masking by man-made noise. In: Effects of anthropogenic noise on animals. Springer, New York, pp 23–46
Dooling RJ, Okanoya K (1995) Psychophysical methods for assessing perceptual categories. In: Klump GM, Dooling RJ, Fay RR, Stebbins WC (eds) Methods in comparative psychoacoustics. Birkhäuser Basel, Basel, pp 307–318. https://doi.org/10.1007/978-3-0348-7463-2_26
Dooling RJ, Saunders JC (1975) Hearing in the parakeet (Melopsittacus undulatus): absolute thresholds, critical ratios, frequency difference limens, and vocalizations. J Comp Physiol Psychol 88(1):1–20
Dooling RJ, Haskell RJ (1978) Auditory duration discrimination in the parakeet (Melopsittacus undulatus). The Journal of the Acoustical Society of America, 63(5):1640–1643.
Dooling R, Searcy M (1980) Early perceptual selectivity in the swamp sparrow. Dev Psychobiol 13(5):499–506
Dooling RJ, Searcy MH (1981) Amplitude modulation thresholds for the parakeet (*Melopsittacus undulatus*). J Comp Physiol 143(3):383–388
Dooling RJ, Ryals BM, Manabe K (1997) Recovery of hearing and vocal behavior after hair-cell regeneration. Proc Natl Acad Sci 94(25):14206–14210
Dooling RJ, Lohr B, Dent ML (2000) Hearing in birds and reptiles. In: Comparative hearing: birds and reptiles. Springer, New York, pp 308–359
Dooling RJ, Leek MR, Gleich O, Dent ML (2002) Auditory temporal resolution in birds: discrimination of harmonic complexes. J Acoust Soc Am 112(2):748–759
Dooling RJ, Ryals BM, Dent ML, Reid TL (2006) Perception of complex sounds in budgerigars (*Melopsittacus undulatus*) with temporary hearing loss. J Acoust Soc Am 119(4):2524–2532
Dooling RJ, Dent ML, Lauer AM, Ryals BM (2008) Functional recovery after hair cell regeneration in birds. In: Salvi R, Popper A, Fay R (eds) Hair cell regeneration, repair, and protection. Springer, New York, pp 117–140
Dyson ML, Klump GM, Gauger B (1998) Absolute hearing thresholds and critical masking ratios in the European barn owl: a comparison with other owls. J Comp Physiol A 182(5):695–702
Edwards EP (1943) Hearing ranges of four species of birds. The Auk 60(2):239–241
Emlen ST (1972) An experimental analysis of the parameters of bird song eliciting species recognition. Behaviour 41(1–2):130–171
Engelmann W (1928) Untersuchungen über die Schallokalisation bei Tieren. Zeitschrift für Psychologie und Physiologie der Sinnesorgane: 1. Abteilung. Zeit für Psychol 105:317–370
Fastl H, Zwicker E (2007) Psychoacoustics. Springer, Berlin
Fay RR (1988) Hearing in vertebrates: a psychophysics databook. Hill-Fay Associates, Winnetka

Feinkohl A, Klump GM (2013) Azimuthal sound localization in the European starling (Sturnus vulgaris): II. Psychophysical results. J Comp Physiol A 199(2):127–138. https://doi.org/10.1007/s00359-012-0774-6

Feinkohl A, Borzeszkowski KM, Klump GM (2016) Azimuthal sound localization in the European starling (*Sturnus vulgaris*): III. Comparison of sound localization measures. Hear Res 332:238–248. https://doi.org/10.1016/j.heares.2015.04.001

Fishbein AR, Idsardi WJ, Ball GF, Dooling RJ (2020) Sound sequences in birdsong: how much do birds really care? Philos Trans R Soc Lond B Biol Sci 375(1789):20190044. https://doi.org/10.1098/rstb.2019.0044

Fishbein AR, Prior NH, Brown JA, Ball GF, Dooling RJ (2021) Discrimination of natural acoustic variation in vocal signals. Sci Rep 11(1):916. https://doi.org/10.1038/s41598-020-79641-z

Gatehouse RW, Shelton BR (1978) Sound localization in bobwhite quail (Colinus virginianus). Behavioral Biology 22(4):533–540.

Gleich O, Langemann U (2011) Auditory capabilities of birds in relation to the structural diversity of the basilar papilla. Hear Res 273(1–2):80–88. https://doi.org/10.1016/j.heares.2010.01.009

Gleich O, Manley GA (2000) The hearing organ of birds and crocodilia. In: Dooling RJ, Fay RR, Popper AN (eds) Comparative hearing: birds and reptiles. Springer, New York, pp 70–138. https://doi.org/10.1007/978-1-4612-1182-2_3

Gleich O, Klump GM, Dooling RJ (1994) Hereditary sensorineural hearing loss in a bird. Naturwissenschaften 81(7):320–323. https://doi.org/10.1007/BF01131950

Gleich O, Dooling RJ, Presson JC, Strutz J (1996) Continuous proliferation of supporting cells and indications for hair cell differentiation in the inner ear of adult song birds with genetic cochlear hearing loss. HNO 44(6):307–312

Gleich O, Dooling RJ, Manley GA (2005) Audiogram, body mass, and basilar papilla length: correlations in birds and predictions for extinct archosaurs. Naturwissenschaften 92(12):595–598. https://doi.org/10.1007/s00114-005-0050-5

Granit O (1941) Beiträge zur Kenntnis des Gehörsinnes der Vögel. Ornis Fennica 18:49–71

Hausberger M, Bigot E, Clergeau P (2008) Dialect use in large assemblies: a study in European starling Sturnus vulgaris roosts. J Avian Biol 39(6):672–682. https://doi.org/10.1111/j.1600-048X.2008.04307.x

Hawkins Jr, JE, Stevens SS (1950) The masking of pure tones and of speech by white noise. The Journal of the Acoustical Society of America 22(1):6–13

Heise GA (1953) Auditory thresholds in the pigeon. Am J Psychol 66(1):1–19. https://doi.org/10.2307/1417964

Hienz RD, Sachs MB (1987) Effects of noise on pure-tone thresholds in blackbirds (*Agelaius* phoeniceus and *Molothrus ater*) and pigeons (*Columba livia*). J Comp Psychol 101(1):16–24. https://doi.org/10.1037/0735-7036.101.1.16

Hill EM, Koay G, Heffner RS, Heffner HE (2014) Audiogram of the chicken (*Gallus domesticus*) from 2 Hz to 9 kHz. J Comp Physiol A 200(10):863–870. https://doi.org/10.1007/s00359-014-0929-8

Hulse SH (2002) Auditory scene analysis in animal communication. In: Advances in the study of behavior, vol 31. Academic, pp 163–200. https://doi.org/10.1016/S0065-3454(02)80008-0

Hulse SH, Cynx J (1985) Relative pitch perception is constrained by absolute pitch in songbirds (*Mimus, Molothrus*, and *Sturnus*). J Comp Psychol 99(2):176–196. https://doi.org/10.1037/0735-7036.99.2.176

Hulse SH, MacDougall-Shackleton SA, Wisniewski AB (1997) Auditory scene analysis by songbirds: stream segregation of birdsong by European starlings (*Sturnus vulgaris*). J Comp Psychol 111(1):3–13. https://doi.org/10.1037/0735-7036.111.1.3

Itatani N, Klump GM (2009) Auditory streaming of amplitude-modulated sounds in the songbird forebrain. J Neurophysiol 101(6):3212–3225. https://doi.org/10.1152/jn.91333.2008

Itatani N, Klump GM (2011) Neural correlates of auditory streaming of harmonic complex sounds with different phase relations in the songbird forebrain. J Neurophysiol 105(1):188–199. https://doi.org/10.1152/jn.00496.2010

Itatani N, Klump GM (2014) Neural correlates of auditory streaming in an objective behavioral task. Proc Natl Acad Sci U S A 111(29):10738–10743. https://doi.org/10.1073/pnas.1321487111
Janesick AS, Heller S (2018) Stem cells and the bird cochlea—where is everybody? Cold Spring Harb Perspect Med. https://doi.org/10.1101/cshperspect.a033183
Jeffress LA (1948) A place theory of sound localization. J Comp Physiol Psychol 41(1):35–39
Kettler L, Christensen-Dalsgaard J, Larsen ON, Wagner H (2016) Low frequency eardrum directionality in the barn owl induced by sound transmission through the interaural canal. Biol Cybern 110(4–5):333–343. https://doi.org/10.1007/s00422-016-0689-3
Kim DO, Moiseff A, Bradley Turner J, Gull J (2008) Acoustic cues underlying auditory distance in barn owls. Acta Otolaryngol 128(4):382–387. https://doi.org/10.1080/00016480701840114
Klump GM (1996) Bird communication in the noisy world. In: Kroodsma DE, Miller EH (eds) Ecology and evolution of acoustic communication in birds. Cornell University Press, Ithaca, pp 321–338
Klump GM (2016) Perceptual and neural mechanisms of auditory scene analysis in the European Starling. In: Bee MA, Miller CT (eds) Psychological mechanisms in animal communication. Springer International Publishing, Cham, pp 57–88. https://doi.org/10.1007/978-3-319-48690-1_3
Klump GM, Langemann U (1995) Comodulation masking release in a songbird. Hear Res 87(1):157–164. https://doi.org/10.1016/0378-5955(95)00087-K
Klump GM, Larsen ON (1992) Azimuthal sound localization in the European starling (*Sturnus vulgaris*). J Comp Physiol A 170(2):243–251
Klump GM, Windt W, Curio E (1986) The great tit's (*Parus major*) auditory resolution in azimuth. J Comp Physiol A 158(3):383–390
Klump GM, Dooling RJ, Fay RR, Stebbins W (1995) Methods in comparative psychoacoustics. Birkhäuser, Basel
Knecht S (1939) Über den Gehörsinn und die Musikalität der Vögel. Zeitschr Vergl Physiol 27(1):169–232. https://doi.org/10.1007/BF00340527
Knudsen EI, Konishi M (1978) A neural map of auditory space in the owl. Science 200(4343):795–797. https://doi.org/10.1126/science.644324
Knudsen EI, Konishi M (1979) Mechanisms of sound localization in the barn owl (*Tyto alba*). J Comp Physiol 133(1):13–21. https://doi.org/10.1007/BF00663106
Knudsen EI, Blasdel GG, Konishi M (1979) Sound localization by the barn owl (*Tyto alba*) measured with the search coil technique. J Comp Physiol 133(1):13–21. https://doi.org/10.1007/BF00663105
Konishi M (1965) The role of auditory feedback in the control of vocalization in the White-Crowned Sparrow. Zeitschrift fur Tierpsychologie 22(7):770–783
Konishi M (1973) How the owl tracks its prey: experiments with trained barn owls reveal how their acute sense of hearing enables them to catch prey in the dark. Am Sci 61(4):414–424
Konishi M (2003) Coding of auditory space. Annu Rev Neurosci 26(1):31–55. https://doi.org/10.1146/annurev.neuro.26.041002.131123
Köppl C (2019) Internally coupled middle ears enhance the range of interaural time differences heard by the chicken. J Exp Biol 222(12). https://doi.org/10.1242/jeb.199232
Kroodsma DE, Miller EH (eds) (1996) Ecology and evolution of acoustic communication in birds. Cornell University Press, Ithaca
Krumm B, Klump G, Köppl C, Langemann U (2017) Barn owls have ageless ears. Proc R Soc B Biol Sci 284(1863):20171584. https://doi.org/10.1098/rspb.2017.1584
Krumm B, Klump GM, Köppl C, Langemann U (2019) The barn owls' minimum audible angle. PLoS One 14(8):e0220652. https://doi.org/10.1371/journal.pone.0220652
Krumm B, Klump GM, Köppl C, Beutelmann R, Langemann U (2022) Chickens have excellent sound localization ability. J Exp Biol 225:jeb243601. https://doi.org/10.1242/jeb.243601
Kuhl PK, Conboy BT, Coffey-Corina S, Padden D, Rivera-Gaxiola M, Nelson T (2008) Phonetic learning as a pathway to language: new data and native language magnet theory expanded (NLM-e). Philos Trans R Soc Lond Ser B Biol Sci 363(1493):979–1000. https://doi.org/10.1098/rstb.2007.2154

Langemann U, Klump GM (1992) Frequency discrimination in the European starling (Sturnus vulgaris): a comparison of different measures. Hear Res 63(1):43–51. https://doi.org/10.1016/0378-5955(92)90072-U

Langemann U, Gauger B, Klump GM (1998) Auditory sensitivity in the great tit: perception of signals in the presence and absence of noise. Anim Behav 56(3):763–769. https://doi.org/10.1006/anbe.1998.0879

Langemann U, Hamann I, Friebe A (1999) A behavioral test of presbycusis in the bird auditory system. Hear Res 137(1):68–76. https://doi.org/10.1016/S0378-5955(99)00139-2

Larsen ON, Dooling RJ, Ryals BM (1996) Roles of intracranial air pressure in bird audition. In: Diversity in auditory mechanics. World Scientific, pp 11–17

Larsen ON, Dooling RJ, Michelsen A (2006) The role of pressure difference reception in the directional hearing of budgerigars (Melopsittacus undulatus). J Comp Physiol A 192:1063–1072

Larsen ON, Christensen-Dalsgaard J, Jensen KK (2016) Role of intracranial cavities in avian directional hearing. Biol Cybern 110(4):319–331. https://doi.org/10.1007/s00422-016-0688-4

Lewald J (1987) The acuity of sound localization in the pigeon (Columba livia). Naturwissenschaften 74(6):296–297.

Lohr B, Dooling RJ (1998) Detection of changes in timbre and harmonicity in complex sounds by zebra finches (*Taeniopygia guttata*) and budgerigars (*Melopsittacus undulatus*). J Comp Psychol 112(1):36. https://doi.org/10.1037/0735-7036.112.1.36

Lohr B, Dooling RJ (2004) Hearing in the Ruby-throated Hummingbird (*Archilochus colubris: Trochilidae*). Paper presented at the Association for Research in Otolaryngology, Daytona, FL

MacDougall-Shackleton SA, Hulse SH, Gentner TQ, White W (1998) Auditory scene analysis by European starlings (*Sturnus vulgaris*): perceptual segregation of tone sequences. J Acoust Soc Am 103(6):3581–3587. https://doi.org/10.1121/1.423063

Maier EH, Klump GM (1990) Auditory duration discrimination in the European starling (*Sturnus vulgaris*). J Acoust Soc Am 88(2):616–621. https://doi.org/10.1121/1.399765

Marean GC, Burt JM, Beecher MD, Rubel EW (1998) Auditory perception following hair cell regeneration in European starling (*Sturnus vulgaris*): frequency and temporal resolution. J Acoust Soc Am 103(6):3567–3580. https://doi.org/10.1121/1.423085

Marler P (1970) Birdsong and speech development: could there be parallels? There may be basic rules governing vocal learning to which many species conform, including man. Am Sci 6:669–673

Marler P (1997) Three models of son learning: evidence from behavior. J Neurobiol 33:501–516

Marler P, Peters S (1977) Selective vocal learning in a sparrow. Science 198(4316):519–521. https://doi.org/10.1126/science.198.4316.519

Marler P, Peters S (1989) Species differences in auditory responsiveness in early vocal learning. In: Dooling RJ, Hulse SH (eds) The comparative psychology of audition: perceiving complex sounds. Lawrence Erlbaum Assoc., Hillsdale, pp 243–273

McAlpine D, Grothe B (2003) Sound localization and delay lines – do mammals fit the model? Trends Neurosci 26(7):347–350. https://doi.org/10.1016/S0166-2236(03)00140-1

Micheyl C, Oxenham AJ (2010) Objective and subjective psychophysical measures of auditory stream integration and segregation. J Assoc Res Otolaryngol 11(4):709–724. https://doi.org/10.1007/s10162-010-0227-2

Miller JD (1974) Effects of noise on people. J Acoust Soc Am 56(3):729–764. https://doi.org/10.1121/1.1903322

Mills AW (1958) On the minimum audible angle. J Acoust Soc Am 30(4):237–246. https://doi.org/10.1121/1.1909553

Moiseff A (1989) Bi-coordinate sound localization by the barn owl. J Comp Physiol A 164(5):637–644

Moiseff A, Konishi M (1981) Neuronal and behavioral sensitivity to binaural time differences in the owl. J Neurosci 1(1):40–48. https://doi.org/10.1523/JNEUROSCI.01-01-00040.1981

Naguib M, Wiley RH (2001) Estimating the distance to a source of sound: mechanisms and adaptations for long-range communication. Animal Behaviour, 62(5):825–837

Nemeth E, Brumm H (2010) Birds and anthropogenic noise: are urban songs adaptive? Am Nat 176(4):465–475

Okanoya K (2004) Song syntax in Bengalese finches: proximate and ultimate analyses. Adv Study Behav 34:297–345

Okanoya K, Dooling RJ (1987) Strain differences in auditory thresholds in the canary (Serinus canarius). J Comp Psychol 101(2):213–215. https://doi.org/10.1037/0735-7036.101.2.213

Osmanski MS, Dooling RJ (2009) The effect of altered auditory feedback on control of vocal production in budgerigars (*Melopsittacus undulatus*). J Acoust Soc Am 126(2):911–919. https://doi.org/10.1121/1.3158928

Payne RS, Drury WH (1958) Marksman of the darkness. Natural History New York 67: 316–323.

Park TJ, Dooling RJ (1991) Sound localization in small birds: absolute localization in azimuth. Journal of Comparative Psychology 105(2):125.

Payne RS (1971) Acoustic location of prey by barn owls (*Tyto alba*). J Exp Biol 54(3):535–573. https://doi.org/10.1242/jeb.54.3.535

Peña JL, Cazettes F, Beckert MV, Fischer BJ (2019) Synthesis of hemispheric ITD tuning from the readout of a neural map: commonalities of proposed coding schemes in birds and mammals. J Neurosci 39(46):9053–9061. https://doi.org/10.1523/JNEUROSCI.0873-19.2019

Pohl NU, Slabbekoorn H, Klump GM, Langemann U (2009) Effects of signal features and environmental noise on signal detection in the great tit, *Parus major*. Anim Behav 78:1293–1300

Prior NH, Smith E, Lawson S, Ball GF, Dooling RJ (2018) Acoustic fine structure may encode biologically relevant information for zebra finches. Sci Rep 8. https://doi.org/10.1038/s41598-018-24307-0

Pumphrey RJ (1961) Part II. Sensory organs: hearing. In: Marshall AJ (ed) Biology and comparative physiology of birds, vol II. Academic, New York, pp 69–86

Retzius G (1884) Das Gehörorgan der Wirbelthiere II: Das Gehörorgan der Reptilien, der Vögel und der Säugethiere. Samson & Wallin, Stockholm

Rice WR (1982) Acoustical location of prey by the Marsh Hawk: adaptation to concealed prey. Auk 99(3):403–413. https://doi.org/10.1093/auk/99.3.403

Roccio M, Senn P, Heller S (2020) Novel insights into inner ear development and regeneration for targeted hearing loss therapies. Hear Res 397:107859. https://doi.org/10.1016/j.heares.2019.107859

Rosowski JJ, Saunders JC (1980) Sound transmission through the avian interaural pathways. J Comp Physiol 136(3):183–190

Rouse AA, Patel AD, Kao MH (2021) Vocal learning and flexible rhythm pattern perception are linked: evidence from songbirds. Proc Natl Acad Sci 118(29):e2026130118. https://doi.org/10.1073/pnas.2026130118

Ryals BM, Rubel EW (1988) Hair cell regeneration after acoustic trauma in adult Coturnix quail. Science 240(4860):1774–1776. https://doi.org/10.1126/science.3381101

Ryals BM, Dooling RJ, Westbrook E, Dent ML, MacKenzie A, Larsen ON (1999) Avian species differences in susceptibility to noise exposure. Hear Res 131(1):71–88. https://doi.org/10.1016/S0378-5955(99)00022-2

Ryals BM, Dent ML, Dooling RJ (2013) Return of function after hair cell regeneration. Hear Res 297:113–120. https://doi.org/10.1016/j.heares.2012.11.019

Salvi RJ, Popper AN, Fay RR (eds) (2008) Hair cell regeneration, repair, and protection springer handbook of auditory research. Springer, New York. https://doi.org/10.1007/978-0-387-73364-7

Saunders JC (1985) Auditory structure and function in the bird middle ear: an evaluation by SEM and capacitive probe. Hear Res 18(3):253–268. https://doi.org/10.1016/0378-5955(85)90042-5

Saunders JC (2010) The role of hair cell regeneration in an avian model of inner ear injury and repair from acoustic trauma. ILAR J 51(4):326–337. https://doi.org/10.1093/ilar.51.4.326

Saunders J, Dooling R (1974) Noise-induced threshold shift in the parakeet (*Melopsittacus undulatus*). Proc Natl Acad Sci 71(5):1962–1965. https://doi.org/10.1073/pnas.71.5.1962

Saunders JC, Dooling RJ (2018) Characteristics of temporary and permanent threshold shifts in vertebrates. In: Slabbekoorn H, Dooling RJ, Popper AN, Fay RR (eds) Effects of anthropogenic noise on animals. Springer, New York, pp 83–107. https://doi.org/10.1007/978-1-4939-8574-6_4

Saunders JC, Rintelmann WF, Bock GR (1979) Frequency selectivity in bird and man: a comparison among critical ratios, critical bands and psychophysical tuning curves. Hear Res 4:303–323
Schooneveldt GP, Moore BC (1989) Comodulation masking release (CMR) as a function of masker bandwidth, modulator bandwidth, and signal duration. J Acoust Soc Am 85(1):273–281. https://doi.org/10.1121/1.397734
Schwartzkopff J (1949) Über Sitz und Leistung von Gehör und Vibrationssinn bei Vögeln. Z Vgl Physiol 31(5):527–608
Schwartzkopff J (1950) Beitrag zum Problem des Richtungshörens bei Vögeln. Z Vgl Physiol 32(4):319–327
Schwartzkopff J (1952) Über den Gehörsinn der Vögel. J Ornithol 93(2):91–103
Slabbekoorn H, den Boer-Visser A (2006) Cities change the songs of birds. Curr Biol 16:2326–2331
Slabbekoorn H, Peet M (2003) Birds sing at a higher pitch in urban noise. Nature 424:267
Starck JM (1995) Comparative anatomy of the external and middle ear of palaeognathous birds. Adv Anat Embry Cell Biol 131:1–137
Takahashi T, Moiseff A, Konishi M (1984) Time and intensity cues are processed independently in the auditory system of the owl. J Neurosci 4(7):1781–1786. https://doi.org/10.1523/JNEUROSCI.04-07-01781.1984
Trainer JE (1946) The auditory acuity of certain birds. Cornell University, Ithaca
Tu H-W, Dooling RJ (2012) Perception of warble song in budgerigars (Melopsittacus undulatus): evidence for special processing. Anim Cogn 15(6):1151–1159. https://doi.org/10.1007/s10071-012-0539-1
Van Noorden LS (1975) Temporal coherence in the perception of tone sequences. PhD thesis, Eindhoven University of Technology
Wada Y (1923) Beiträge zur vergleichenden Physiologie des Gehörgangs. Pflugers Arch Gesamte Physiol Menschen Tiere 202(1):46–69
Ware HE, McClure CJ, Carlisle JD, Barber JR (2015) A phantom road experiment reveals traffic noise is an invisible source of habitat degradation. Proc Natl Acad Sci U S A 112(39):12105–12109
Welch TE, Dent ML (2011) Lateralization of acoustic signals by dichotically listening budgerigars (*Melopsittacus undulatus*). J Acoust Soc Am 130(4):2293–2301. https://doi.org/10.1121/1.3628335
Wiley RH (2015) Noise matters. Harvard University Press. https://doi.org/10.4159/9780674287044
Wisniewski AB, Hulse SH (1997) Auditory scene analysis in European Starlings (*Sturnus vulgaris*): discrimination of song segments, their segregation from multiple and reversed conspecific songs, and evidence for conspecific song categorization. J Comp Psychol 111(4):337–350
Wright TF, Brittan–Powell EF, Dooling RJ, Mundinger PC (2004) Sex-linked inheritance of hearing and song in the Belgian Waterslager canary. Proc R Soc Lond B Biol Sci 271(suppl_6):S409–S412. https://doi.org/10.1098/rsbl.2004.0204
Zann RA (1996) The zebra finch: a synthesis of field and laboratory studies. Oxford University Press, Oxford

Chapter 7
Discoveries in Marine Mammal Hearing

Douglas Wartzok and Darlene R. Ketten

7.1 Introduction

Humans have long been fascinated by mammals that transitioned from terrestrial to aquatic life. "Marine mammal" is a term that covers approximately 150 species that have one major feature in common: the ability to conduct life functions in water. Some, like the pinnipeds (seals, sea lions, and walruses) and fissipeds (otters), are amphibious, meaning they are able to function both on land and in water. Others, like the cetaceans (dolphins and whales) and sirenians (manatees and dugongs), are "obligate" fully aquatic mammals that are unable to survive on land for any protracted period of time. They conduct all major functions, including breeding, feeding, and communicating, in water. Several odontocete species have extraordinary abilities to make prolonged, deep dives, even though they are air-breathing mammals.

Marine mammals did not evolve from a single ancestor. Rather, the major groups evolved from different ancestral terrestrial species at different times and in different regions, with a number of intermediate stages. Pinnipedia (taxonomic order Carnivora) includes the Otariidae (sea lions), Odobenidae (walrus), Phocidae (true seals), and Fissipedia (sea otters). Most live in cooler ocean waters. They evolved in Pacific waters approximately 30–20 million years ago. By contrast, the earliest cetacean and sirenian fossils indicate that they evolved in warmer regions bordering the

D. Wartzok (✉)
Department of Biological Sciences, Institute of Environment, Florida International University, Miami, FL, USA
e-mail: wartzok@fiu.edu

D. R. Ketten
Biology Department, Woods Hole Oceanographic Institution, Woods Hole, MA, USA
e-mail: dketten@whoi.edu

D. R. Ketten et al. (eds.), *A History of Discoveries on Hearing*, Springer Handbook of Auditory Research 77,
https://doi.org/10.1007/978-3-031-41320-9_7

Tethys Sea, located between the prehistoric continents of Gondwana and Laurasia (Domning 1982; Barnes et al. 1985).

There is good fossil evidence that cetacean aquatic transitions began 50–60 million years ago. Mesonychid condylarths, hooved carnivores related to artiodactyls that existed from the Paleocene to Eocene, are the ancestral line of the Archaeoceti. They included an amphibious creature called *Ambulocetus*—the walking whale. This line evolved into the two whale suborders of Odontoceti (toothed whales) and Mysticeti ("moustached" baleen whales) (Fordyce 1980; Gingerich et al. 1983; Thewissen et al. 1994). Currently, there are 78 extant species of odontocetes and 15 extant species of mysticetes, including new species recognized as recently as 2021 using morphometric and genetic techniques (Rosel et al. 2021).

Mysticetes are large, pelagic animals ranging from 6 m to over 30 m that filter feed by skimming or lunge feeding (Kawamura 1980). Odontocete species range from 1 to 20 m, with species that inhabit every aquatic habitat, including river dolphins, coastal porpoises, pelagic, open ocean dolphins, and larger beaked whales and sperm whales. Like bats, odontocetes operate in dusky, dark environments, in some cases foraging in profoundly dark, abyssal depths near the seafloor. Although microchiropteran bats and odontocetes evolved independently, dolphins rival bats in their abilities to image and interpret fine details of their environment purely through sound. Both developed ears specialized for echolocation, a biosonar system for acoustically probing the environment.

7.1.1 Transition from Land to Water

Specimens that represent the progressive stages of terrestrial to amphibious to aquatic mammals show a wide range of physical adaptations to accommodate the demands of functioning in water. These transitions to water required not only hydrodynamic changes in body shape but also adaptations consistent with the differences that the physics of sound in water imposes on the systems for sound production, reception, transduction, and processing by these animals. Critical features of sound in any medium that form acoustic cues for survival are frequency, speed, wavelength, and intensity. Frequency, measured in cycles per second or Hertz (Hz), is the ratio of the speed of sound (meters per second) divided by the wavelength (meters per cycle):

$$f = c / \lambda \tag{7.1}$$

Sound speed differs according to the density of the medium. Sound speed in water (1530 m/s) is approximately 4.5 times faster than in air (340 m/s), although it varies with any factor affecting density such as salinity and temperature. To operate in water, marine mammal ears must be "tuned" to these factors. Indeed, the lack of this aquatic tuning is one reason human ears are unreliable underwater for tasks like frequency and directionality (Sørensen et al. 2022). Therefore, the ability of amphibious pinniped species to hear to some degree in both media is all the more remarkable.

Substantial structural changes in the ears of marine mammals parallel their degree of aquatic adaptation, from minor in amphibious, littoral species like otters and sea lions, to more extensive in seals, to extreme in mysticetes and odontocetes. This chapter focuses on the most exceptional auditory adaptations among marine mammals. These are found primarily in cetaceans and some pinnipeds.

7.2 Historical Views of Marine Mammals

Curiosity about marine mammals is as old as hominids, with cave and rock wall paintings in Cueva de la Pileta in Spain picturing basking seals and whales in Atacama in Chile, as well as evidence of active Stone Age hunting and harvesting of marine mammals (Clarke 1946). We see evidence of ancient observations today also in indigenous knowledge of the natural history, behavior, attention to sounds, and vocalizations of marine mammals that predate the classical era and continue to be essential for success in their traditional hunts.

Observations have continued over the centuries, and theories abound about how these animals function in water, many of which centered on how they use sound and what sounds they produce. However, despite our great curiosity, substantial progress was not made in understanding their exceptional acoustic abilities and anatomical specializations until the mid to late twentieth century, when technology was sufficient to document, analyze, and even duplicate their sonar capabilities.

7.2.1 Ancient Natural History

As evidenced by statues, pottery, and coinage showing dolphins with human dolphin riders in Greek and Roman societies, scholars, and naturalists were keenly interested in dolphins and even integrated them into myths and legends. Arion (c 600 BCE), returning home from his triumph in Sicily, convinced his mutinous crew to allow him to play his lyre and sing prior to being thrown into the sea. His playing attracted dolphins, one of which, after the mutinous crew sailed away, was reported to carry him to safety on its back (Bullfinch 1903). Similarly, a fragment of a Pindar poem (c 500 BCE) reads, "…the singing of flutes gently animates the dolphin in the depths of the waveless sea…" Today, we might dismiss these as acts of curiosity or annoyance, but we do know that dolphins are sometimes drawn to acoustic signals in the water, in some cases mimicking them, or, in some cases, coming to the surface in the presence of depth sounders and fish finders during fisheries operations (Payne 1995).

The most prolific classical observations of marine mammals are found in Aristotle's treatise on animals (350 BCE, translation by D'Arcy Wentworth

Thompson in 1910). Aristotle made laborious studies of both live, swimming animals and stranded ones. He noted, "Now, the seal has the passages visible whereby it hears; but the dolphin can hear, but has no ears, nor yet any passages visible." Aristotle noted that dolphins produce audible sounds when out of the water, and he concluded that dolphins could hear in water based on a traditional fishing method that used "…a loud and alarming resonance… to induce the creatures to run in a shoal high and dry up on the beach and so to catch them while stupefied with the noise." He further stated that dolphins had acute hearing because they could only be surrounded by boats if the hunters made no noise with their oars or nets and that sounds from surging boats would cause the pod to scatter. Modern artisanal fisheries continue today to use loud percussive underwater sounds, like those of the Taiji drive fisheries in Japan and smaller whaling operations in the Faroe Islands.

Gaius Plinius Secundus (Pliny the Elder, 24–79 CE) in his *Historia Naturalis* adhered to *most* of the observations of Aristotle on marine creatures. Yet, curiously, Pliny mixed descriptions of whales not only with seals but also with fishes, and especially sharks (Romero 2012). Although Pliny clearly recognized the mammalian features of suckling young and the absence of gills, he characterized the bones as cartilaginous, an error that persisted for over 1500 years and led to the misclassification of many marine mammals, particularly whales, until well into the seventeenth century.

7.2.2 Middle Ages to Modern Era

Myths about marine mammals flourished in the Middle Ages as did industries that grew by harvesting whale products, especially from strandings. The commercial importance of marine mammals is underscored by an English law in 1324, making all stranded or caught cetaceans ("Great Fish") in Britain property of the Crown (CERWG 2006). Drawings of stranded whales, some rather fancifully rendered, are common in Europe from the fifteenth century onward, many depicting masses of harvesters next to a stranded whale on the beach, while others featured fearsome, monstrous leviathans.

By contrast, the Renaissance produced the first anatomically reliable drawings of marine mammals by artists and anatomists such as Belon and Rondelet who produced images of whole animals, skulls, and organs, particularly from smaller dolphins. These were useful illustrations but did little to advance functional knowledge except to establish Latin names for many species and to provide imagery that aided later scientists in classification.

Peter Artredi, a student of Linnaeus, established the fundamental classification framework for Linnaeus's work (Romero 2012). Artredi's work led to Linnaeus and his students re-establishing the "Cete" as Mammalia, setting the stage for the return of scientists to the question of how these mammals sense their environment and how they hear. After all, the sea is not silent, and as Aristotle said, neither are they.

7.3 Anatomy of Underwater Hearing

7.3.1 Cetacean Ears

The anatomy of pinniped ears is essentially like that of land mammals. Some pinniped species lack a detectable external ear flap (pinna), but all have a well-defined external auditory canal, which on land is air filled. By contrast, the ears of all cetaceans are highly modified and are the focus of this section.

Not surprisingly, the earliest research and observations on cetaceans involved anatomy, often derived from stranded specimens. Several millennia after Aristotle, there was a rebirth of scientific interest in marine mammal anatomy. Hunter (1787) noted that cetaceans have all the basic components of the auditory system of a terrestrial mammal: an auditory meatus of varying sizes and widths, a tympanic membrane (eardrum), a Eustachian tube (middle ear pressure control mechanism), and ossicles (three small bones of the middle ear connecting the eardrum to the inner ear).

In all mammals, the middle and inner ear components are located within a bony case referred to as the tympanic bulla or bone (houses the middle ear) and the periotic bulla (houses the inner ear) (Fig. 7.1). In most mammals, these bones form part of the skull or are fused to the skull on at least one of their surfaces. Hunter (1787) noted that the "bony part of the organ" was basically separate from the head bones, in contrast to the situation in land mammals. Remington Kellogg (1928) noted that this was true only in the odontocete cetaceans, whereas in baleen whales, the periotic bone has a long process that is wedged between the exoccipital and squamosal bones.

Hunter also established that all bones associated with hearing in cetaceans were more massive and denser than the corresponding bones in land mammals. He further observed that the skull cavity (fossa) in which the bullae are situated is lined with a plexus of blood vessels. Hunter also described the unusual eardrum of the minke whale, *Balaenoptera acutorostrata*. This is possibly the earliest scientific report of the exceptionally long, hollow, everted tympanic membrane common to baleen whales, later termed a "glove finger."

As in other mammals, cetacean inner ears have multiple bony and fluid-filled membranous labyrinths forming the vestibular system (with looped semicircular canals) and the cochlea (a spiral tri-chambered canal with a basilar membrane that supports hair cells that respond to acoustic stimuli) (Figs. 7.2 and 7.3). Hunter stated that the cochlea was rather large in most cetaceans, increasing with their body mass, which is consistent with cochlear length:body mass scaling in land mammals (West 1985; Ketten 2000). Hunter reported however that the semicircular canals were anomalously small in comparison to the cochlear dimensions in all the whales and dolphins he examined, a feature "rediscovered" by many later researchers, but, as yet, not fully understood. Hunter dissected harbor porpoises (*Phocoena phocoena*, the smallest odontocete), several species of larger odontocetes, and several baleen whales, reporting that the number of turns in the cochlear spiral ranges from 1.5 to 2.5 and varies by species.

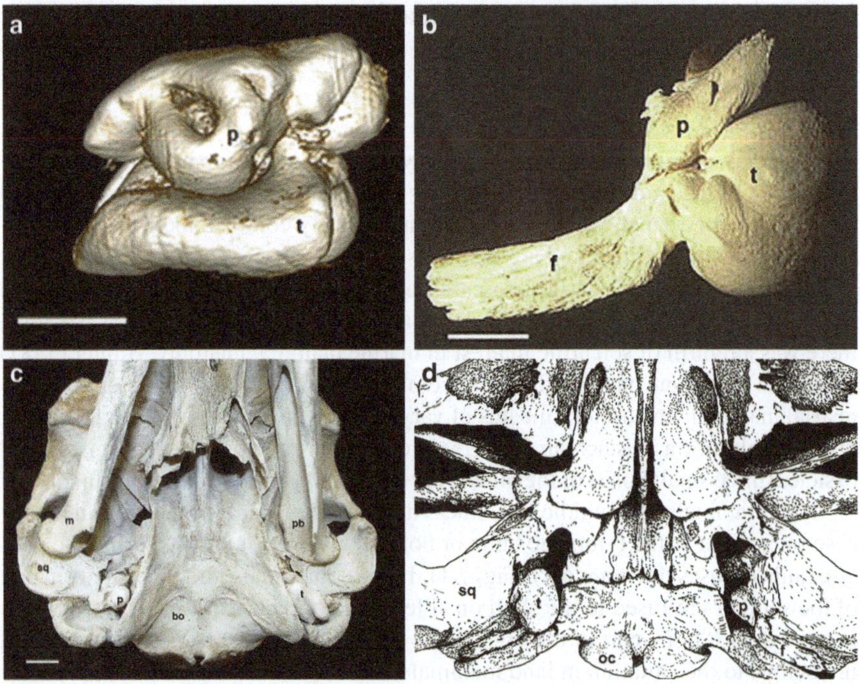

Fig. 7.1 Cetacean ear bones. bo basioccipital; eo exoccipital; f flange; m mandible; oc occipital condyle; p periotic bulla; pb pan bone region of mandible; sq. squamosal; t tympanic bulla. (Figure 7.1 ©2023 Darlene R. Ketten, all rights reserved.) (**a**) Medial view of right tympanic and periotic bullae of harbor porpoise (*Phocoena phocoena*). Scale = 1 cm. (**b**) Lateral view of periotic and tympanic of humpback whale with flange (*Megaptera novaeangliae*). Scale = 2 cm. (**c**) Ventral view of a bottlenose dolphin skull (*Tursiops truncatus*). Scale = 3 cm. (**d**) Ventral view of humpback whale skull. Right tympanic (t). Left tympanic bulla has been removed to show periotic and periotic flange (f) wedged between occipital and squamosal bones. (Drawing by I. Milde)

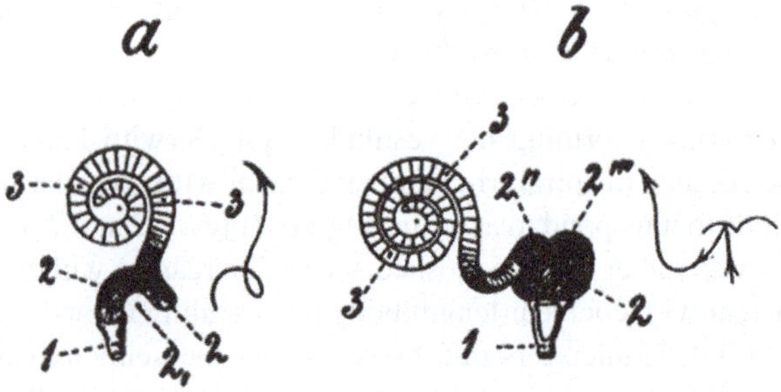

Fig. 7.2 Section through stapes, vestibule, and scala vestibuli. Right side, ventral view. (**a**) *Phocoena* (**b**) human. (Arrows indicate the main direction of molecular movement in labyrinth.) 1 Stapes; 2 Vestibule; 2″ Axis, around which one can imagine the whale tubular vestibule rotated; 3 Scala vestibuli. Only in the case of *Phocoena* are all the structures seen in a natural horizontal plane, but in humans, the scala vestibuli in the figure first had to be turned into the horizontal plane. Note: No scale was provided in the original; however, the drawings accurately represent the comparative sizes of the two ears. Porpoise basilar membranes are 22–24 mm in length; human membranes are 30–34 mm. (Image and caption (trans.) Boenninghaus (1904))

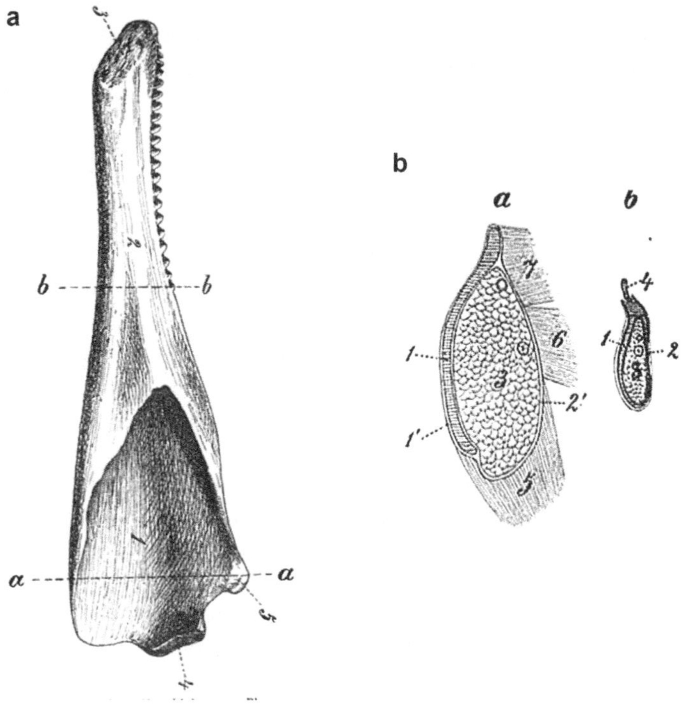

Fig. 7.3 Right mandible of *Phocoena*, internal view (**a**) and cross-section (**b**). (**a**) 1 Outer wall, 2 Inner wall, 3 Symphysis, 4 Glenoid fossa, 5 Coronoid process. Lines *a–a* and *b–b* refer to posterior and anterior transverse sections of the jaw as shown in 7.3**b** *a,b*: 1 Outer bony wall; 1′ Periosteum; 2 Inner bony wall; 2′ Inner wall periosteum; 3 Bone marrow (fatty tissue filling the canal); 4 Lateral ridge; 5 Mylohyoid ridge. (Caption translated from original text Boenninghaus (1904))

A potential reason for the few detailed anatomical studies on marine mammal middle and inner ears in this period is the size and extreme density of cetacean ear bones. The fact that Hunter undertook opening whale and dolphin periotics to observe the cochlear shapes is remarkable. Wever et al. (1971a) remarked that "the dolphin ear is a formidable object for histology…," noting it required 1 year to prepare for histology. Furthermore, cochlear soft tissues are seldom preserved in stranded animals. These issues make it difficult, even with modern tools, to study the middle and inner ears of cetaceans.

It was not until the late nineteenth and early twentieth centuries that highly detailed illustrations of whale ears appeared. Georg Boenninghaus (1904) produced a massive anatomical work on marine mammal ears with illustrations and theories on how underwater hearing works—albeit some of these were more fanciful than factual. He clearly showed differences in size and shape among cochleae of cetaceans and compared them to human ears (Fig. 7.2). Also, Boenninghaus was able to make some observations of inner ear membranes in harbor porpoises, noting that the basilar membrane was thicker and stiffer than in calves and humans. Still, he freely admitted that the specimen was in poor condition and details could no longer be recognized. This continues to be an issue with the ears of any stranded animal. Much later, exceptional illustrations based on histologic sections of dolphin inner ears were published by Wever et al. (1971a, b, c, 1972) from experiments discussed in Sects. 7.3.2 and 7.3.3.

7.3.2 *Sound Conduction to the Middle Ear*

A key question has been how underwater sound energy is conveyed to the middle ear and inner ear and whether these transduce sound by the same processes as in land mammals. The middle ear in cetaceans is housed in the shell-like tympanic bulla connected at one edge to the extremely dense, compact periotic bone, which contains the inner ear (Fig. 7.1). Hunter (1787) hypothesized that the tympanic bulla operated as a sounding box. The odontocete auditory canal is narrow and occluded with debris and wax. Consequently, several alternative sound paths were proposed in place of the external auditory canal to the inner ear.

Boenninghaus (1904) concluded "…because of its ivory-like hardness it is more likely to come close to iron than to tubular bone" and "…if now sound waves from the air fall through water, total reflection already occurs at an angle of incidence…". This is one of the earliest statements against bone conduction in cetaceans.

Kellogg (1928) agreed with Hunter, concluding that vibrations must be transmitted through the ossicular chain to the cochlea. Fraser and Purves (1954) speculated later that the bullae sat in an air-filled cavity and theorized that external vibrations reached the ossicular chain only via longitudinal vibrations of the auditory meatus and fibers attached to the malleus. However, their assumption of air in the cavity is not quite correct; in life, it is filled with a spongy soft tissue, the peribullar plexus.

Alternative paths to the middle and inner ear were proposed by Reysenbach de Haan (1957) and specifically one at or near the bullae by Dudok van Heel (1959, 1962). Since the acoustic transmission characteristics of blubber and seawater are similar, Reysenbach de Haan theorized that a soft tissue conductor would provide more fidelity than an occluded canal or bone conduction. Based on live animal behavioral studies, Dudok van Heel (1959, 1962) also obtained data demonstrating that the minimum audible angle (MAA) in bottlenose dolphins and porpoises was not consistent with intermeatal distances between the canals but was consistent with interbullar distances (see Sect. 7.4.9.2 for additional detail). Because the tympanic and periotic bullae are extracranial and are suspended in a space adjacent to the rear edge of the mandible, Dudok van Heel's results suggested that there was a bone or soft tissue path to the ear in that vicinity.

Earlier studies made observations on relevant soft tissues near the bullae. Hunter (1787, p. 387) stated the fat "…of this order of anmals, except the Spermaceti, is what we gerally term oil….and is probably the most fluid of animal fats…" and likened the fat of Spermaceti to "ruminant tallow." Boenninghaus (1904) very accurately described specialized fatty tissues aligned with and filling the mandible. He refers to them as bone marrow but notes their different character. Boenninghaus stated that the tissues are "…often colossal deposits of fat… present in all toothed whales…" and that they are lighter than blubber as well as "liquid and oily," which is true at normal room temperatures. He primarily discussed how the shape differed from that of land mammals and how it related to "…the spindle-shaped, fish-like transformation of the body." Boeninghaus also noted the fragility of the posterior region of the mandible, which, because it has no opposing wall, appears less dense (Fig. 7.3). Boenninghaus did not speculate on the function of this complex of jaw and fats, but did note other theories related to its function. One attributed to Rapp and Claudius was to amplify sound; another theory by Monro was that the fats made the head more buoyant to "rest easily on the surface." Neither theory was persuasive to Boenninghaus.

Only in the latter half of the twentieth century was substantial progress made in understanding the unique chemistry, anatomy, and acoustic properties of these fatty tissues. Kenneth Norris (1964, 1968) proposed that sounds reached the bulla by way of an "acoustic window" (also called the "pan bone" by whalers) after examining a dolphin skull on the beach. This refers to an area of the outer wall near the rear section of the mandible (position 1 in Fig. 7.3). Because there is no inner wall in that area and the bone is very thin, the area appears translucent on a dried skull. Varanasi and Malins (1971) demonstrated that fats surrounding and filling the mandible and the core of the melon are wax esters, which are very different from blubber and other body tissues with acoustic impedances close to seawater (Ketten 1992; Koopman et al. 2006; Ketten et al. 2021). Using implanted hydrophones in postmortem dolphin heads, Norris and Harvey (1974) reported that the best acoustic signal transmission occurred with sound sources placed along the jaw. They proposed that the arriving sound was conducted through the overlying fatty tissue and through the pan bone, to the intramandibular fats (Fig. 7.3), which acted as a wave guide, conveying sound to the middle ear. Norris did not, however, address how the transmission of incoming signals may be affected by the impedance differences between soft tissues and bone and potential reflections at a bony surface.

Around this time, several experiments were conducted by Theodore Holmes Bullock et al. (1968) to record the evoked potentials of auditory centers in live dolphins. Bullock et al. used four species of cetaceans (striped dolphin, *Stenella coeruleoalba*; pantropical spotted dolphin, *S. attenuate*; rough-toothed dolphin, *Steno bredanensis*; and the bottlenose dolphin, *Tursiops truncatus*). Among the variables they tested was whether different positions of the sound stimulus affected response levels. They found the largest responses from placements along a small area of the jaw for sounds above 20 kHz. There were no significant differences in auditory responses among the species.

James McCormick, a student in the laboratory of Ernest Glen Wever, led experiments that measured cochlear potentials in live dolphins. The experiments by McCormack et al. (1970) were conducted on bottlenose dolphins and Pacific white-sided dolphins (*Lagenorhynchus obliquidens*). These experiments also measured the electrical potentials of the hair cells for different sound stimuli placements. McCormack et al. (1970) found that responses were strongest when the stimulus was placed over the lower jaw, consistent with Bullock et al. (1968). However, Bullock et al. (1968), and later Alexander Supin and Vladimir Popov (1990), found the best thresholds for frequencies below 20 kHz for stimuli placed nearer the external meatus. The combined results suggest odontocetes may have two parallel systems, one for the generation and reception of ultrasonics and one for lower frequency communication signals. However, this possibility remains to be investigated.

McCormack et al. (1970) also conducted studies leading to the hypothesis that the external auditory meatus was dysfunctional. They opened the tympanic bulla and removed the ligamental connection between the tympanic membrane and the ossicular chain and removed the malleus, causing a 4 dB loss in hearing as measured by the cochlear potential. They concluded therefore that the middle ear ossicles have little function and that sound stimuli are transmitted by bone conduction with acoustic stimuli causing differential motion between the ossicular chain and the cochlear capsule. This, they hypothesized, would produce stapes motions at the oval window, creating basilar membrane displacements in the cochlea, stimulating the stereocilia of the hair cells, and generating a neural impulse conveyed to the auditory brainstem. This intracochlear process is consistent with the inner ear hair cell mechanisms known to occur in terrestrial mammals. However, it must be noted that by opening the bulla, these experiments immediately altered the attachment of the middle ear ossicles to the tympanic bone and may have disrupted middle ear soft tissues (e.g., ligaments, epithelium, and corpus cavernosum) that may have affected ossicular motions. Furthermore, bone conduction is known to worsen thresholds, especially at high frequencies, and to diminish frequency resolution and localization (von Békésy 1948; Kim et al. 2011).

These experiments were a major step forward, not only because of the data on hearing but also because of advances in surgical treatments and anesthesia with halothane that continue to be invaluable methods for treating marine mammals (Fig. 7.4, Ridgway 2008). The work was possible because of the collaborations that Bullock, Wever, and McCormick formed with Sam Ridgway, a coauthor on the resulting research papers and one of the most productive marine mammal scientists of the time.

These pioneering studies provided some answers about dolphin hearing, but the major questions of the exact paths formed by which tissues in which species are still not completely resolved. It is unlikely such experiments will be replicated because of global recognition and concern for conservation of threatened and endangered species, and many countries have regulations like the US Marine Mammal Protection Act of 1972, which set a high bar for invasive research.

Improved methods for imaging anatomy in vivo and testing hearing in marine mammals are now possible, which offer alternatives to invasive experiments. R. Stuart Mackay was a physiologist involved in dolphin research and an early advocate of the use of radiologic imaging (Mackay 1987). In recent years, advances in noninvasive imaging and modeling have proven useful in addressing questions about marine mammal head and ear structures and responses at even microscopic scales (Ketten 1992, 2000; Ketten et al. 2021). To really solve this puzzle, newer approaches, such as noninvasive imaging, finite element modeling, and tissue vector analyses, coupled with improved delivery of sound stimuli, and auditory evoked potential methods for marine mammals may settle continuing debates about underwater sound reception pathways.

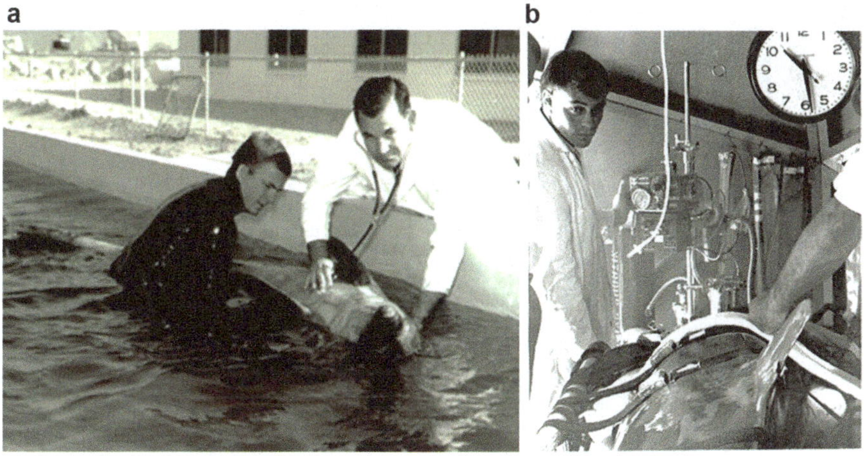

Fig. 7.4 (**a**) Dr. Sam Ridgway (right) and an assistant conducting a medical exam on a trained bottlenose dolphin in the US Navy's Marine Mammal Program in 1967. (US Navy photo) (**b**) Dr. James McCormick monitoring a dolphin anesthetized using a modified Bird-Respirator. (reprinted with permission from Dr. James G. McCormick)

7.3.3 Biosonar Adaptations of Cetacean Inner Ears

Yamada and Yoshizaki (1959) examined the cochlea of odontocetes histologically and found the basilar membrane was supported by the secondary spiral lamina for 68–82% of the cochlear canal. They also noted that the width of the basilar membrane increased relatively little over most of the basal turn, where the highest frequencies are encoded. They attributed the high frequency capabilities of odontocetes to these features. Because wavelengths in water are 4.5 times larger than in air (Eq. 7.1), odontocetes require much higher frequencies to obtain acoustic information for similarly sized targets as bats. Consequently, most odontocete echolocation signals and peak sensitivities range from 40 to 110 kHz (Southall et al. 2019; see also Sect. 7.4.1). In contrast to high frequency odontocete ears, Yamada and Yoshizaki (1959) noted that the thin, wide membranes and limited support structures of mysticete cochleae resembled that of larger terrestrial mammals, especially elephants.

Wever et al. (1971a, b, c) produced a series of papers addressing odontocete cochlear anatomy (Fig. 7.5). They noted that throughout the inner ear, the organ of Corti shows cellular hypertrophy and auxiliary membrane support, which stiffens cochlear elements, which is consistent with cochlear structure in other high frequency ears. Similar to Yamada and Yoshizaki (1959), Wever et al. (1971b, 1972) reported that the basilar membrane is suspended between bony laminae in the majority of the basal turn.

Changes that affect membrane stiffness (e.g., width, thickness, and rigidity of supports) determine the frequency range over which a mammal can hear. Wever and colleagues found that the rigidity of the basilar membrane attachment decreased

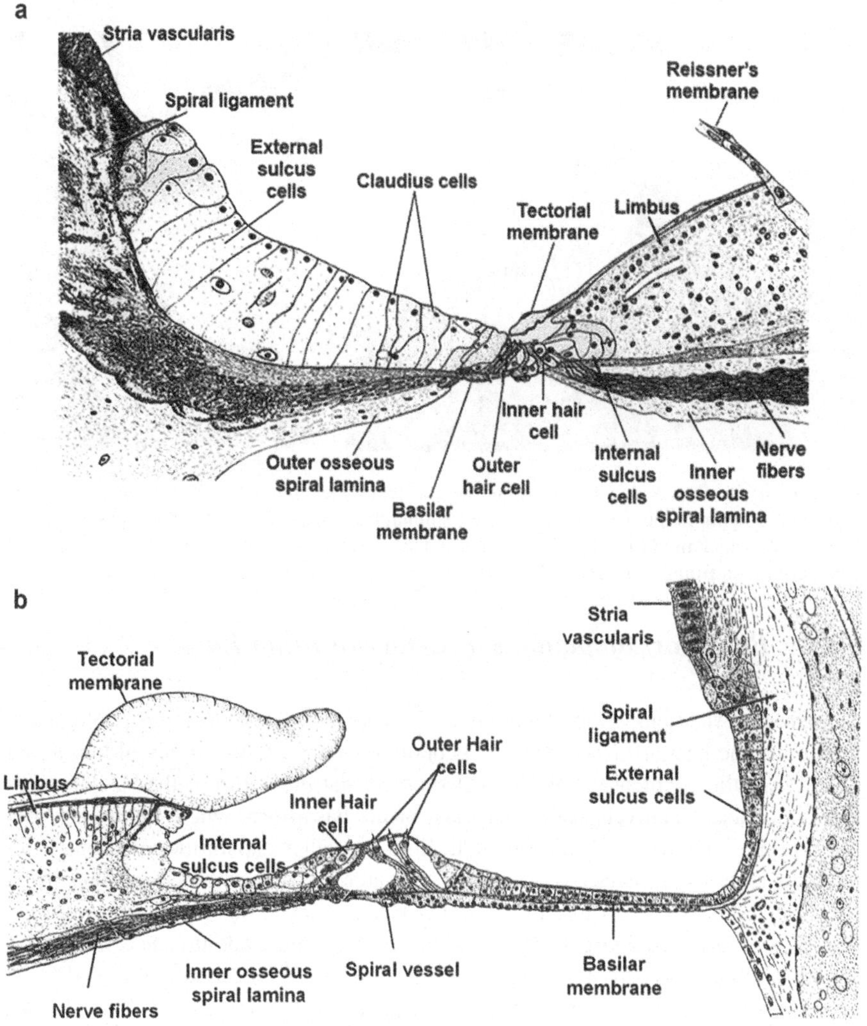

Fig. 7.5 Cochlear duct anatomy of bottlenose dolphin (Wever et al. 1971a). (**a**) Basal turn cross-section of scala media showing inner and outer bony laminae supporting a thick, narrow, basilar membrane consistent with high frequency tuning. (**b**) Apical section showing thin inner lamina and a thin, wide basilar membrane consistent with lower frequency tuning. (reprinted with permission from Dr. James G. McCormick.)

over the length of the cochlear duct, and at the apical end, it was attached only by connective tissue. The membrane width increased 14-fold, from 25 μm at the base to 350 μm at the apex, compared with a 6.25-fold width increase for human basilar membranes (Wever et al. 1971b). The narrow, firmly attached basal portion suggested that dolphins hear higher frequencies than many other mammals, and the 14-fold increase in width implied a broad range of hearing. Bottlenose dolphins have significantly higher neural densities per millimeter of basilar membrane that are threefold to sixfold the density in humans (approximately 95,000 ganglion cells in the bottlenose dolphin versus 30,000 ganglion cells in an adult human ear) (Wever et al. 1971c). This suggests more extensive auditory processing power in the dolphin and is consistent with superior frequency discriminations.

7.4 Discovering Ultrasonic Hearing and Echolocation

7.4.1 Recordings

Much of the early audiometric and recording work on cetacean hearing was done by William Schevill and Barbara Lawrence (1949, 1953) and Winthrop Kellogg et al. (1953). Schevill and Lawrence (1949) reported the first underwater recordings of identified cetacean vocalizations, in this case from beluga whales, *Delphinapterus leucas*, in the Saguenay River in Quebec, Canada. The fact that these whales vocalize, and their underwater vocalizations could be heard in air, was likely known to natives for centuries. The phenomenon was first reported by Parry (1821) who stated that beluga whales "emit a shrill, ringing sound, not unlike that of musical glasses when badly played." Beddard (1900) noted that "their [sic. beluga whales] vocal capabilities have earned for them the name of 'Sea Canary.'" Schevill and Lawrence went to the Saguenay River because there were no other marine mammals in the vicinity, and thus they could be certain the vocalizations they recorded were from beluga whales.

Schevill and Lawrence's equipment consisted of a Rochelle salt crystal hydrophone, a three-tube amplifier (both constructed at Woods Hole Oceanographic Institution), and a Gray Audiograph dictating machine using plastic disks (Watkins and Wartzok 1985). The combined frequency response of this equipment was about 100–4000 Hz. In addition to whistles and squeals, Schevill and Lawrence recorded some sounds that were reminiscent of an echo sounder. The totality of vocalizations was summarized by Schevill and Lawrence (1949, p. 144) as "...This loquaciousness contrasts markedly with most terrestrial herd mammals and compares with such chatterboxes as monkeys and men." They immediately noted that recordings needed to be extended to higher frequencies.

By 1953, Schevill and Lawrence had acquired equipment capable of recording with a flat response between 50 Hz and 196 kHz and equipment capable of presenting tones between 100 Hz and 153 kHz. They tested a recently captured female bottlenose dolphin that was held in a turbid, mud-bottomed tidal inlet, which provided a quasi-anechoic test environment. One drawback, however, was the presence of snapping shrimp, which provided background crackling noises at higher frequencies. The investigators taught the animal to go to a feeding station when it heard a projected sound and demonstrated consistent responses to tones between 150 Hz and 120 kHz. The response rate dropped off after 120 kHz (down to 50% at 126 kHz and to 30% at 130 kHz to 13% at 153 kHz, the limit of their equipment). Although Schevill and Lawrence (1953) detected no echolocation pulses, possibly due to high levels of high frequency background noise, they did note some energy in the dolphin's vocalizations up to the 196 kHz limit of their equipment.

After reviewing literature on the auditory anatomy of bottlenose dolphins, Schevill and Lawrence concluded that signals reached the cochlea via bone or body conduction (Sect. 7.3.2). They noted that human divers can detect ultrasonic frequencies but are unable to show pitch discrimination at frequencies above their in-air hearing limit. Schevill and Lawrence noted the need for pitch discrimination studies in dolphins to further understand the functional upper frequency hearing limits.

Kellogg et al. (1953) presented the first recordings of bottlenose dolphins making ultrasonic clicks. They showed that the clicks had frequencies above 120 kHz and repetition rates of 5–100 s^{-1}. They discussed the advantages of a high frequency echolocation system in that the high frequencies would be much less likely to be masked and the high frequencies would greatly enhance the resolving power of an echolocation system. When Kellogg and colleagues played back their recordings at 1/16 and 1/32 of recorded speed, listening humans could detect the echoes from the clicks. They noted that this evidence, combined with Schevill and Lawrence's (1953) ultrasonic hearing results, showed that dolphins had the components of an echolocation system, but the proof would be if they were able to avoid objects in water in the absence of vision.

7.4.2 Behavioral Observations

The first notation of odontocete behavior consistent with echolocation was made by McBride in 1947. After McBride's death, Schevill published those notes as McBride (1956). McBride showed that under conditions where vision was likely inoperable (in murky water at night), bottlenose dolphins never ran into fine mesh fishing nets and would jump over the cork line to escape. However, dolphins could be captured in nets with a 10-inch square weave. The nets typically surrounded a pod of dolphins, and if dolphins entangled in the net ended up pulling the cork line below the water, a dolphin not entangled in the net would leap over the cork line to escape. These observations, along with his knowledge of the expanded size of the dolphin auditory cortex, caused McBride to query whether "the above described behavior is associated with some highly specialized mechanism enabling the porpoise to learn a great deal about his environment through sound?" (p. 154).

An interesting footnote to this first paper suggesting echolocation by a marine mammal comes from Forrest G. Wood, onetime head of the US Navy's Marine Mammal Program. Wood quoted from a Woods Hole Oceanographic Institution report to the Office of Naval Research written by Schevill: "In 1956, we finally got published a posthumous note (McBride 1956) by Arthur McBride of Marine Studios, in an effort to establish his priority in observational evidence of echolocation of obstacle avoidance by porpoises. Shortly thereafter, we published our own note on the food-finding behavior of our captive of the previous summer; this included our evidence that the animal used echo-ranging for this purpose" (recounted by Dourdeville 2019, p. 32). Not too many scientists in any era prepare a posthumous publication for a colleague to assure his priority over their own groundbreaking work!

Schevill and Lawrence (1956) showed that a bottlenose dolphin could consistently find fish when there were no external clues in an exceedingly murky environment where vision was essentially inoperative. They recorded echolocation signals from the dolphin and noted that they became a sequence of "creaks" associated with horizontal movements of the head as the dolphin closed in on the fish.

7.4.3 Behavioral Experiments

Kellogg (1958) reported on two bottlenose dolphins, an adolescent male and a mature female, which were studied for their echolocation abilities in a dredged pool connected to a natural harbor at the Marine Laboratories of Florida State University (St. Teresa, FL). The soft bottom of the pool made for a low-reflective environment for the acoustic studies and the brown, turbid water limited vision. For humans, visibility was as low as 25 cm and routinely no more than 50 cm. Kellogg and assistants first threw fish into the water when the dolphins were 30 feet away. The animals raced to the spot and found the sinking fish. If the fish were lowered into the water quietly, the dolphins continued normal swimming, and only after 10–15 s, when the dolphins next made an exploratory ultrasonic click, did they begin rapidly clicking and retrieve the fish.

Kellogg then formalized the experiment by lowering either a 15-cm spot croaker fish, *Leiostomus xanthurus*, or a 30-cm flathead gray mullet, *Mugil cephalus*, from behind a plywood screen. The spot croaker was the preferred prey and the positive target. In the first set of 16 trials, there were four errors, but after a few days, the final 140 trials including some sessions in the dark of night produced zero errors.

In the next experiment, two identical fish, both spot croakers, were presented, with the variable being a plexiglass plate placed randomly between one of the fish and the dolphin. The two fish were equally visible, but in 202 trials, the dolphin never attempted to retrieve the fish behind the plexiglass. To remove any fish-associated gustatory clues, a net was spread across the tank with two openings in it that could be blocked randomly with a sheet of plexiglass. Of 100 trials, there were only two errors when the dolphin tried to pass through the blocked opening. When 36, 5-cm triangular metal poles were placed in the pond with a 2.4-m separation, the dolphins were able to swim between them with no contact during sessions conducted at night in the dark phase of the moon.

To confirm the ability to locate targets and avoid obstacles when vision was ruled out completely, Norris et al. (1961) trained a young adult female bottlenose dolphin to accept eye cups, thereby completely blocking any visual input. The blindfolded animal was able to unerringly distinguish between small pieces of fish and a water-filled capsule weighted to sink at the same rate as the fish. If the fish were dropped within inches of a hydrophone or a suspended pole, the dolphin could pick up the fish without hitting the nearby object. During the approach to the fish pieces, the click emission rate increased typically from 16 to 190 clicks/s. The blindfolded dolphin was also able to navigate through a variable maze of suspended 2.5-cm-diameter poles.

With echolocation firmly established for bottlenose dolphins, and subsequently for all odontocetes tested, the pace of research on all aspects of echolocation increased rapidly. Whitlow W.L. Au, Paul Nachtigall, and Alexander Supin were notable contributors to what we currently know of echolocation in marine mammals through their collaborative research at Hawaii Institute of Marine Biology (Honolulu). (See Au (2015) for a comprehensive summary of later echolocation research)

7.4.4 Audiograms

The first marine mammal audiogram was obtained by C. Scott Johnson (1967)[1] for an 8- to 9-year-old male bottlenose dolphin (Fig. 7.6). Johnson used operant conditioning and tested over a range from 75 Hz to 150 kHz. The great care Johnson took with all aspects of this experiment produced results that have not been improved upon for this species in the subsequent decades.

To do the work, Johnson constructed, as best he could, an anechoic chamber for the bottlenose dolphin. The signal presentation chamber was constructed of redwood and lined with rubberized pig and horse hair packing material. The sound level in the chamber was below zero sea state for 40 Hz to 40 kHz. An alerting light started a trial. The animal was trained to press a lever, which put his head in a specific position that turned out the light, and a tone was presented or not. If the tone was presented and detected, the animal swam to the other side of the tank and pressed a second lever to receive a fish reward. Errors resulted in a 90-s timeout. Thresholds were determined using an up-down staircase method with step sizes of 1, 2, or 3 dB. Three different transducers were used based on the frequency tested.

Results showed best sensitivity near 50 kHz at a threshold of approximately 45 dB re 1 μPa. As Johnson (1967) points out, this may not be the absolute threshold but rather the threshold at which the dolphin would commit to having heard the signal. The threshold increased as the frequency decreased, reaching about 137 dB re 1 μPa at 75 Hz. The threshold rose slowly to 55 dB re 1 μPa at 100 kHz and then more rapidly to 135 dB re 1 μPa at 150 kHz. When comparing thresholds in dB re 1 μW/cm^2, the proper metric to compare air and water thresholds (Tavolga and Wodinsky 1963), the best hearing thresholds of dolphins and humans are similar although separated by almost 50 kHz. Near threshold, dolphins can detect level changes of 1 dB. This compares with minimum intensity differences of 5–6 dB in

[1] First published in 1966 as a technical report: US Naval Ordnance Test Station Report T. P. 4178, 1966, 28 pp.

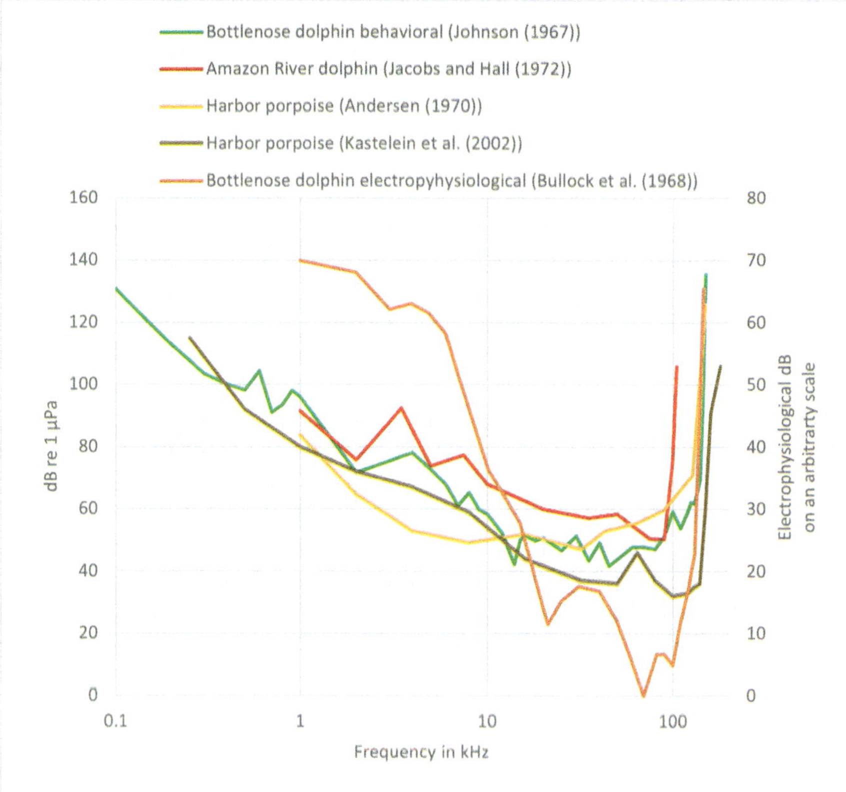

Fig. 7.6 Audiograms of bottlenose dolphins, Amazon River dolphin, and harbor porpoises. (Modified from Johnson (1967), Bullock et al. (1968), Jacobs and Hall (1972), Andersen (1970), and Kastelein et al. (2002))

humans. Johnson noted that this enhanced intensity discrimination (ID) is advantageous for echolocating.

Bullock et al. (1968) used electrical recordings of the evoked potential from the inferior colliculi, medullary auditory centers, and the medial geniculate in four species of anesthetized dolphins to determine the neural response to acoustic signals (see also Sect. 7.3.2). The stimuli were 0.3- to 10-ms pulses of various frequencies having rise and decay times of at least 0.1 ms. Striped dolphins were the most common subjects, although a few pantropical spotted dolphins, rough-tooth dolphins, and bottlenose dolphins were also tested. The evoked potential audiograms (Fig. 7.6) were similar across all four species. Regardless of where the evoked potential was recorded, the audiogram generated was in close agreement with that determined by operant conditioning (Johnson 1967). Because the absolute intensity of the stimuli could not be determined accurately, thresholds were determined from the attenuation of the maximum intensity. The best hearing was about 60 kHz, with the highest frequencies eliciting a response between 120 and 140 kHz.

Jacobs and Hall (1972) reported that a subadult male Amazon River dolphin, *Inia geoffrensis*, had its lowest threshold between 75 and 90 kHz (Fig. 7.6), frequencies well above those of the bottlenose dolphin. Jacobs and Hall used 5-s pure tones for stimuli and a staircase method with 5 dB steps to determine the 50% thresholds.

Andersen (1970) obtained an audiogram from a 3.5-year-old female harbor porpoise in response to 1.5-s pure tones using the constant stimuli method with a minimum separation of 5 dB. The hearing range was approximately 40–60 kHz (Fig. 7.6). Andersen (1970) noted he could not accurately measure background noise in the tank. This audiogram of this harbor porpoise is significantly less sensitive to higher frequencies than a later audiogram of a young male harbor porpoise (Kastelein et al. 2002). Kastelein et al. (2002) used an up-down staircase method with 4 dB steps of frequency-modulated (FM) tones and found the best sensitivity at 110 kHz. Hearing loss from age, noise exposure, disease, and equipment and experimental design can result in different audiograms for animals of the same species. Harbor porpoises have been classified as Type I very high frequency cetaceans in contrast to the larger class of Type II high frequency cetaceans, which includes bottlenose dolphins (see marine mammals grouped by hearing sensitivity, Southall et al. 2019).

Bertl Møhl (1968), working in Aarhus, Denmark, conducted hearing experiments with a 3- to 4-year-old male common seal (*Phoca vitulina*) to determine both in air and in water audiograms. Møhl used an innovative approach in which the subject controlled when, for how long, and the number of times the stimulus was presented before making a presence–absence decision. The psychophysical method employed was constant stimuli with sufficient measurements around the threshold to give a standard error of threshold of 2 dB.

Møhl found the seal's best hearing frequency in air was 16 kHz; in water, it was 32 kHz (Fig. 7.7). As with the dolphin, the seal's sensitivity underwater at frequency of best hearing was essentially the same (within 1 dB) as a human's best sensitivity in air. Thus, Møhl concluded that the seal's hearing was primarily water adapted. Later, Terhune and Ronald (1972, 1975b) described a similar underwater audiogram for a harp seal, *Pagophilus groenlandicus*, and a ringed seal, *Phoca hispida*. Ronald Schusterman et al. (1972) showed an otariid, a California sea lion (*Zalophus californianus*), had a lower underwater high frequency cutoff at 28 kHz and approximately 15 dB lower sensitivity throughout its range of best hearing (Fig. 7.7).

Møhl noted that humans lose about 30 dB of sensitivity when they move from the air to underwater, which is equal to the calculated difference based on the difference between air and water impedances. If the water-adapted seal ear responded similarly, he expected a 30 dB decrease in sensitivity when he tested the seal in air. What he found was that between 1 and 16 kHz, the air and water audiograms declined from the best frequency at the same rate, 5 dB/octave, and were separated by only 15 dB re 1 μW/cm^2 (Fig. 7.8). The seal in air had an anomalous increase in sensitivity at 2 kHz that Møhl suggested could have been caused by resonance in the 50- to 60-mm-long auditory meatus. At frequencies above that of best hearing (16 kHz (air) and 32 kHz (water)), the audiogram slopes were also similar at 65 dB/octave up to 22.5 kHz (air) and 64 kHz (water).

Subsequent work (Møhl and Ronald 1975) showed that this difference was attributable to differences in the pathways of sound to the ear in water (via tissue

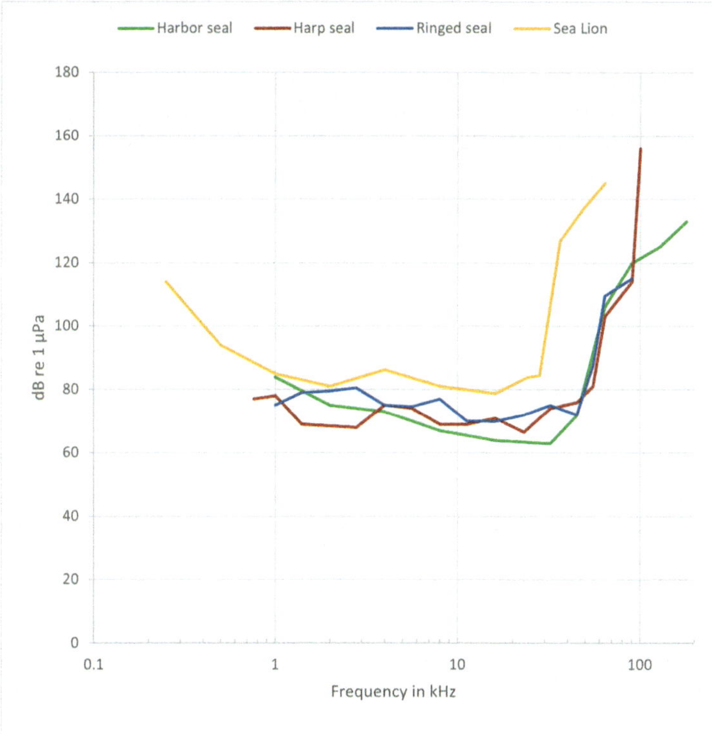

Fig. 7.7 Underwater audiograms of harbor seal, harp seal, ringed seal, and California sea lion. (Modified from Møhl (1968), Terhune and Ronald (1972, 1975b), and Schusterman et al. (1972))

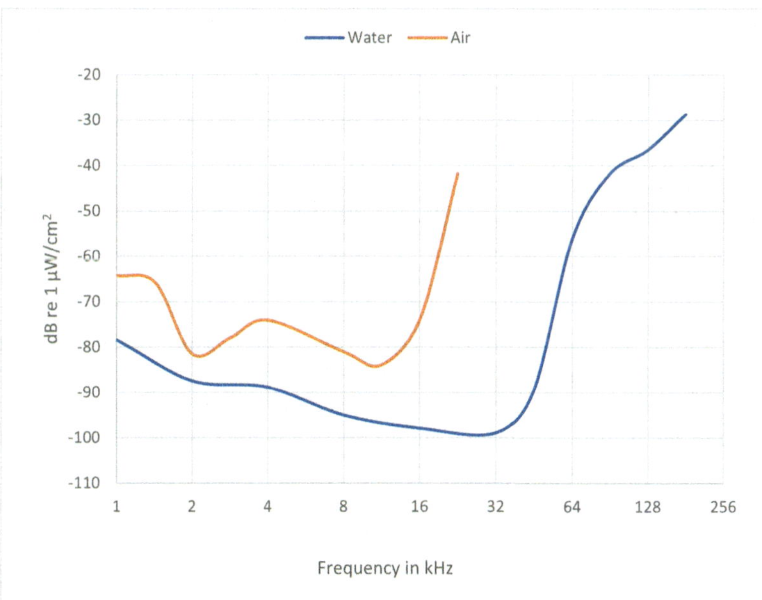

Fig. 7.8 Audiograms of harbor seal in air and water plotted as dB re 1 µW/cm^2 to compare thresholds in the two media. (From Møhl (1968). Used with permission)

conduction ventral to the auditory meatus) and in air (via the auditory meatus with accompanying midfrequency amplification). For an updated review of pinniped amphibious hearing, see Reichmuth et al. (2013), in which they not only review more recent pinniped studies but also demonstrate that harbor seals hear as well underwater as fully water-adapted cetaceans, as described above, but are also able to hear as well in air as terrestrial carnivores when in-air thresholds are not noise masked.

7.4.5 Critical Ratios and Critical Bands

Johnson (1968) investigated critical ratios and critical bands in the same bottlenose dolphin tested in his audiogram experiments. He measured critical ratios of just detectable tone intensity compared with the power spectral density of wideband noise. Following Fletcher (1940), Johnson (1968) then calculated critical bands as $CB = 10^{CR/10}$. When he plotted critical ratios against frequency, he found that the dolphin data overlapped with the human curve where tested frequencies overlapped and followed an extrapolation of the human data at higher frequencies (Fig. 7.9). At 100 kHz, the critical bandwidth was about 10 kHz. Johnson (1968, p. 967) concluded that "Information on the auditory capabilities of *Tursiops truncatus* accumulated from behavior experiments indicates thus far that the animal's inner ear functions in very much the same way as the human inner ear. The porpoise's ability to detect much higher frequencies appears to be the major difference discovered so far in behavioral experiments." More recent experiments, while confirming Johnson's critical ratio measurements, have shown that directly measured critical bands range from 11 times wider at 30 kHz to 2.2 times wider at 120 kHz compared with the calculated critical bands (Au and Moore 2000).

Terhune and Ronald (1971) reported the first in-air critical ratios for a pinniped, the harp seal. Later, Terhune and Ronald (1975a) reported the first underwater critical ratios for two ringed seals. The critical ratios for the harp seal in air ranged from 23 dB at 4 kHz to 35 dB at 8.6 kHz. For the ringed seals underwater, the ratios ranged from 30 dB at 4 kHz to 35 dB at 32 kHz. They concluded the basic pattern of an increase in critical ratio with increasing frequency held for animals in air and water.

Bullock et al. (1968) investigated critical bandwidth using neurophysiological techniques. They demonstrated the existence of critical bandwidth by showing that masking was confined to a relatively narrow band of frequencies surrounding that of the stimulus. However, they used a single frequency tone as the masking stimulus and hence the results are not comparable to more typical white noise masking stimuli. They did show, however, that as the masking tone moved away from the stimulus tone, the change in stimulus threshold was about 2 dB for 1% change in frequency.

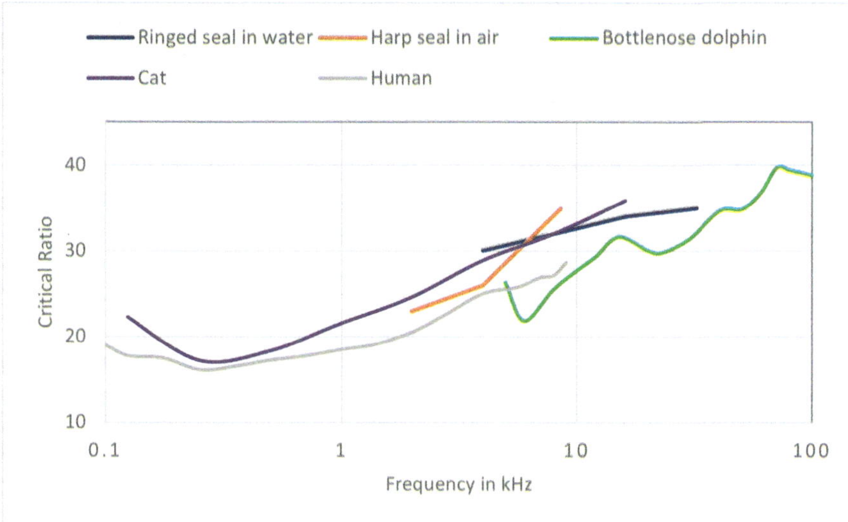

Fig. 7.9 Critical ratio for ringed seal in water, harp seal in air, bottlenose dolphin in water, and cat and human. (Modified from Terhune and Ronald (1975a) using data from Hawkins and Stevens (1950), Watson (1963), Johnson (1968), and Terhune and Ronald (1971))

7.4.6 Electrophysiological Studies of Auditory Signal Processing

Bullock and Ridgway (1972) repeated many of the experiments conducted by Bullock et al. (1968) but without heavily anesthetized dolphins. They found, to record an evoked potential in the midbrain, the signal had to have a fast rise time and be above 10 kHz. A tone at the optimal frequency of 40 kHz would evoke no response if the rise time was longer than 4 ms. The response increased with a faster rise time, showing a clear difference between a rise time of 50 μs and one of 20 μs.

An advantage of using unanesthetized dolphins was that the animals could be trained to give strings of echolocating clicks on command (and were willing to par ticipate!). Looking at the evoked potential recovery after a click, the primary wave reached 100% in response to an artificial click occurring only 0.5 ms after the dolphin click. It took twice as long to reach 100% recovery if the first and second signals were both artificial clicks. In the anesthetized dolphin (Bullock et al. 1968), some recordings were able to follow artificial click trains at rates up to 2000 s^{-1}.

While Bullock and Ridgway (1972) were unable to record evoked potentials from either low frequency or slowly rising signals in the midbrain, a few electrode placements in the cerebrum did show responses to stimuli in the frequency–time domain of dolphin whistles. The cells that responded to these stimuli were unresponsive to the click-like stimuli.

Bullock et al. (1971) conducted similar experiments on 12 sea lions. Most of the experiments were conducted in air, but some were done with the ears submerged, while the nostrils were out of the water. Most of the recording electrodes were placed in the inferior colliculus of the brain. The sea lion responded to signals with rise times as long as 50 ms, whereas the dolphin response falls to zero if the rise time was greater than 4 ms. The sea lion evoked potential showed greatest amplitude at 4–6 kHz, little response at 30 kHz, and no response at 35 kHz. Sea lion evoked potentials follow click trains up to 300 to a maximum of 800 s^{-1}. This compares with the dolphin following at 2000 s^{-1}.

Bullock et al. (1971) acknowledged that looking at the electrophysiological responses could not conclusively answer a question in a great debate at that time, which was whether pinnipeds can echolocate. Poulter (1966) thought the evidence supported sea lion echolocation, whereas Schusterman (1967) refuted the idea that sea lions echolocated. Bullock et al. (1968) noted that the evoked potentials recorded in the midbrain of pinnipeds indicated that they were not capable of the time and frequency responses shown in dolphin echolocators. Bullock et al. (1971, p. 385) concluded, based on the electrophysiology data, that "the sea lion is in these respects much like a cat gone to sea."

7.4.7 Intensity Discrimination

Several lines of evidence placed the bottlenose dolphin ID ability at about 1 dB. Johnson (1967) used an up-down staircase method to determine ID thresholds, concluding that the consistency of performance at the threshold implied that the ID was about 1 dB. Bullock et al. (1968) also showed that for some electrode positions the recorded evoked potential change in response to stimulus intensity was 1 dB.

Evans (1973) studied the ability of blindfolded bottlenose and Amazon River dolphins to detect changes in echolocation target strength. He used a series of 18-cm-long, cork-neoprene cylinders, each of a different radius, creating a series of objects with target strengths differing by 1 dB. He showed that the bottlenose dolphins were able to distinguish between cylinders differing in target strength by 1 dB and that the Amazon River dolphin could distinguish cylinders differing in target strength by 2 dB.

Burdin et al. (1973) took a different approach with Black Sea bottlenose dolphins, *Delphinus delphis ponticus*. They trained the animals to distinguish between unmodulated white noise and white noise modulated at 3 Hz. The just-distinguishable depth of modulation was approximately 5%. They noted that with similar stimuli, humans had a similar depth of modulation detection. Given that human ID is about 1 dB, this result implied that these bottlenose dolphins had a similar ID ability.

7.4.8 *Frequency Discrimination*

Thompson and Herman (1975) presented the first study of a bottlenose dolphin frequency discrimination ability across its full range of hearing sensitivity. They trained an adult female to discriminate between a constant frequency (CF) signal and an FM (2 Hz) signal where the center frequency was the same as the CF signal. The CF and FM signals were 40–100 dB above threshold values based on Johnson's (1967) audiogram. A staircase method was used to determine the threshold difference limens (DLs) (Fig. 7.10). Between 2 and 53 kHz, DL values were 0.002–0.003 and never exceeded 0.008 up to 130 kHz. Humans listening underwater in the same experimental setup had DLs of 0.0035–0.0041 at 1, 2, and 4 kHz, the range of human best sensitivity, and 0.0056 at 8 kHz, consistent with human hearing in air.

Møhl (1968) tested the frequency discrimination ability of a male harbor seal. The animal was trained to distinguish between pulses of CF and pulses in which the frequency alternated at 1 Hz between that of the CF and a test frequency of variable ΔF/F above or below the CF. Møhl found that over the range of 1–57 kHz, the difference limen was relatively constant around 0.013. At frequencies above 60 kHz the seal was incapable of making frequency discrimination (Fig. 7.10). Møhl noted

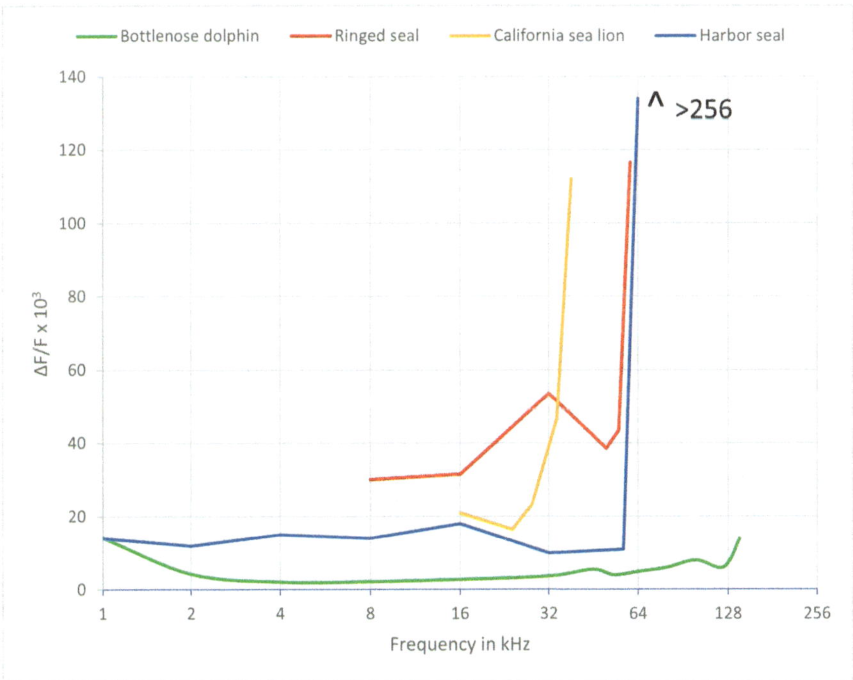

Fig. 7.10 Threshold frequency difference limens for bottlenose dolphin, ringed seal, California sea lion, and harbor seal tested underwater. (Modified from Møhl (1968), Thompson and Herman (1975), Terhune and Ronald (1976), and Schusterman and Moore (1978))

that seals in water had responses to signals as high as 160 kHz but that a reasonable upper frequency hearing limit, based on an ability to discriminate, was at 60 kHz.

Terhune and Ronald (1976) trained two ringed seals to distinguish between pulses of a CF and FM pulses. The frequency discrimination of ringed seals is very similar to that of harbor seals (Fig. 7.10). The sharp decrease in frequency discrimination corresponds to the high frequency cutoff in their audiograms (Fig. 7.6).

Ronald Schusterman and Patrick Moore (1978) provided the first frequency discrimination information on an otariid, a California sea lion. The experimental subject was an 11- to 12-year-old male that was the subject of the audiogram discussed above (Schusterman et al. 1972). The test stimuli were two pure tones of 1 s duration and 100 ms rise and fall times. Because the investigators were primarily interested in the upper frequency limit of discrimination, they chose frequencies of 16, 24, 28, 34, and 38 kHz. The test signal was either at the tested standard frequency or at a frequency below the test frequency. The signals were 20–30 dB above threshold at each frequency. The sea lion indicated that it detected the second signal as different from the first with an underwater bark. If the sea lion did not detect a difference, it remained silent. In either case, the trial ended when the seal pressed an "end of trial" paddle. A staircase method was used to establish the thresholds. Between 16 and 28 kHz, the DL was basically constant at 0.02 but more than doubled at 34 kHz and rose to more than 0.11 at 38 kHz. In all cases, the subjects had good frequency discrimination only over the range of best hearing as determined by their audiograms.

7.4.9 *Directional Discrimination*

7.4.9.1 Pinnipeds

Møhl (1964) used directional discrimination to assess the underwater hearing reception pathways of a juvenile harbor seal. He conducted the same directional discrimination test in water at a frequency of 2 kHz and in air at a frequency of 500 Hz to compensate for the differences in the speed of sound in the two media, which resulted in identical phase differences between the two ears if the reception pathways were the same. The seal was required to place its head in a basket to press the signal-initiation paddle. The signal was 0.58 s in duration, which meant that the location of the seal's head was fixed for the entire signal duration. The seal's responses were described as "somewhat desultory."

Møhl did not use either a staircase or constant stimuli method to determine the MAA. He plotted the percentage of responses to the right transducer as a function of the angle from the center line between the transducers and the seal's head in the test position. Angles to the right were positive, and angles to the left were negative. He then determined a best-fit straight line between 6° to the left and 6° to the right. The MAA was defined as the numerical value halfway between the azimuth for 25% correct and 75% correct. He concluded that there was no statistically significant difference in minimal angle in the two media (approximately 3.1° ± 1.9° in water;

4.2° ± 2.4° in air). Assuming that there was no difference in neural processing of phase differences in the two media and that in air the baseline distance separating the ears for directional discrimination was the spacing between the external openings of the auditory meatus, these results indicated that in water and air, the sound pathway to the cochlea started at the surface of the head at the distal end of the external auditory meatus. Although Møhl (1964, p. 289) cautioned that "the minimum audible angle found in this way is a figure of calculation only and is not meant to have any special biological significance," some subsequent researchers have quoted his MMA results as if they were comparable to MMAs determined in more standard psychophysical experimental settings.

Terhune (1974) studied the sound localization ability of a 3-year-old harbor seal both underwater and in the air using a click train of 30 pulses per second where each pulse had a rise time of 10 μs and signal duration of 0.2 ms. The MAAs were 9 ± 4° underwater and 3 ± 4° in air. The difference in MMAs was consistent with the location discrimination occurring via the interaural time difference and the reception path in both media originating on the surface of the head. This result agreed with the directional discrimination studies of Møhl (1964) and the cochlear microphonic study of Møhl and Ronald (1975) showing that the lowest threshold sound pathway in a harp seal was via the auditory meatus orifice in air and an area just below the orifice underwater.

Moore (1975) measured the directional discrimination ability of a 3-year-old male California sea lion to a click-like tone generated by driving the transducer to ring with one cycle of a 1 kHz tone with a repetition rate of 30 Hz and a total duration of 2.5 s. A modified method of constants was used to determine the 63% and 75% correct responses. For this click stimulus, the MAAs were 6° and 9°, respectively, for 63% and 75% correct responses.

Moore and Au (1975) measured a California sea lion's directional discrimination for pure tones. The experimental subject and experimental setup were the same as in Moore (1975). In this case, the stimuli were pure tones with a rise and fall time of 5 ms and a duration of 20 ms presented 30 times a second. A modified method of constants was employed to determine the MAAs using a 70% correct response criterion. The results showed 12° at 500 Hz, 4° at 1 kHz, 42.4° at 4 kHz, 13.5° at 8 kHz, and 18° at 16 kHz. They could not determine a threshold at 2 kHz. When they combined data from this experiment with data from Moore (1975) when he used pure tones, including data at 250 Hz and looking specifically at the percent correct responses when the separation was 20°, they obtained the data shown in Fig. 7.11.

The overall pattern in Fig. 7.11 is like that described for humans (Mills 1972). Moore and Au (1975) similarly attribute directional discrimination at frequencies below 1 kHz to time cues based on interaural phase disparities, while interaural intensity differences are the primary cue for the improved discrimination at frequencies above 4 kHz. Between these frequencies, phase differences are less pronounced, and the wavelength is too long for the head to provide shadowing to produce intensity differences.

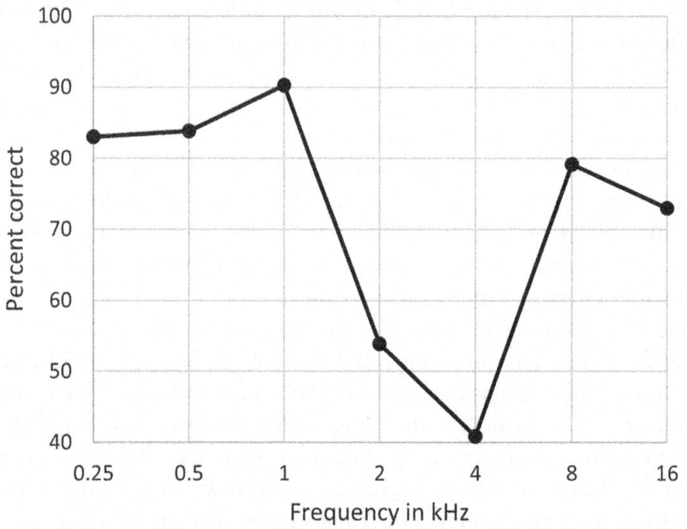

Fig. 7.11 Percent correct responses for directional discrimination by a California sea lion when the audible angle was set to 20°. (Reproduced from Moore and Au (1975) with permission of the Acoustical Society of America)

Moore and Au (1975) used hydrophones mounted on a sea lion skull to determine the intensity differences for various frequencies and various azimuths. When the behaviorally determined MAA was plotted on the intensity–azimuth curves for 4, 8, and 16 kHz, the skull-measured intensity difference was 3 ± 0.4 dB.

Moore and Schusterman (1976) used psychophysical techniques to determine the minimum detectable intensity difference at 16 kHz using a different California sea lion subject. This animal was able to detect intensity differences of 3.2 dB consistent with the intensity-based directional discrimination hypothesis for higher frequencies.

7.4.9.2 Dolphins

Dudok van Heel (1962) reported on the first attempt to determine the ability of a male harbor porpoise to discriminate between identical sound sources separated by varying amounts. The animal died before too many data points were obtained, but Dudok van Heel was able to determine that at 6 kHz the animal's threshold was 7.9° as measured from the longitudinal axis of the animal and the midpoint between the transducers. At 3.5 kHz, the threshold was 11°.

Dudok van Heel compared these values with those obtained with a human in air exposed to the same type of stimuli but with the frequency adjusted to compensate for the air–water speed of sound difference. If he assumed the point of reception was at the opening of the external auditory meatus, the porpoise would have about half the directional discrimination ability of a human. However, if he assumed the point of reception was at the bulla, the directional discrimination was comparable. Dudok van Heel (1962, p. 458) concluded, “It seems highly improbable that the precision of directional hearing of a porpoise would be less compared with that of man, taking into account...the highly developed sense of hearing of these animals.” He concluded that this result lent weight to Reysenbach de Haan’s (1957) contention that the effective distance between the ears is the distance between the lateral surfaces of the bulla in contrast to Fraser and Purves’ (1954) proposition that it was the distance between the surface points of the external auditory meatus.

The first thorough study of directional discrimination in cetaceans was conducted by Donna Renaud and Arthur Popper (1975). The subadult male subject, Heptuna (see Moore and Popper 2019 for Heptuna’s history as a researcher), was positioned on a bite bar 18 m from the sound sources, which allowed for the investigation of small angles. Localization test signals were at least 40 dB above the frequency-dependent threshold found by Johnson (1967). The animal was trained to swim to the corresponding response paddle to indicate whether the pulsed-tone signal had been heard from the left or right transducer. A modified up-down staircase method in which two correct responses would decrease the angle and one incorrect response would increase the angle yielded a threshold at the 70% level. When the stimulus was a click train, the method of constant stimuli was used to determine the MAA.

The first set of experiments had the animal bite a horizontal bar and position in a straight line to the buzzer. Under these circumstances, the MAA was somewhat frequency dependent in that the values for 6 kHz (3.6°), 90 kHz (3.2°), and 100 kHz (3.8°) were significantly different from the 2° to 3° MAAs found in the 10–80 kHz range. Using a 40 kHz signal as the standard, Renaud and Popper tested the MAAs at different azimuths. The bite plate was rotated in the horizontal plane so that the stationed animal’s body was at 15°, 345°, 30° and 330° relative to a line from the buzzer to the animal. At 40 kHz the localization acuity was 2.5° with an azimuth of 0°. This decreased to 1.7° and 1.4° at 15° and 345° respectively. This result was anomalous compared with results from humans and other animals but was noted to be consistent with the hypothesis that maximum sensitivity is along the lateral side of the mandible. The MAAs increased to 5.3° at 30° and 5.2° at 330°.

Next Renaud and Popper investigated directional ability in the vertical plane by using a vertical bite bar and having the animal on its side. Testing at 30, 60, and 90 kHz there was no significant difference in MAAs from those determined in the horizontal plane.

7.5 Cold War Science and a Thaw

A critical period in our understanding of marine mammal bioacoustics arose through the US Navy’s Marine Mammal Program, which began in 1960 with a goal of duplicating the sonar abilities of dolphins. The research was also designed to determine whether dolphins, as well as beluga whales and sea lions, were able to perform tasks such as underwater surveillance, mine detection, and equipment deliveries to divers.

These same tasks were also of interest to navies worldwide, and especially to the USSR naval force. The Soviet Union was in an arms race with the USA at that time, and both countries were in an underwater dolphin sonar race as well. Because these programs were classified, it was unclear to the scientific world exactly what had been discovered by either side. However, when the Berlin Wall came down in 1989, the Cold War was on its last legs. The USA declassified its marine mammal program in 1992, and scientists in both countries began to publicly publish the results of their research for the last 30 years. It became apparent that initial investigations for military applications led to even broader questions such as odontocete hydrodynamics, limits to the types of objects and materials they could detect, dive physiology, and, above all, how did dolphin biosonar work.

From a scientific perspective, one quick way to exchange information and ideas is by direct communication at conferences. A series of international conferences on biosonar began in the mid-1960s (Busnel and Fish 1966) in Frascati, repeating in 1978 (Jersey Islands), 1986 (Helsingor), 1989 (Rome), and most notably in Moscow in 1991 (see Nachtigall and Moore 1988; Thomas and Kastelein 1990; Thomas et al. 1992). These were extremely productive meetings, which increased steadily in participants, including notable scientists such as Whitlow Au, Nicoli Dubrovsky, Ronald Kastelein, Bertl Møhl, Paul Nachtigall, Vladimir Popov, Karen Pryor, Ronald Schusterman, Galina Solntseva, and Alexander Supin. The success of the later meetings was due in large part to the skills and enthusiasm of the chief organizer, Jeanette Thomas. The meetings also fostered collaborations that continue to this day.

Among the significant new scientific findings reported at the meetings was a demonstration that interaural time distances vs. high frequency limits are consistent for land and aquatic mammals *if* intermeatal distances are used for in-air experiments, but intercochlear distances are used for hearing in water, building on the earlier work of Reyenbach de Haan and Dudok van Heel (Supin and Popov 1993; Moore et al. 1995; see also Sect. 7.4.9.1). There was also a dramatic increase in auditory evoked potential data (Supin and Popov 1990, 1993; Bibikov 1992; Dolphin and Mountain 1992). Cetacean auditory brainstem responses were found to have distinct species-dependent peak, amplitude, and latency characteristics that differ from many land mammals, such as three positive peaks with increasing amplitudes for harbor porpoises and bottlenose dolphins but longer latencies in porpoises. The results show that bottlenose dolphins and false killer whales (*Pseudorca* spp.) can extract envelope following response features at higher modulation frequencies than other mammals, implying that dolphins may have central nervous system adaptations for both low and high frequencies, consistent with their ability to process whistles and echolocation signals simultaneously.

There is another important contributor to Cold War sonar and dolphin hearing research, Heptuna, mentioned in Sect. 7.4.9.2, which had a scientific "career" that stretched from 1970 to 2010. His responses and task performance data were the basis for at least 30 papers during that time (Moore and Popper 2019). Indeed, the unsung heroes of this research area, in every country, are the animals and their trainers who provided data at the time and who continue to do so to this day. Such studies laid the foundation for what are now standard research tools and broarder biosonar research directions.

7.6 Acoustic Communication

The primary focus of this chapter has been on hearing. However, much of the inspiration for work on hearing was based on earlier studies of sounds produced by marine mammals. The role of echolocation clicks in stimulating work on high frequency reception has been presented above. We thus conclude this chapter with a brief overview of two areas of sound production that initiated much more work on acoustic communication in marine mammals.

7.6.1 Songs

Broughton (1963, p. 883) defined a song as "a series of notes, generally of more than one type, uttered in succession and so related as to form a recognizable sequence or pattern in time." Under this criterion, several marine mammal species have been reported to have song patterns among their vocalizations. The first marine mammal so identified was the bearded seal, *Erignathus barbatus* (Ray et al. 1969). Alaskan natives had long recognized the song of the bearded seal naming the animal "aveloouk" (in the Upik dialect) or "ayuktuk" (in the Lnupik dialect) translated as "the one that sings" or "the singer," respectively. Ray et al. (1969) described the song as variations on five themes that were stereotyped and repetitive. They hypothesized that the song was related to breeding because it was heard at the time of the year when the seals were known to breed. The hypothesis was strengthened when some of the singers were collected, and all were found to be sexually mature males.

A few years later, Payne and McVay (1971) reported similar results from humpback whales (*Megaptera novaeangliae*). As with the Alaskan native hunters of bearded seals, hunters of large mysticete whales long knew of their vocal abilities. The vocalizations studied by Payne and McVay had previously been identified as coming from humpback whales, but Payne and McVay were the first to carefully analyze the structure and realize that these vocalizations also fit the definition of song. They demonstrated a six-tier structure to the vocalizations: subunit < unit < phrase < theme < song < song session. While the number of phrases in a theme varies, the sequence of themes in a song is consistent. They were unable to identify the sex of the vocalizing whales or to ascribe a particular function to the song although mating, pair formation, or maintenance of a migrating group through song were suggested. Subsequent research (e.g., Herman et al. 2013) has shown that only male humpback whales sing and that the singing is done primarily by sexually mature males within a lek breeding structure.

7.6.2 Signature Whistles and Codas

Bottlenose dolphins and several other dolphin species produce individual-specific whistles that are termed signature whistles. Probably the earliest report of signature whistles, although not recognized as individual-specific, was McBride's (1940) note about vocalizations of dolphins when first placed in captivity, particularly vocalizations between a female and her calf.

Signature whistles were first described in the scientific literature by Caldwell and Caldwell (1965), who observed five recently captured bottlenose dolphins from the same wild school. Each animal had a characteristic whistle that was distinct in its frequency modulation pattern from the others. While the basic contour of each signature whistle was relatively constant, the number and intensity were related to stress levels of the individual. With mild stress, the whistles would increase in intensity and number, whereas in high-stress situations, the whistles would cease. The signature whistle made up well over 90% of all whistles during the 3 weeks following capture. Later studies have shown that in wild populations, signature whistles make up 40–70% of the whistles. Dolphins learn and refine their signature whistles during the first 3 months of life and then maintain that signature whistle with few modifications for life. Dolphins can copy other individuals' signature whistles but with subtle changes, so it is clear that the copy is indeed a copy. Fifty years of research on dolphin whistles are reviewed by Janik and Sayigh (2013).

Sperm whales, *Physeter macrocephalus*, use echolocation clicks in foraging and also use the same structured clicks for communication via time-structured sequencing of the clicks termed a coda (Watkins and Schevill 1977). The coda lasts from 0.5 to 1.5 s and has 3–40 clicks. The coda can be repeated up to 60 times over a period of several minutes. Each coda is individual specific, at least over the relatively short periods of time the whales can be tracked. The codas are produced only when the whales are submerged. Codas appear to be used when conspecifics are in the same general area. When multiple whales are producing codas, each coda is different from the others in terms of number of clicks and temporal sequencing of clicks in the coda. Watkins and Schevill (1977, p. 1487) reported an instance of two acoustically tracked sperm whales exchanging codas. One had a nine-click coda, and the other one had a seven-click coda. "The nine-click coda was answered by the seven-click coda for five alternations. But during the fifth exchange, the nine-coda was superimposed on the seven-coda, and the seven-coda continued (at the seven-click-per-sec rate) for two more clicks, now also nine clicks. Then there follows a fast (about 35 per sec) series of 14 clicks apparently from the seven-coda whale, judging by relative frequency spectra. This fast 14-click series was answered by 14 clicks from the nine-coda whale at the nine-click-per-sec rate. And finally the whales returned to three more alternations of seven- and nine-click codas, respectively."

7.7 Conclusions

The history of marine mammal hearing discoveries began with field and anatomical observations. To these foundations were added progressively more detailed investigations of the anatomical underpinnings of hearing and great leaps in the sophistication of technologies that opened our eyes—and ears—to the extreme frequency ranges these animals both produce and detect.

Marine mammals are acoustically diverse. Their hearing is as acute as that of land mammals, but on average they have wider frequency ranges and better temporal and spatial sensitivity. Most marine mammal species have some ultrasonic capability, and all exhibit some aquatic auditory adaptations. Both mysticetes and odontocetes appear to have soft tissue channels for sound conduction to the ear. Pinnipeds appear to use an air-filled external canal for sound reception, at least in air. Comparisons of the hearing characteristics of otariids and phocids suggest that there are at least two types of pinniped ears, with phocids being better adapted for underwater hearing.

Throughout the ages, from ancient to modern, a hallmark of research on marine mammals has been to overcome the barrier of obtaining accurate observations and measurements of structures, responses, and functions in these creatures. In some cases, this work was hazardous, while in others, it was painstaking, perhaps even tedious. In many cases, it required ingenuity and careful planning to adapt existing resources and technologies to conduct the work on and within the Ocean. The scientists whose research is summarized in this chapter often compensated for equipment limitations with extensive thought and careful, innovative experimental designs.

As noted in several places in this chapter, great progress was made on how and what these animals hear, but at the same time, these efforts only scratched the surface of the proposed questions. It is remarkable that observations as old as those of Aristotle are still relevant today (such as the way sound is used in fisheries), but it is also a testament to the complexity and sophistication of the anatomical and physiological adaptations of marine mammals. The field is wide open for new insights and approaches to enrich our knowledge of mammalian aquatic life.

Acknowledgments There are too many exceptional scientists who advanced our knowledge of marine mammals to properly acknowledge all. This chapter is dedicated to a few whom we had the privilege to know and were fortunate to have as mentors, colleagues, and friends. Bill Watkins and Bill Schevill, assisted in his early work by his wife, Barbara Lawrence, an outstanding scientist in her own right, laid the foundation for and influenced marine mammal acoustic work over the subsequent 70 years. In addition to their many publications and archived recordings, they set the standard for inventing and using state-of-the-art equipment coupled with insightful natural history observations to unravel crucial questions about these fascinating mammals. They also imbued the field with their high integrity.

Sam Ridgway was the epitome of a dedicated veterinarian and innovative scientist who approached his work with great intellect and his friends with great humor. He is rightly considered to have created the field of marine mammal medicine, and his work on dolphin hearing set critical benchmarks for that research. He is arguably one of the giants upon whose shoulders all future marine mammal researchers stand.

Jeanette Thomas was a pioneer, who bravely entered a field in which women were rare. To that, add her dedication to mentoring young scientists coupled with a sparkling personality. These qualities made her not only a great scientist but also a successful organizer of multiple international meetings that turned marine mammal science into a truly global discipline.

Dr. Doris Meyerdiercks and Dr. Dorothe Poggel of the Hanse-Wissenschaftskolleg (HWK) provided invaluable advice, support, and encouragement throughout this entire project. This is publication #1581 from The Institute of Environment at Florida International University.

References

Andersen S (1970) Auditory sensitivity of the harbour porpoise, *Phocoena phocoena*. In: Pilleri G (ed) Investigations on Cetacea, vol 2. Institute for Brain Research, Bern, pp 255–259

Aristotle 350 BCE (1910) The history of animals. Book 1, Section 11 (trans: Thompson DW). John Bell, London. http://www.esp.org/books/aristotle/history-of-animals/html/

Au WWL (2015) History of dolphin biosonar research. Acoust Today 11(4):10–17

Au WWL, Moore PB (2000) Critical ratio and critical bandwidth for Atlantic bottlenose dolphin. J Acoust Soc Am 88:1635–1638

Barnes LG, Domning DP and Ray CE (1985) Status of studies on fossil marine mammals. Mar Mamm Sci 1:15–53

Beddard FE (1900) A book of whales. G.P. Putnam's Sons, New York

Bibikov NG (1992) Auditory brainstem responses in the harbor porpoise (*Phocoena phocoena*). In: Thomas J, Kastelein R, Supin AY (eds) Marine mammal sensory systems. Plenum Press, New York, pp 197–211

Boenninghaus G (1904) Das Ohr des Zahnwales zugleich ein Beitrag zur Theorie der Schalleitung. Eine biologische Studie Zool Jahrb (Anatomie) 19:13–360

Broughton WB (1963) Glossarial index. In: Busnel RG (ed) *Acoustic behavior of animals*. Elsevier, London, pp 824–910

Bullfinch T (1903) The age of fables or beauties of mythology. 1855 edition: Klapp WH. Henry Altemus Co., Philadelphia. https://www.google.com/books/edition/_/606CAAAAIAAJ?hl=en&gbpv=1

Bullock TH, Ridgway SH (1972) Evoked potentials in the central auditory system of alert porpoises to their own and artificial sounds. J Neurobiol 3:79–99

Bullock TH, Grinnell AD, Ikezono E, Kameda K, Katsuki Y, Nomoto M, Sato O, Suga N, Yanagisawa K (1968) Electrophysiological studies of central auditory mechanisms in cetaceans. Z Vgl Physiol 59:117–156

Bullock TH, Ridgway SH, Suga N (1971) Acoustically evoked potentials in midbrain auditory structures in sea lions (Pinnipedia). Z Vgl Physiol 74:372–387

Burdin VI, Markov VI, Reznik AM, Skornyakov VM, Chupakov AG (1973) Determination of the just noticeable difference for white noise in the Black Sea bottlenose dolphin (*Tursiops truncatus ponticus barabasch*). In: Mills H (ed and trans) Morphology and ecology of marine mammals: seals, dolphins, and porpoises. Wiley, New York, pp 169–173

Busnel R-G, Fish JF (1966) Animal sonar systems, NATO Advanced Study Institute (ASIAS), vol 28. Springer

Caldwell MC, Caldwell DK (1965) Individualized whistle contours in bottlenose dolphins (*Tursiops truncatus*). Nature 207:434–435

CERWG (2006) Annex 9: The crown right to whales. Final report. 8 pp

Clarke JGD (1946) Seal-hunting in the Stone Age of North-Western Europe: a study in economic prehistory. Proc Prehist Soc 12:12–48

Dolphin WF, Mountain DC (1992) The envelope following response: scalp potential elicited in the Mongolian gerbil using SAM acoustic stimuli. Hear Res 58:70–78

Domning DP (1982) Evolution of manatees: a speculative history. J. Paleontol 56:599–619
Dourdeville KM (2019) The William E. Schevill and Barbara Lawrence legacy in marine mammal science. In: Brophy CC (ed) *The voices of marine mammals*. Old Dartmouth Historical Society/ New Bedford Whaling Museum, New Bedford, pp 31–39
Dudok van Heel W (1959) Audio-direction finding in the porpoise (*Phocaena phocaena*). Nature 183:1063. https://doi.org/10.1038/1831063a0
Dudok van Heel WH (1962) Sound and Cetacea. Neth J Sea Res 1:407–507
Evans WE (1973) Echolocation by marine delphinids and one species of fresh-water dolphin. J Acoust Soc Am 54:191–199
Fletcher H (1940) Auditory patterns. Rev Mod Phys 12:47–65
Fordyce RE (1980) Whale evolution and Oligocene southern ocean environments. Palaeogeogr Palaeoclimatol Palaeoecol 31:319–336
Fraser FC, Purves PE (1954) Hearing in cetaceans. Bull Br Mus 2:103–116
Gingerich PD, Wells NA, Russell DE, Shah SM (1983) Origin of whales in epicontinental remnant seas: new evidence from the early Eocene of Pakistan. Science 220:403–406
Hawkins Jr, JE, Stevens SS (1950) The masking of pure tones and of speech by white noise. J Acoust Soc Am 22:6–13
Herman LM, Pack AA, Spitz SS, Herman EYK, Rose K, Hakala S, Deakos MH (2013) Humpback whale song: who sings? Behav Ecol Sociobiol 67:1653–1663
Hunter J (1787) Observations on the structure and oeconomy of whales. Philos Trans R Soc Lond 77:371–450
Jacobs DW, Hall JD (1972) Auditory thresholds of a fresh water dolphin, *Inia geoffrensis* Blainville. J Acoust Soc Am 51:530–533
Janik VM, Sayigh LS (2013) Communication in bottlenose dolphins: 50 years of signature whistle research. J Comp Physiol A 199:479–489
Johnson CS (1967) Sound detection thresholds in marine mammals. In: Tavolga WN (ed) *Marine bioacoustics II*. Pregamon, Oxford, pp 247–260
Johnson CS (1968) Masked tonal thresholds in the bottlenosed porpoise. J Acoust Soc Am 44:965–967
Kastelein RA, Bunskoek P, Hagedoorn ML Au, WWL, de Haan D (2002) Audiogram of a harbor porpoise (*Phocoena phocoena*) measured with narrow-band frequency-modulated signals. J Acoust Soc Am 112:334–344
Kawamura A (1980) A review of food of balaenopterid whales. Sci Rep Whales Res Inst 32:155–197
Kellogg R (1928) The history of whales: their adaptation to life in the water (concluded). Q Rev Biol 3:174–208
Kellogg WN (1958) Echo ranging in the porpoise. Science 128:982–988
Kellogg WN, Kohler R, Morris HN (1953) Porpoise sounds as sonar signals. Science 117:239–243
Ketten DR (1992) The marine mammal ear: specializations for aquatic audition and echolocation. In: Webster D, Fay R, Popper A (eds) The evolutionary biology of hearing. Springer, New York, pp 717–750
Ketten DR (2000) Cetacean ears. In: Au WWL, Fay RRR, Popper AN (eds) Hearing by whales and dolphins. Springer, New York, pp 43–108
Ketten DR, Simmons JA, Riquimaroux H, Simmons AM (2021) Functional analyses of peripheral auditory adaptations related to echolocation in air vs. water. Front Ecol Evol 9:661216. https://www.frontiersin.org/article/10.3389/fevo.2021.661216
Kim N, Homma K, Puria S (2011) Inertial bone conduction: symmetric and anti-symmetric components. J Assoc Res Otolaryngol 12(3):261–279
Koopman HN, Budge SM, Ketten DR, Iverson SJ (2006) The topographical distribution of lipids inside the mandibular fat bodies of odontocetes: remarkable complexity and consistency. IEEE J Ocean Eng 31(1):95–106
Mackay RS (1987) Whale heads and ray diagrams. Mar Mamm Sci 3(3):283–285
McBride AF (1940) Meet mister porpoise. Nat Hist 45:16–29

McBride AF (1956) Evidence for echolocation by cetaceans. Deep-Sea Res 3:153–154
McCormack JG, Wever EG, Palin J, Ridgway SH (1970) Sound conduction in the dolphin ear. J Acoust Soc Am 48:1418–1428
Møhl B (1964) Preliminary studies on hearing in seals. Videnskabelige meddelelser fra Dansk naturhistorisk forening i Kjöbenhavn 127:283–294
Møhl B (1968) Auditory sensitivity of the common seal in air and water. J Aud Res 8:27–38
Møhl B, Ronald K (1975) The peripheral auditory system of the harp seal, *Pagophilus groenlandicus* (Erxleben, 1777). Rapports et Procès-Verbaux des Réunions 169:516–523
Moore PWB (1975) Underwater localization of click and pulsed pure-tone signals by the California sea lion (*Zalophus californianus*). J Acoust Soc Am 57:406–410
Moore PWB, Au WL (1975) Underwater localization of pulsed pure tones by the California sea lion (*Zalophus californianus*). J Acoust Soc Am 58:721–727
Moore P, Popper AN (2019) Heptuna's contributions to biosonar. Acoust Today 15(1):44–52
Moore PWB, Schusterman RJ (1976) Discrimination of pure-tone intensities by the California sea lion. J Acoust Soc Am 60:1405–1407
Moore PWB, Pawloski DA, Dankiewicz L (1995) Interaural time and intensity difference thresholds in the bottlenose dolphin (*Tursiops truncatus*). In: Kastelein RA, Thomas JA, Nachtigall PE (eds) Sensory systems of aquatic mammals. DeSpil Publishers, Woerden, pp 11–25
Mills AW (1972) Auditory localization. In: Tobias, JV (ed) Foundations of modern auditory theory, Vol. II. Academic Press, New York, pp 303–348
Nachtigall PE, Moore PWB (eds) (1988) Animal sonar: processes and performance NATO advanced study institute on animal sonar systems. Plenum Press
Norris KS (1964) Some problems of echolocation in cetaceans. In: Tavolga WN (ed) Marine bioacoustics. Pergamon Press, New York, pp 317–336
Norris KS (1968) The evolution of acoustic mechanisms in odontocete cetaceans. In: Drake ET (ed) Evolution and environment, a symposium presented on the occasion of the one hundredth anniversary of the foundation of Peabody Museum of Natural History at Yale University. Yale University Press, New Haven, pp 297–324
Norris KS, Harvey GW (1974) Sound transmission in the porpoise head. J Acoust Soc Am 56:659–664
Norris KS, Prescott JH, Asa-Dorian PV, Perkins P (1961) An experimental demonstration of echolocation behavior in the porpoise, *Tursiops truncatus* (Montagu). Biol Bull 120:163–176
Parry WE (1821) Journal of a voyage for the discovery of the north-west passage from the Atlantic to the Pacific. Adam Small, Philadelphia
Payne R (1995) Among whales. Scribner. 431 pp
Payne RS, McVay S (1971) Songs of humpback whales. Science 173:585–597
Poulter TC (1966) The use of active sonar by the California sea lion. J Aud Res 6:165–173
Ray GC, Watkins WA, Burns JJ (1969) The underwater song of *Erignathus* (Bearded seal). Zoologica 54:79–82
Reichmuth C, Holt MM, Mulsow J, Sills JM, Southall BL (2013) Comparative assessment of amphibious hearing in pinnipeds. J Comp Physiol A 199:491–507
Renaud DL, Popper AN (1975) Sound localization by the bottlenose porpoise *Tursiops truncatus*. J Exp Biol 63:569–585
Reysenbach de Haan FW (1957) Hearing in whales. Acta Otolaryngol Suppl 134:1–114
Ridgway S (2008) History of veterinary medicine and marine mammals: a personal perspective. Aquat Mamm 34:471–513. https://doi.org/10.1578/AM.34.3.2008.471
Romero A (2012) When whales became mammals: the scientific journey of cetaceans from fish to mammals in the history of science. In: New approaches to the study of marine mammals. InTech. https://doi.org/10.5772/50811
Rosel PE, Wilcox LA, Yamada TK, Mullin KD (2021) A new species of baleen whale (*Balaenoptera*) from the Gulf of Mexico, with a review of its geographic distribution. Mar Mamm Sci 37:577–610

Schevill WE, Lawrence B (1949) Underwater listening to the white porpoise (*Delphinapterus leucas*). Science 109:143–144

Schevill WE, Lawrence B (1953) High-frequency auditory response of a bottlenosed porpoise, *Tursiops truncatus* (Montagu). J Acoust Soc Am 25:1016–1017

Schevill WE, Lawrence B (1956) Food-finding by a captive porpoise (*Tursiops truncatus*). Beviora 53:1–15

Schusterman RJ (1967) Perception and determinants of underwater localization in the California sea lion. In: Busnel R-G (ed) Animal sonar systems. Laboratoire de Physiologie Acoustique, Jouy-en-Josas, pp 535–617

Schusterman RJ, Moore PW (1978) The upper limit of underwater frequency discrimination in the California sea lion. J Acoust Soc Am 63:1591–1595

Schusterman RJ, Balliet RF, Nixon J (1972) Underwater audiogram of the California sea lion by the conditioned vocalization technique. J Exp Anal Behav 17:339–350

Sørensen K, Christensen-Dalsgaard J, Wahlberg M (2022) Is human underwater hearing mediated by bone conduction? Hear Res 420:108484., ISSN 0378-5955. https://doi.org/10.1016/j.heares.2022.108484

Southall BL, Finneran JJ, Reichmuth C, Nachtigall PE, Ketten DR, Bowles AE, Ellison WT, Nowacek DP, Tyack PL (2019) Marine mammal noise exposure criteria: updated scientific recommendations for residual hearing effects. Aquat Mamm 45:125–232

Supin AY, Popov VV (1990) Frequency selectivity of the auditory system of the bottlenosed dolphin *Tursiops truncatus*. In: Thomas JA, Kastelein RA (eds) Sensory abilities of cetaceans: laboratory and field evidence. Plenum Press, New York, pp 385–393

Supin AY, Popov VV (1993) Direction-dependent spectral sensitivity and interaural spectral difference in a dolphin: evoked potential study. J Acoust Soc Am 93:3490–3495

Tavolga WN, Wodinsky J (1963) Pure tone thresholds in nine species of marine teleosts. Am Mus Nat Hist Bull 126:179–239

Terhune JM (1974) Directional hearing of a harbor seal in air and water. J Acoust Soc Am 56:1862–1865

Terhune JM, Ronald K (1971) The harp seal, *Pagophilus groenlandicus* (Erxleben, 1777). X. The air audiogram. Can J Zool 49:385–390

Terhune JM, Ronald K (1972) The harp seal, *Pagophilus groenlandicus* (Erxleben, 1777). III. The underwater audiogram. Can J Zool 50:565–569

Terhune JM, Ronald K (1975a) Masked hearing thresholds of ringed seals. J Acoust Soc Am 58:515–516

Terhune JM, Ronald K (1975b) Underwater hearing sensitivity of two ringed seals (*Pusa hispida*). Can J Zool 53:227–231

Terhune JM, Ronald K (1976) The upper frequency limit of ringed seal hearing. Can J Zool 54:1226–1229

Thewissen JGM, Hussain ST, Arif M (1994) Fossil evidence for the origin of aquatic locomotion in archaeocete whales. Science 263:210–212

Thomas J, Kastelein R (eds) (1990) Sensory abilities of cetaceans: field and laboratory evidence, NATO science series A: life science, vol 196. Plenum Press. 714 pp

Thomas J, Kastelein R, Supin A (eds) (1992) Marine mammal sensory systems. Plenum Press, New York

Thompson RKR, Herman LM (1975) Underwater frequency discrimination in the bottlenosed dolphin (1-140 kHz) and the human (1-8 kHz). J Acoust Soc Am 57:943–948

Varnassi U, Malins DG (1971) Unique lipids of the porpoise (*Tursiops gilli*): Differences in triacyl glycerols and wax esters of acoustic (mandibular canal and melon) and blubber tissues. Biochem Biophys Acta 231:415

von Békésy G (1948) Vibrations of the head in a sound field and its role in hearing by bone conduction. J Acoust Soc Am 20:749–760

Watkins WA, Schevill WE (1977) Sperm whale codas. J Acoust Soc Am 62:1485–1490

Watkins WA, Wartzok D (1985) Sensory biophysics of marine mammals. Mar Mamm Sci 1:219–260
Watson CS (1963) Masking of tones by noise for the cat. J Acoust Soc Am 35:167–172
West CD (1985) The relationship of the spiral turns of the cochlea and the length of the basilar membrane to the range of audible frequencies in ground dwelling mammals. J Acoust Soc Am 77:1091–1101
Wever EG, McCormick JG, Palin J, Ridgway SH (1971a) The cochlea of the dolphin, *Tursiops truncatus*: general morphology. Proc Natl Acad Sci U S A 68:2381–2385
Wever EG, McCormick JG, Palin J, Ridgway SH (1971b) Cochlea of the dolphin, *Tursiops truncatus*: the basilar membrane. Proc Natl Acad Sci U S A 68:2708–2711
Wever EG, McCormick JG, Palin J, Ridgway SH (1971c) The cochlea of the dolphin, *Tursiops truncatus*: hair cells and ganglion cells. Proc Natl Acad Sci U S A 68:2908–2912
Wever EG, McCormick JG, Palin J, Ridgway SH (1972) Cochlear structure in the dolphin *Lagenorhynchus obliquidens*. Proc Natl Acad Sci U S A 69:657–661
Yamada M, Yoshizaki F (1959) Osseous labyrinth of Cetacea. Sci Rep Whales Res Inst 14:291–304

Chapter 8
Development of Models for Bat Echolocation

James A. Simmons and Andrea Megela Simmons

8.1 Introduction

The history of bat echolocation began with the foundational observations of the Italian physiologist Lazaro Spallanzani and the Swiss naturalist Louis Jurine in 1793–1794 (MacArthur 2000) and continued to the direct experimental demonstration of echolocation in action by Donald R. Griffin and Robert Galambos at Harvard University in 1938–1942 (Griffin 1958). The discoveries of Griffin and Galambos stimulated an enormous body of research on the behavior, biology, and mechanisms of echolocation (Fenton et al. 2016). Prior reviews of this body of research (Grinnell et al. 2016; Grinnell 2018) have focused primarily on the biology of echolocation. This chapter discusses, instead, the historical development of conceptual models of echolocation. These models were informed by contemporaneous technological developments of radar and sonar systems, which are relevant to understanding how the bat's auditory system might exploit the wide ultrasonic bandwidths of biosonar signals using much lower-frequency neural systems. Echolocation evolved in bats as a perceptual system based on the acoustics of echoes (Teeling et al. 2016), and it is to be expected that aspects of auditory responses might reflect common parameters that also govern the operation of radar (Simmons et al. 2017).

The chapter begins by acknowledging two early ideas (1912–1920) that echolocation itself was initially a theory, not a demonstrated mode of sensing, but with

J. A. Simmons
Department of Neuroscience, Carney Institute for Brain Science, Brown University, Providence, RI, USA
e-mail: James_Simmons@brown.edu

A. M. Simmons (✉)
Department of Cognitive, Linguistic, & Psychological Sciences, Carney Institute for Brain Science, Brown University, Providence, RI, USA
e-mail: Andrea_Simmons@brown.edu

D. R. Ketten et al. (eds.), *A History of Discoveries on Hearing*, Springer Handbook of Auditory Research 77,
https://doi.org/10.1007/978-3-031-41320-9_8

significant technological motivation derived from the wartime need to improve radar and radio communication technology. The chapter then turns to examining the proliferation of hypotheses proposed during the roughly two decades that followed the actual discovery of echolocation and the publication of Griffin's *Listening in the Dark* (1958), a masterpiece detailing what was then known about this sensing system. These hypotheses drew largely from what was then understood about human perception of complex sounds. The first international meeting on Animal Sonar Systems in Frascati, Italy (Busnel 1966) marked the transition from a discovery and speculation phase of research on echolocation to a prolonged period of intense, highly diverse experimentation and modeling that has continued to the present. The influential Frascati meeting led almost immediately to discoveries about different types of echolocation, to critical new experiments on how target range is determined from echo delay, how target velocity is determined from echo Doppler shifts, and how target shape and fluttering motion are conveyed by echoes (Simmons 2017; Grinnell 2018). In particular, that meeting inaugurated intensive study of how echolocation might work on perceptual and neural levels (Pollak and Casseday 1989; Grinnell 1995). Early theories, viewed in the light of modern evidence, anticipate a degree of incorporation of external acoustic factors into auditory response patterns that amounts to an internal auditory model of the propagation and scattering of sounds (Simmons 2012).

8.2 Early Theorizing (1912–1920)

On the basis of observations of bats being able to fly without the use of vision, Spallanzani and Jurine concluded that bats used their ears and hearing for orientation in the dark, but they offered no speculations on how this might operate (Griffin 1958; MacArthur 2000). Indeed, the sounds bats used, being inaudible to humans, were unknown to both Spallanzani and Jurine and remained unknown for a further 150 years until proper instrumentation was invented (Pierce and Griffin 1938). The alternative idea that bats rely on a tactile sense to detect and avoid obstacles was still considered up until the time of the discovery that bats really do emit ultrasounds (Griffin and Galambos 1941). The use of touch did seem somewhat reasonable based on aspects of bat morphology—"nerve centers" on the wings and unusual facial structures sparsely covered with hairs, with large bare surfaces of skin to act as sensing arrays (Fig. 8.1). Still, Spallanzani's original experiments demonstrated that bats with their wings and fur covered with varnish or library paste were able to avoid obstacles successfully. We now know that tactile sensing using wing hairs distributed over the bare skin of the wings is involved in flight control and not echolocation (Sterbing-D'Angelo et al. 2011).

Fig. 8.1 Illustrations from the 1912 Scientific American interview with Sir Hiram Maxim. https://archive.org/details/newsystemforprev00maxirich/page/13/mode/1up. Left is an illustration of Welwitsch's mouse-eared bat (*Myotis welwitschii*): "this bat furnishes us with a very good illustration of the sensitive wing that enables a bat to send out vibrations and to receive the echo. The spots on the wing probably represent nerve centers" (p 149). Welwitsch's bat is an FM bat native to Africa. Right is an illustration of the head of Blainville's leaf-chinned bat (*Mormoops blainvillii*), "which shows the highest development of the organ of the sixth sense to be found anywhere in animated nature. The whole face, including the ears, is covered with this organ; the nose, ears and chin, are all occupied and covered with sensitive hairs. The eyes are small and of very little use" (p 149). Blainville's leaf-chinned bat emits short constant frequency signals followed by a shallow frequency-modulated sweep. This bat is found in the Greater Antilles islands. By "sixth sense," the article means a sensory system in addition to hearing, seeing, smelling, tasting, and feeling. It is interesting to note that bat external morphology was well known at this time, and so the characteristics of the face and wings could reasonably be applied to explain the "sixth sense"

8.2.1 The Role of Bats in Inspiring Maxim's Acoustical Iceberg Detector

The sinking of the RMS Titanic in 1912 led to efforts to redesign collision warning systems for ships. Sir Hiram Maxim, the inventor of the first fully automatic machine gun and an early aircraft engine, described in a 1912 interview in *Scientific American* an acoustical iceberg detector based on what he thought bats did. Maxim proposed that bats avoided obstacles in the dark by using sounds inaudible to humans. He concluded that these sounds were at low frequencies (<20 Hz) generated by the bat's wingbeats, not by their vocalization system. These low-frequency sounds traveled outward, were modified and reflected back by objects in the vicinity, and were then detected by tactile sense organs in the wing membranes or on the face (Fig. 8.1). These reflections—echoes—indicate the "character and size" of objects, and they are "analyzed exactly as light waves would be by our eyes under similar conditions" (Scientific American 1912, p 148). Maxim's theory was surprisingly comprehensive: There is a mechanism for producing the sounds, the physical fact of reflected echoes, a receiving sense organ, a statement to the effect that echoes must be subjected to analysis, and an apparatus specifically for displaying the objects. In the sense that the display shows time for distance on a plot, the display creates what would be recognizable as "sonar images." Little is known about how Maxim conceived his idea, but it was grounded in known properties of sound propagation in air—low frequencies undergo less atmospheric attenuation than high frequencies and thus are more suited for long-range detection.

Maxim's acoustical detector included a siren at the front of the ship that broadcasts intense, low-frequency sounds outside the human audible range. These sounds would travel outward to strike any large obstacles in the ship's course and then return as echoes. The echoes would be picked up by a receiving device consisting of a flexible diaphragm stretched over the back end of a cylinder, with the front end acting as the receiving aperture. Maxim called the receiver an "artificial ear." The diaphragm's movements operated electrical switches that controlled bells of different sizes. Very slight deflections of the diaphragm rang a small bell, while stronger deflections rang a larger bell, thus signaling object size and potentially the urgency of the collision hazard. The diaphragm's deflections also moved a pencil, which made a recording of the waveform of the received echoes on a moving strip of paper. From this permanent record, the arrival time of echoes, hence the distance to the object, could be estimated. Maxim had, in effect, invented a practical sonar system consisting of a transmitter, a receiver, and a sonar display. A similar airborne-sound echo-ranging device was described in a patent by the British mathematician Lewis Fry Richardson in 1912 (Charnock 1981).

8.2.2 Hartridge and the Use of High-Frequency Sound

From his observations of bats flying in rooms and his knowledge of sound propagation and scattering, the British physiologist Hamilton Hartridge proposed that bats relied on hearing actual sounds and not on tactile sensitivity to detect and avoid obstacles in darkness (Hartridge 1920). In particular, he suggested, contrary to Maxim, that bats emit high-frequency sounds to produce "sound pictures" and to recognize objects from returning echoes. Hartridge did not have the technology to measure bats' acoustic emissions, but he reasoned that frequencies above the human audible range could produce echoes from small objects due to their short wavelengths. Moreover, the small heads of bats and their short interaural spacing indicated that they probably detected and localized high-frequency sounds, which explains the inaudibility of these sounds to humans without technical aid. Interestingly, from a technological perspective, Hartridge drew a further analogy between the bats' orientation abilities and the sound-ranging apparatus used to listen for aircraft during World War I.

Some years later, Hartridge (1945) expanded on his earlier ideas in light of the new facts about echolocation discovered by Griffin and Galambos (Griffin 1958). He described the similarity between the basic principles of what he called the "audio-location" of bats and of "radio-location" of what we now call radar. The major difference is the use of ultrasonic (in Hartridge's time described as "supersonic") sound waves by bats compared with radar's dependence on short electromagnetic waves. Radar had already been developed more or less independently by the main belligerents in World War II, and there was considerable convergence on common receiver principles across different designs. Much early work in the USA was conducted at the "Rad Lab" at the Massachusetts Institute of Technology (Buderi 1996). Griffin and Galambos were both close by at Harvard University at that time, and there was some local cross-fertilization of ideas between the radar and the biological communities. For example, an important early technical problem in radar was to suppress the effects of the intense outgoing broadcast on the sensitive receiver, so it could register very weak echoes arriving very shortly after the broadcast. The engineering solution was a transmit-receive switch and a blanking mechanism that altered the mode of reception to discriminate against broadcasts and in favor of echoes.

Hartridge went so far as to suggest that middle ear muscle contractions, which had already been observed in bats by Galambos (1942), could attenuate the bat's intense outgoing sound and restore hearing sensitivity for echoes, similar to the role of a radar set's transmit-receive switch. We now know that not only do the bat's middle ear muscles contract to reduce sensitivity to broadcasts but also that they relax in a graded fashion over time to improve sensitivity to echoes smoothly at progressively longer delays (Kick and Simmons 1984).

Hartridge recognized that the resolving power of the echolocating bat's system might be similar to that of radar since both used wavelengths in the centimeters and tens of centimeter bands, with the bat having the added advantage of "stereophonic" perception by the two ears. Binaural interactions comprise the foundation of sound localization, which explicitly treats the two ears as elements in an array to create directional sensitivity (Blauert 1996). Modern radars use arrays of receiving elements and exploit their combinations to form narrow beams (Skolnik 1980).

8.3 Target Ranging from Echo Delay

Coming now to the main concern of early theories of echolocation, Hartridge (1945) suggested that bats could determine the distance to objects by the time delay of their echoes. Target ranging is, of course, a basic function of radar or sonar, and a large part of engineering design is devoted to estimating delay (Woodward 1953; Skolnik 1980). Several different methods were proposed for the bat's auditory system to achieve the same goal. The background came from an important problem in radar combined with the wide gap between the ultrasonic frequencies of echolocation sounds and the significantly slower response properties of neurons as processing elements compared with electronic components: Most radar systems transmit at frequencies of hundreds of megahertz using pulses with frequency content covering a small fraction of the band around the actual, high center (i.e., average) broadcast frequency (Skolnik 1980). At such high frequencies, the short wavelengths confer advantages for echo formation when they impinge on aircraft or other targets, but the frequencies are too high to process directly into displays. The spectrum of radar echoes is several megahertz, which is narrow relative to the magnitude of their actual center frequencies. It is elevated far above the zero origin of the frequency scale, being centered at hundreds of megahertz. The solution is to shift echo frequencies down to a workable range while preserving their all-important bandwidth. To do this, radar receivers borrowed from commercial radio technology. The method, called heterodyning, mixes the radar echoes with an internally generated fixed reference frequency, close to the center frequency of the original transmission, and sorts out the frequency difference between the echoes and the reference as the stand-in for the echoes themselves. Through heterodyning, the echoes, which originally varied by at most a few megahertz around the very high center frequency of hundreds of megahertz, still vary by the same few megahertz but are now shifted down to fit close to zero frequency as the origin of their spectrum. The system thus retains the advantages of short wavelengths for echo formation, while the lower-frequency signals to be displayed retain their original wide bandwidth so vital for achieving high delay acuity.

Much like a radar set, early theories of echolocation considered the possibility that the high ultrasonic transmitted frequencies might have to be shifted down to lower frequencies to accommodate limitations imposed by the slow responses of neurons. Bats were known then to emit ultrasonic frequencies up to at least 100 kHz (Griffin and Galambos 1941; Galambos and Griffin 1942; Möhres 1953). Moreover, most of the sounds were known to be frequency modulated (FM), starting at a high frequency and sweeping down to lower frequencies. This information was imprecise and somewhat sketchy, being largely dependent on visual inspection of oscilloscope screen photographs of sample signals (Griffin 1958). At the time, the receptor cells of bats' inner ears had been demonstrated to respond to these high frequencies (Galambos 1942), but transduction from high-frequency receptor activation into impulses in auditory neurons stood as a bottleneck to understanding how biosonar "works." Neurons were widely understood to be able to produce repetitive responses no faster than 1 or 2 kHz, and then only for brief bursts. How could the waves in ultrasonic echoes be registered well enough to determine echo delay if the very elements that do the work are unable to respond at sufficiently high frequencies? It was not until the groundbreaking neurophysiological work of Griffin's graduate student Alan Grinnell (1963a, b) that direct experimental verification that single neurons in the bat's brain respond to ultrasound was available. (For the contemporaneous perspective, a useful review of the state of knowledge about central auditory mechanisms of echolocation at that time was published by O'Dell Henson in 1966).

8.3.1 Beat Frequency and Pitch Theories

In 1958, Griffin published *Listening in the Dark*, which recounted the discovery of echolocation in bats. This superb book aroused scientific curiosity and enthusiasm about echolocation to such an extent that within a few years, several theories were advanced to account for bat's ability to perceive echoes (Grinnell 1995). The broadcasts emitted by the few species of bats studied at that time consisted of ultrasonic frequencies between 20 and 100 kHz, with cycle-by-cycle periods of 10–50 μs. In contrast, the time intervals between broadcasts and their echoes are much longer. For objects at distances of 15 cm to 3 m, for example, echo delays range from roughly 1 to 20 ms. These time intervals coincide with periods at frequencies from about 50 to 1000 Hz, frequencies that humans can hear and perceive as the corresponding pitches. Thus, there was no doubt that these lower frequencies are effectively processed by the auditory system. The theoretical approach was for the bat to extract the longer intervals between broadcasts and echoes and then fit them into known lower (i.e., human) auditory sensitivity to determine delay.

8.3.2 The Beat Frequency Theory of FM Echolocation

If, as Hartridge (1945) suggested, bats could determine the distance to a target from the arrival times of echoes, then bats must be able to hear individual echoes as discrete, identifiable sounds and must be able to resolve small differences in echo arrival time relative to the broadcast. But no mechanism had as yet been proposed to allow the detection of echoes at the short arrival times encountered by bats. Furthermore, as Hartridge noted, for targets at close ranges, echoes might return before emissions had ceased. Overlap of emissions with echoes would complicate the bat's task of perceiving echoes as discrete events since the strong emissions could mask the presence of simultaneously occurring weaker echoes. This limitation was turned into a means of solving the problem of shifting high ultrasonic frequencies into low, auditorily friendly frequencies to determine delay. It capitalized on the finding that bats emitted FM sounds. J. David Pye, a biologist at the University of London, proposed that the overlap of emissions and echoes produces a beat frequency, much like a musical beat, in the bat's auditory system that could then be used to estimate target distance and direction (Pye 1960, 1961, 1966). A key feature is that the beat frequency is the difference between the broadcast and the echo, which is lower than the actual ultrasonic frequencies.

Figure 8.2 illustrates how the beat frequency theory operates. If a bat emitting FM sounds sweeping linearly from high to low is near enough to a target so that echoes return before the emissions have ended, there will be a period of time when both signals overlap to stimulate the ear simultaneously. When two frequencies stimulate the ear at the same time, one hears not only these frequencies but also a third frequency (a beat note or beat frequency) corresponding to the frequency difference between the two signals. Thus, the bat should hear the difference (beat) frequency as a separate sensation, with a frequency lower than that of either the emission or the echo. If the FM sweep is linear, the beat frequency is proportional to the separation in arrival time between emission and echo. If echoes arrive earlier, the beat frequency is low; if they arrive later, the beat frequency increases. Thus, a bat emitting FM sounds and receiving overlapping echoes can detect the presence of the target by perceiving the beat frequency, and it can determine the distance to

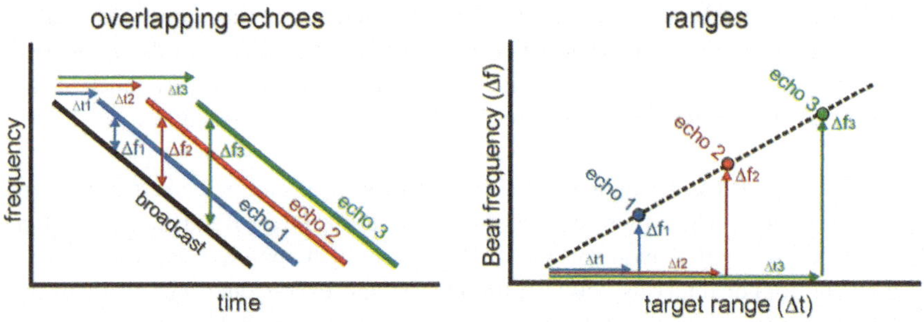

Fig. 8.2 Pye's beat frequency theory for frequency-modulated (FM) echolocation. (Left) FM broadcast (black) is followed by three echoes at different time delays (Δt; blue, red, green). Overlap of echoes with the broadcast creates frequency differences (Δf) that are registered as beat notes (beat frequencies) to represent delay for displaying the target range (Right)

the target from that beat's perceived frequency or pitch. This scheme accommodates multiple targets because they yield echoes that overlap with the emission, but at different time delays. Each delay generates a different beat frequency, and since the ear can discriminate among multiple frequencies by their perceived pitches, the bat should be able to perceive one target embedded within a cluster of several targets by perceiving one beat frequency among several.

Pye (1961) pointed out that electronic models producing beats from "bat-like" sounds improve human performance in sound localization tasks (see below). There are, however, two important limitations to the beat frequency hypothesis as applied to echolocation. First, it assumes that the bat's FM emissions sweep linearly. This allows each target range to be associated with a discrete beat note that is constant throughout the period of overlap. At the time, the linear-sweep assumption seemed plausible. It was soon discovered, however, that many bat FM echolocation sounds are curvilinear rather than linear in shape (Fenton et al. 2014), so the beat will shift in frequency during the period of overlap and each target range will not simply be associated with a single beat frequency. More seriously, it was also soon learned that bats shorten their FM broadcasts as they approach a target or an obstacle, keeping the duration slightly shorter than the echo delay (Griffin 1958). Consequently, overlap of broadcasts and their echoes is mostly avoided, at least for the object of the bat's immediate interest and in acoustically uncluttered environments (Schnitzler and Henson Jr 1980). Finally, at the time, the beat frequency theory seemed only to apply in situations where bats are close enough to their targets that echoes return above a certain intensity so they can interact with the broadcast; otherwise, no audible beats could be formed. If the echo is significantly weaker, the resulting beat would go unnoticed.

Leslie Kay (1961, 1962), a British engineer, also proposed a theory of echolocation based on the generation of beats. His idea incorporated two additional features over Pye's (1961) theory—one to get around the limitation that bats might produce sounds short enough to largely prevent overlap with echoes, and another to accommodate not only echo-ranging but also horizontal localization. Kay's interest in echolocation derived from his work in the design of navigational aids for the visually impaired (Kay 1966). From his studies of human performance, he concluded that FM signals are more effective than steady tones for judging the presence of an object, so his device emitted FM sounds. His device was a wearable acoustic aid that resembled eyeglasses, but with a miniature ultrasonic loudspeaker for emitting broadcasts and two miniature microphones for receiving the echoes. To accommodate his concern with localization in azimuth, Kay's device employed separate binaural presentation of beat frequencies to the two ears.

In its most important contribution toward understanding echolocation, Kay's theory (Fig. 8.3) proposed that the original FM emission is not necessarily the signal being mixed with echoes to produce beats. For the shortest echo delays, the emitted sound overlaps the echoes, but for longer delays, the bat regenerates its emission to itself, perhaps by "humming" another FM sound that then mixes with echoes to produce beats. For longer delays, it is the interaction between the echo and this later, regenerated, internal sound that is important. No direct overlap between the emission and the echo is necessary. This hypothesis extends the bat's ability to determine target range to longer distances, first because the regenerated emission could be weaker than the broadcast so its mixing with weaker echoes from objects at longer ranges would still produce beats, and, second, because the delayed regeneration of the broadcast FM sweep would be available to mix with echoes that arrive after the accrual broadcast has ended.

Significantly, Kay's (1962) theory goes into perceptual details: It was proposed that echolocation most likely provides the bat with precise estimations of range only at close distances. In this condition of "focused" echolocation, the actual emission and not a subsequently regenerated replica serve as the reference signal for producing a beat with the echo. Observations of bats turning away from obstacles only when quite near (Griffin 1958; Schnitzler and Henson Jr 1980), or turning to reject

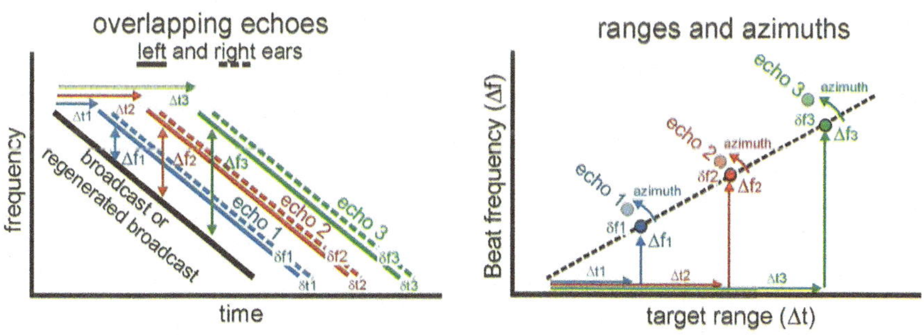

Fig. 8.3 Kay's beat frequency theory with regenerated broadcast. (Left) Frequency-modulated (FM) broadcast or its regenerated (i.e., "hummed") equivalent (black) is followed by three echoes at different time delays (Δt; blue, red, green). Overlap of echoes with the broadcast creates frequency differences (Δf) that are registered as beat notes to represent delay for displaying the target range (Right). Delays (Δt) differ slightly between left and right ears (δt) due to interaural spacing (~1 cm in many bats), so beat frequencies (δf) differ slightly, too. Using the slight binaural beat frequency differences, azimuth is perceptible along with range

capture of a pebble thrown into the air after approaching to within a fraction of a meter (Griffin 1958), suggest that the bat gathers more information about targets at close range when in this focused mode. Research has now broadened the concept of focused echolocation to include how bats select individual objects for paying attention (Simmons 2014).

Kay's acoustic aid for the visually impaired is an effective device based on the beat frequency principle. The question here is whether echolocating bats actually use beat frequencies created by adding ultrasonic reference frequencies to ultrasonic echo frequencies to determine the target range by reducing the effective frequency of stimuli. The first aspect of beat frequency theories to be examined experimentally was the requirement for overlap between FM broadcasts and echoes. Cahlander et al. (1964) analyzed tape recordings and high-speed motion-picture films of aerial captures of airborne mealworms by the little brown bat, *Myotis lucifugus*. These recordings were made to assess the bat's acoustic behavior in detail during interception of targets (Griffin 1958). During laboratory experiments that kept the bat close enough to the ultrasonic microphone for the echolocation sounds to be registered quite well, mealworms were projected into the air for the flying bat to chase and capture. During its approach to catch the mealworm, the bat shortened its emissions to prevent overlap of outgoing and returning signals even when the bat was so close to the target that it could just about reach the mealworm with its wingtip. Thus, there is no effective overlap of emissions and echoes, which obviates the initial version of the beat frequency hypothesis. Second, somewhat later experiments using electronic or "virtual" echoes established that echo delay is indeed the acoustic cue that bats use to perceive target range (Simmons 1973; Simmons and Grinnell 1988). This result does not directly indicate how bats determine delay, however, so further consideration of various theories of target ranging is still warranted.

In these same psychophysical experiments, done with big brown bats (*Eptesicus fuscus*), microphones were placed very close to the bat to record all its outgoing emissions. No evidence was found of the bat regenerating, or "humming," FM sounds following the broadcasts that might overlap with echoes (Simmons 1973). These findings did not entirely disprove the beat frequency approach, but they encouraged consideration of other types of theories about target ranging by bats (Simmons 2017).

8.3.3 Beat Frequency Theory of Constant Frequency Echolocation

Although bats that emit FM echolocation sounds shorten their broadcasts as they approach a target, thus avoiding overlap of each individual broadcast and with any short-delay echoes from a single target, substantial overlap between emissions and echoes nevertheless does occur in constant frequency (CF) echolocation (Schnitzler and Henson Jr 1980; Fenton et al. 2014). The CF echolocation sounds of the greater horseshoe bat (*Rhinolophus ferrumequinum*), which emits at about 80 kHz, or the mustache bat (*Pteronotus parnellii*), which emits at about 60 kHz, often have durations in the tens of milliseconds, as opposed to the <10 ms duration of FM broadcasts. CF broadcasts are at higher ultrasonic frequencies on average than most FM broadcasts, and they are long enough to overlap with echoes for targets located as far away as several meters, creating the conditions for generating beat notes.

Pye (1960) applied his beat frequency theory to CF echolocation to explain how bats might process CF echoes. The information they contain is not about their delay, however, but about their frequency, which in practice means the relative velocity of the target to the bat. Figure 8.4 illustrates the concept. The bat emits a long CF signal at a high ultrasonic frequency and receives an echo from a target that it is approaching. The echo frequency is raised slightly higher by the amount of the upward (i.e., approaching) Doppler shift (Δf), creating a beat frequency. The bat perceives the beat frequency to represent the act of reception of the echo and the target's approach velocity, instead of the actual ultrasonic frequency of the echo. Changes in the velocity difference would be reflected in changes in the beat, too, so the bat can determine if the target has changed its relative velocity. Moreover, because the beat is much lower in frequency than either the ultrasonic emission itself or the slightly elevated ultrasonic frequency of the echo, it stimulates the auditory system far removed from the frequencies in the original sounds. This feature of the beat frequency hypothesis is a different way of addressing the problem of self-deafening caused by the intense broadcast. The long duration and large overlap of the CF sounds would prevent a transmit-receive switch analogous to radar from separating the echo from the broadcast.

As for determining distance, when the German zoologist Franz Peter Möhres (1953) described the long-duration CF emissions of horseshoe bats (*Rhinolophus*), he suggested that these bats might perceive the distance to targets from the intensity rather than from the arrival time of echoes. CF signals contain only a very narrow band of frequencies, so they are poor for registering timing directly. However,

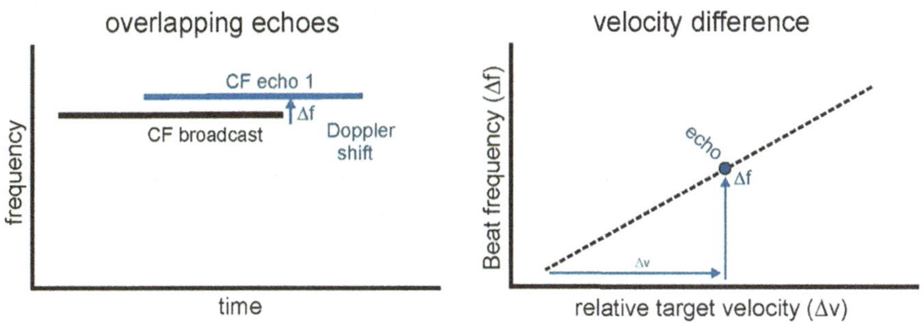

Fig. 8.4 Pye's beat frequency theory for constant frequency (CF) echolocation. (Left) Long-duration CF ultrasonic broadcast (black) followed by echo (blue) from the target being approached, causing echo to be Doppler shifted to a slightly higher frequency (Δf). Overlap of the echo with the broadcast creates frequency beats (Δf) that register the velocity difference of the approach (Right)

echoes are weaker for targets at long range, and as the bat approaches, they become more intense, which makes Möhres' suggestion a viable alternative. Pye (1960) adopted this suggestion when he proposed that the intensity of the beat note could indicate target distance as effectively as the intensity of the echo itself. We now know that both horseshoe bats and mustache bats emit short FM signals attached to their long CF signals and that they perceive target range from the delay of the FM component in echoes (Simmons 1974; Grinnell 1995).

The true role of the CF signals themselves was clarified when it was discovered that horseshoe bats make fine adjustments to the frequency of their broadcasts to compensate for slight upward Doppler shifts as they approach an object (Schnitzler 1973; Schuller et al. 1974). The specific behavior is to shift the frequency of CF broadcasts downward to compensate for upward Doppler shifts in echo frequency that occur because the bat is flying toward the object it is facing. Doppler shift compensation keeps the frequency of echoes locked onto an internally determined reference frequency even though the bat and the object are in relative motion. The bat's exceptional ability to resolve very small changes in ultrasonic frequencies is based on the fine frequency tuning of receptors in the inner ear (Neuweiler 2000). These receptors are very sharply tuned to the reference frequency, and the Doppler shift compensation response is controlled by adjusting echoes to fall into the narrow zone of this tuned frequency. As a result of the sharp frequency selectivity of receptor and neural responses required to control the response, the emitted CF sound and returning Doppler-shifted echoes probably do not mix together within the tuning range of the same sensory cells, and beats would not be produced to any significant extent. Where emissions are long in duration, some degree of overlap may even be necessary for the Doppler compensation response to occur (Schuller et al. 1974).

8.3.4 *Time-Difference Pitch Theory*

The beat frequency theories were conceived to translate ultrasonic frequencies down to lower frequencies where known auditory processes, identified in humans, might be plausibly invoked to explain how bats analyzed echoes. The Swedish biologist Jan Nordmark (1960, 1961) proposed a different means for the bat to detect echoes and determine target range from echo delay, also by exploiting low frequencies implicitly built into the sequences of broadcasts and echoes (Fig. 8.5). Nordmark observed from Griffin's recordings that big brown bats emit trains of pulses at regular repetition rates and receive corresponding trains of echoes at times corresponding to echo delay. The critical feature of Nordmark's theory is a perceived pitch derived from the two separate pulse trains—the broadcasts and their echoes set apart in time by the delay of the echoes. When humans listen to two pulse trains with the same fixed repetition rate but offset in time from each other, they perceive a time-difference pitch having a frequency equal to the reciprocal of the time offset (Thurlow and Small 1955). The two pulse trains run concurrently, and they evoke a pitch sensation determined by the time offset from one train (the broadcasts) to the

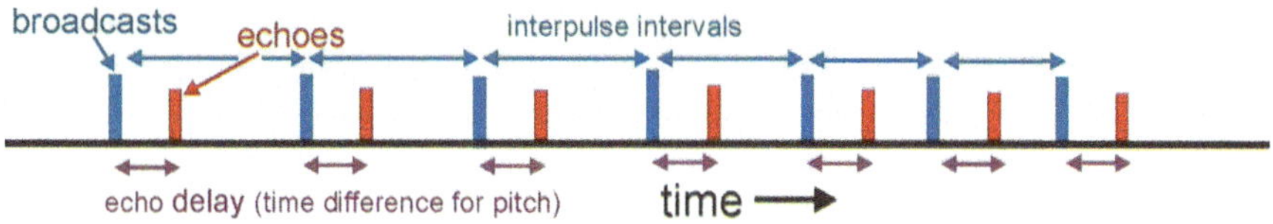

Fig. 8.5 Nordmark's time-difference pitch theory. The bat's broadcasts (blue) and their echoes (red) each form a pulse train defined by the interpulse interval between successive broadcasts (inverse of repetition rate). Broadcasts and echoes each comprise a pulse train with the same interpulse intervals, just offset in time from each other by the delay of the echoes. The bat perceives a pitch corresponding to the frequency related to this offset, a frequency that has echo delay as its period

other (the echoes). Nordmark's hypothesis is that the bat perceives the time-difference pitch corresponding to the reciprocal of the delay of echoes, which is the target's range (Fig. 8.5). The target is detected by the presence of the pitch, and the target's range is determined in part by its strength (weaker pitches for farther targets). The bat can track the target's range from the rate of change of the pitch (progressively rising pitch, progressively shortening distance).

As in beat frequency theories, the time-difference pitch theory has the bat perceiving a tone at a lower frequency than the ultrasonic frequencies actually present in emissions and echoes. Nordmark (1960) suggested that the bat experiences time-separation pitch only if echoes occur within 10 ms of emissions, which sets 100 Hz as the lowest perceived pitch, with progressively higher pitches as the bat approaches closer and the delay shortens. If time-separation pitches are perceived for echo delays from 10 ms down to about 1 ms (target ranges from about 2 m to 20 cm), the corresponding time-difference pitches are at frequencies from 100 Hz up to 1000 Hz. Griffin (1958) estimated the little brown bat's distance of detection for insects and other airborne targets at about 2 m, so the parameters specified by the theory are realistic.

Nordmark's time-difference pitch would come into psychological existence at roughly the same range at which the bat shows signs of having detected the target. If more than one target is present, several different time-separation pitches might arise and be perceived to differentiate objects at different distances. Similar to the beat frequency theories, multiple targets are represented by multiple low frequencies, generated from different "pitches" or from different beat notes. The beat frequency theories produce a beat separately for each broadcast-echo pair. Successive pairs produce progressively decreasing beat notes as the echo delay becomes shorter (Figs. 8.2 and 8.3). A distinctive feature of the time-difference pitch theory is that it considers the whole sequence of emitted sounds and the concurrent sequence of their echoes as the source of the perceived pitch, which rises as the target range shortens. The idea that the bat uses a global aspect of the combined broadcast and echo streams has ecological relevance because the bat's observed behavior suggests a smoothly integrated process of interception, not a series of jerky motions.

Time-separation pitches likely to be experienced by bats would be in the range from perhaps 100 to 1000 Hz, and yet bats do not hear these low frequencies at all

(Pollak and Casseday 1989; Neuweiler 2000). However, it is not necessary that low-frequency sounds be received by the ear and heard in their own right for periodicities conveyed by higher frequencies to be heard as low-frequency pitches. The pitches associated with time delays between the bat's emissions and echoes could be generated within the nervous system by mechanisms having characteristics quite different from the peripheral mechanics of the middle and inner ears, which determine the range of frequencies registered by the cochlea and the shape of the audiogram. The bat could well hear periodicity pitches at low frequencies without being able to hear low-frequency sounds since the periodicity information is conveyed by neural responses evoked by the higher-frequency emissions and echoes. Moreover, the dynamics of neural responses to lower-frequency sounds that evoke pitches resemble the dynamics of responses evoked by ultrasonic sounds, so some intersection of target ranging with pitch remains plausible (Simmons and Simmons 2011).

8.3.5 Envelope Detection Theory

Nordmark (1961) made a curious comment about the FM character of the bat's broadcasts. He stated that FM sweeps are unlikely to be the critical characteristic of the emitted sounds. (This idea assumed that the true frequency spectrum of echoes would not be represented in auditory terms.) Instead, the important feature is their broad bandwidth and short duration, more like clicks. In this view, the FM sweeps are merely the way the bat achieves its wide emitted frequency band. They are only "accidental by-products" of the sound production mechanism. Only the timing of the pulse-echo sequences is used for distance perception. Essentially, the envelopes of emissions and echoes are all that he considered significant for target-range perception using the time-difference pitch sensation.

The role of sound envelopes comes up in a very sophisticated theory proposed by the American engineer John Stewart (1979; Stewart and Kasson 1968) in an auditory model that encompassed target ranging and also the perception of target shape. Stewart suggested bats perceive the envelopes of their broadcasts (which he called "primaries") and the envelopes of echoes to detect targets and determine target features. Stewart's theory (Fig. 8.6) was based on an electronic model of the inner ear that incorporated known nonlinear properties of mechanical activation of the basilar membrane in the cochlea. The lower frequencies present in the envelopes of ultrasonic sounds are extracted by the nonlinear properties of mechanical activation at the basal (high-frequency) end of the cochlea and then travel to the more apical cochlear "places" corresponding to different low sound frequencies present in the envelopes.

The principle underlying Stewart's model is demodulation of the low envelope frequencies by nonlinear mechanics of the inner ear at the places where high ultrasonic frequencies activate the basilar membrane, followed by these low frequencies activating their corresponding places toward more apical locations along the basilar membrane. The lower frequencies then evoke auditory perceptual processes known

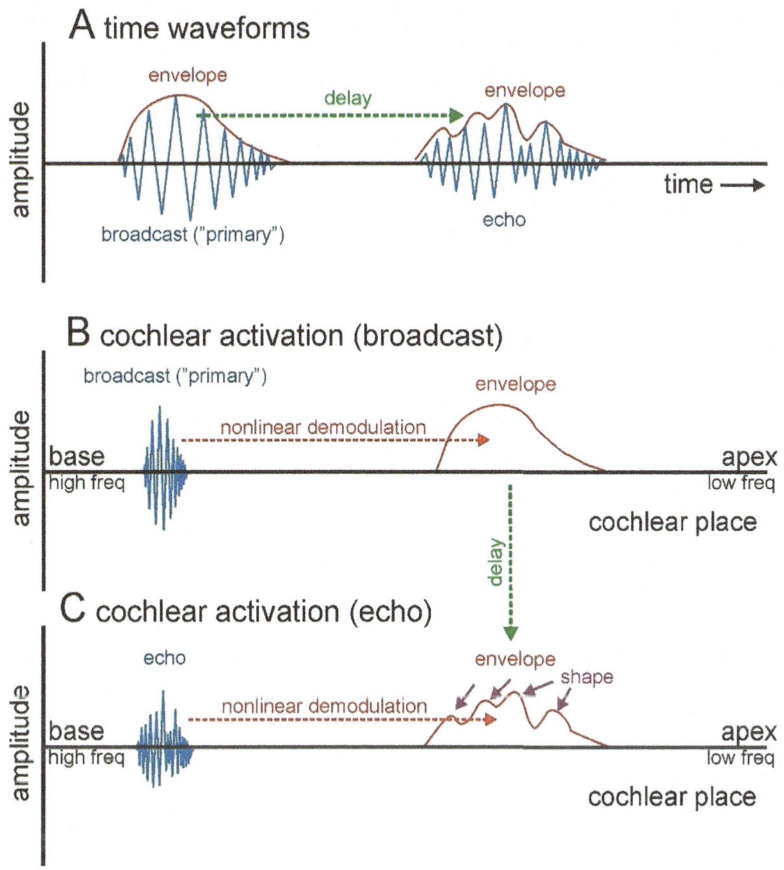

Fig. 8.6 Stewart's nonlinear envelope demodulation theory. (**a**) Frequency-modulated (FM) broadcast ("primary") followed by echo; ultrasonic waveform cycles (blue) and low-frequency envelope (red). The broadcast has a smooth envelope from frequency stability of vocalization; the echo has a rougher envelope due to frequency dependence of target reflectivity (i.e., echo spectrum is sensitive to target shape). Cochlear place representations of the broadcast (**b**) and the echo (**c**). Ultrasonic waveform cycles (blue) activate the basal, high-frequency end of cochlea for both broadcast and echo. Cochlear nonlinearity generates low-frequency envelope cycles at the base that propagate toward the apex and induce activation at places corresponding to the low envelope frequencies. Target range is determined from echo delay using periodicity pitch derived from the interval between cochlear envelope responses to broadcast (**b**) and echo (**c**). Target shape is determined using perceived timbre derived from the spread of low frequencies derived from roughness of the echo envelope

in humans, thus leaving no doubt as to their existence. These low-frequency psychoacoustic effects are mobilized to perceive the target range as a periodicity pitch associated with the time separation between the envelope peak of the emission and the envelope peak of the echo (Fig. 8.6). The envelope of the emission is relatively smooth because the bat's vocal system generates wideband FM sounds with a broad frequency response. In contrast, the envelope of the echo is rougher and fluctuates more rapidly across frequencies because targets have a more complex frequency dependence of reflectivity, which appears as amplitude modulation. These more spread-out envelope frequencies stimulate a wider segment of the low-frequency, more apical region of the basilar membrane, where they evoke a sensation of timbre that the bat uses to perceive the target shape.

Stewart's theory specifically accommodates the contemporaneous finding that, in a typical pursuit sequence, bats shorten their FM broadcasts to avoid overlap of broadcasts and echoes as they approach a target, except possibly at the very end of the interception maneuver when the bat's sounds have shortened as much as they can (Cahlander et al. 1964). The time delay between the envelope of the echo and of the emission decreases, so the frequency of periodicity pitch associated with this time delay increases. The bat could track the target's declining range during approach from the steady increase in this pitch as each broadcast-echo pair occurs. This is not the same as Nordmark's (1960) theory, which considered the sequence of broadcasts as a whole pulse train running in step with the corresponding sequence of echoes that trail it by the delay of the echoes. The time-difference pitch is determined by the offset of the echo pulse train relative to the broadcast pulse train. In Stewart's theory, the perceived pitch is generated by the time between each member of the pair of pulses—one broadcast and its echo. Stewart even considered that eventually the bat will be near enough to the target that the echo will overlap the end of the emitted sound to some extent, and the resulting beat frequency would be demodulated by cochlear nonlinearity, too. The short interval of overlap generates beat frequencies equal to the difference in frequency between the lower frequencies at the end of the FM emission and the higher frequencies at the start of the FM echo. Detected by the same nonlinear mechanism that extracts the envelope from the echo, these beat frequencies would be represented in the mid-frequency region of the cochlea, away from both the ultrasonic frequencies in the emissions themselves and from the lower frequencies in the demodulated envelopes. The onset of these beat frequencies would signify that the bat is close enough to reach out and seize the target. Their occurrence might be analogous to the miniature radar proximity fuse in an antiaircraft shell; it could trigger the bat's grabbing of the target at the moment it is within reach.

8.4 Pulse Compression Theory

The problem with envelope detection is that the wide bandwidth of the bat's ultrasonic sounds is replaced by the lower-frequency, narrower-band signals represented by their envelopes. It destroys much of the original information in the FM signals before they are used to determine echo delay (Sanderson et al. 2003). We now know that echolocating bats perceive exceptionally small changes in the delay of echoes on a fine time scale and with a fine perceptual organization that shows they do not destroy the critical wideband ultrasonic signal information widely assumed to be required by the nominal limits of auditory neurons (Simmons 1993, 2012). Because the accuracy of delay determination depends on having as much bandwidth as is feasible within other constraints, replacing the ultrasonic signals with their envelopes is an irreversible loss of information. The just-described early theories were concerned with one particular other constraint that the bat's inner ear could receive ultrasonic sounds but not then faithfully represent their true frequencies due to limitations in the effective bandwidth of neurons, which nominally is about 1 kHz. The appeal of these early theories is that they did not require the bat's auditory nervous system to operate directly at ultrasonic frequencies. Instead, auditory capabilities familiar to humans who hear at lower frequencies could be the secret of echolocation. But this leads to a conceptual objection to the previously described theories, all of which replace direct processing of the broadcasts and echoes themselves with extraction of lower-frequency signals as their surrogates. Because generating wideband vocalizations is a significant biomechanical challenge (Au and Suthers 2014; Metzner and Müller 2016), bats ought not to emit their FM signals if they do not intend to fully use their broad bandwidth for perceiving targets (Glaser 1974). The ostensible reason for emitting wideband FM signals is to achieve high echo-delay accuracy (Sanderson et al. 2003), but this cannot be done just by emitting FM signals unless the receiver itself actually uses the available bandwidth, not the narrowband envelope. The audiograms of echolocating bats do indeed cover the ultrasonic frequencies they emit (Grinnell 1963a, 1995). Indeed, the discovery of echolocation depended in part on the demonstration that the receptors of the inner ear responded to ultrasonic frequencies (Galambos 1942). The bat's use of a truly wideband receiver, not an envelope receiver, is, thus, implied by their use of FM broadcasts, even if it might seem implausible on neurophysiological grounds.

In 1961, G. K. Strother, a physicist at the Pennsylvania State University, proposed a theory of echolocation based on an analogy between bat FM sounds and a type of radar signal-processing strategy that also exploited FM signals. There were several aspects of the performance of echolocating bats that Strother indicated were in need of theoretical explanation—the ability of bats to detect small targets at considerable distances, even in the presence of other, larger objects; the ability of bats to detect targets in the presence of intense, ultrasonic interfering noise; the continuous variation in parameters of FM echolocation sounds as bats approached targets; and the possibility that bats might be able to detect echoes even when they overlap outgoing sounds. These same features of echolocation performance were cited as challenges for theorizing by Pye, Kay, and Nordmark.

Strother related the FM structure of bat sonar signals to a design for a new type of radar. In 1951, the Bell Telephone Laboratories (Murray Hill, New Jersey) began developing a radar system that transmitted FM signals and incorporated a sonar receiver to achieve both good target-detection and target-ranging performance (Buderi 1996). It used FM-transmitted signals to achieve both wide bandwidth and high energy (Klauder 1960; Klauder et al. 1960). After being kept classified for nearly a decade, the characteristics of this system, which Bell Labs called "chirp radar" after the use of FM signals, were published. Strother realized that the bat might take similar advantage of its own FM signals if it possessed an appropriate sonar receiver.

The essential features of Strother's 1961 application of pulse-compression theory to echolocation are illustrated in Fig. 8.7. The bat transmits an FM signal that sweeps downward across a wide band of ultrasonic frequencies across a fixed total duration with a fixed sweep slope. Built into the receiver's design is a system of frequency-dependent time delays that counteract the direction of the sweep, so that all of the frequencies dispersed across the sweep come out of the receiver at the same time. This specific feature of the arrangement of delays defines a pulse-compression receiver. The chirp radar principle compresses the information strung out along the duration of the FM sweep into a single, very short pulse that corresponds to the cross-correlation function of the echo to the broadcast. The process of compression or dechirping can be viewed as tilting the slope of the FM sweep until it is vertical. The receiver is "matched" to the transmitted signal, and one can describe the receiver and the signal as being designed for each other. There is one disadvantage to having the system of frequency-dependent delays built into the

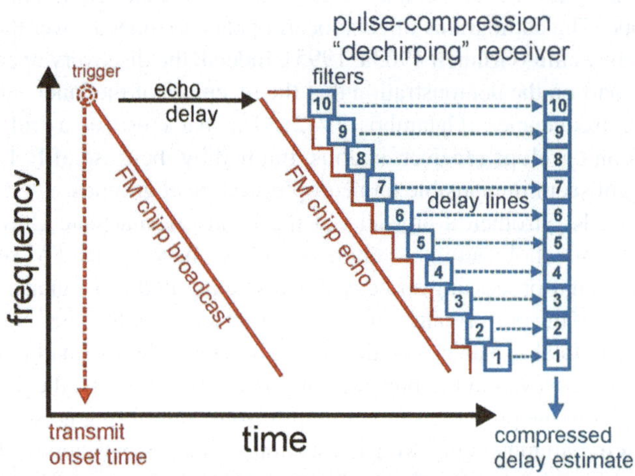

Fig. 8.7 Pulse compression or frequency-modulated (FM) chirp receiver. The downward-sweeping FM chirp broadcast is followed by an FM echo at a particular delay. Because successive frequencies occur at different times along the sweep, the information they contain about delay cannot be extracted simply by summation across frequencies. By segregating the frequencies into multiple sub-bands (1–10) and then delaying each of the segregated signals by an amount of time opposite to the location of that sub-band in the FM sweep, the individual frequency segments are brought into alignment (hence, pulse compression) and can be added together to produce a single, combined delay estimate. Delay accuracy incorporates all of the information relevant to delay (i.e., all frequencies), while the combined estimate sums the energy distributed across the sweep to maximize the probability of detection. When the countervailing delays exactly mirror the FM sweep, the compressed echo is the cross-correlation function between the broadcast and the echo. Note that the chirp receiver is matched to the slope of just one FM sweep, but bats change their sweep slopes from one broadcast to the next

receiver's hardware, however. The hardware, indeed, its actual wiring, has to be modified if the FM transmission is changed to a different duration.

There are two desirable effects of echo compression. First, all of the energy present in the long FM sweep is concentrated together into a single peak. None of the energy in the signal is lost as a result of determining the signal's presence by only looking for the onset or offset of the signal's envelope. (Hartridge (1945) suggested that bats listened only for the echo's offset to mark the echo delay.) Both onset and offset are very obvious features of the echo if the signal is strong, but it is difficult to detect the onset or offset if the signal is weak and buried in noise. The compressed pulse can be viewed as a collective onset and offset marker for all of the frequencies spread out along the FM signal. And since the energy is combined into the compressed pulse, the echo can be detected with maximum efficiency. Second, since the signal has been shortened into a single peak, it is easier to specify exactly when the echo occurs from the sharply defined time of occurrence of the peak. Target range is indicated very accurately by the time interval between the instant the sonar sound is emitted (trigger in Fig. 8.7) and the occurrence of the compressed peak representing the echo.

Strother (1961) outlined the advantages of a pulse compression model for echolocation in terms of the bat's observed performance. The ability of bats to detect small targets using weak echoes could be explained by the increase in the amplitude of the echo signal after it is compressed by the sonar receiver. The bat's ability to detect targets when interfering noise is present could also be attributed to the concentration of signal energy into a shorter pulse (Griffin et al. 1963). It is important to note that the signal-to-noise ratio of the echo is not actually increased by pulse compression, but the energy in the echo is mobilized optimally for detection to occur (McCue 1966). In the absence of pulse compression, performance would be suboptimal. This relates to Glaser's (1974) point that the use of FM sounds implies the use of pulse compression; otherwise, why bother emitting such complicated signals (see below).

The receiver of John Klauder's (1960; Klauder et al. 1960) chirp radar system developed at Bell Labs contains a system of frequency channels and electronic delay lines which impose a slightly different delay on each frequency along the FM sweep in the echo. Because the inner ear does, in fact, separate the frequencies of sounds into different channels for processing by the central auditory system, the pulse compression concept is attractive as a model for echolocation. The problem with using the cochlea as the substrate for pulse compression is that it delays low frequencies more than high frequencies, which is the opposite of pulse compression for downward-sweeping FM chirps (Fig. 8.7); instead, it shows pulse stretching. This was cited by Willem van Bergeijk (1964), then at Bell Labs, as a reason to doubt that pulse compression occurred in the bat's inner ear. J. J. Gerald McCue (1966, 1968), at the Massachusetts Institute of Technology's Lincoln Laboratory, described an attempt to measure propagation delays in the cochlea of little brown bats to determine whether a pulse compression mechanism was feasible. His experiment recorded cochlear microphonic responses, the same physiological signals used by Galambos (1942) to show that the bat's inner ear responded to ultrasonic sounds.

In the resulting measurements, the propagation delays in the cochlear microphonic response to ultrasonic frequencies did exhibit frequency-dependent delays, but in the wrong direction for pulse compression. We now know that cochlear microphonic responses are not primarily responses evoked by receptor cells tuned to the frequency of the stimulus; instead, they originate from a broad region of the basilar membrane, tuned to higher frequencies, that respond with a similar phase. Their in-phase responses summate and thus are more readily recorded, but they do not directly register frequency-dependent delays in the cochlea (Dallos 1972). Using evoked neural responses recorded from the auditory midbrain of big brown bats, the latencies to auditory activation have been estimated as requiring about 1-ms upsweep to match the inner ear's own internal stretching of wideband sounds (Luo et al. 2019). Thus, low frequencies are delayed more than high frequencies, so a pulse compression receiver is not present in the bat's cochlea. van Bergeijk (1964), McCue (1966), and Strother (1966) did not actually rule out pulse compression, or its equivalent, however. If compression exists, it would have to be implemented in the central auditory system instead of in the inner ear.

8.4.1 Spectrogram Correlation Theory

The real importance of the pulse-compression theory of echolocation lies not in the particular mechanism that compresses FM echoes but in the pulse that each FM echo is compressed into. This pulse, or a reasonable facsimile thereof, is manifested directly in what bats perceive (Simmons 1993). The frequency-dependent delay system of Klauder's chirp radar receiver (Fig. 8.7) is only one of a number of different ways to achieve the same end—to convert the FM echo into its corresponding compressed pulse (Skolnik 1980). It is the compressed pulse itself that provides the advantages of chirp radar, not the process used to obtain the compressed pulse. In signal-processing terms, the output of pulse compression is the cross-correlation function of the echo with its corresponding broadcast (Skolnik 1980). In contemporary receivers, digital signal processing methods replace the hard-wired delay lines and frequency segmentation of the original chirp receiver. No one believes that echolocating bats sample their broadcasts and echoes at rates of hundreds of kilohertz and then apply Fourier transforms to estimate echo delay. Instead, the types of theories described above were intended to hew closer to the possibilities offered by the auditory system.

The bat reduces the duration of its FM sweeps while keeping the bandwidth relatively fixed as it approaches a target (Cahlander et al. 1964). The chirp receiver design illustrated in Fig. 8.7 incorporates delay lines that provide different amounts of delay to different frequencies to compensate for the dispersion of frequencies along the FM sweep. The arrangement of the delays is a mirror image of the FM sweep or the spectrogram of the chirp. Changing the duration of the sounds changes the slope of the FM sweeps through the relative timing of their frequencies, which is best portrayed by their spectrograms. In the receiver, it would require different values of delay across frequencies to achieve the same degree of pulse compression for different FM sweep durations. To make such a change, the internal wiring of the delay system would have to be changed for each sound, which is extremely cumbersome. It might be feasible to have a small number of different, parallel delay systems inside the receiver to allow for the use of a correspondingly small number of transmitted signals. The appropriate delay pattern would be switched on to match the particular broadcast being emitted. But the bat changes its broadcasts over a continuous range of durations, and therefore, of FM sweep slopes. To take advantage of pulse compression while accommodating the changing durations of broadcasts, the method for registering the time of the broadcast has to be changed. The design is no longer pulse compression in the classic sense, but spectrogram compression or spectrogram correlation (Saillant et al. 1993). The reference time for conventional pulse compression is the single trigger signal that marks the instant at which the broadcast is emitted (Fig. 8.7). Then, the echo arrives after some delay, and the compressed pulse is used to register that time. To achieve the versatility for using different FM broadcasts, the trigger signal is broken into pieces and distributed across the frequencies in the FM sweep (Fig. 8.8). In this scheme, each frequency in the broadcast marks its own time-of-occurrence as it goes out. Then when the echo returns, the occurrence of each frequency in the echo is marked in the same

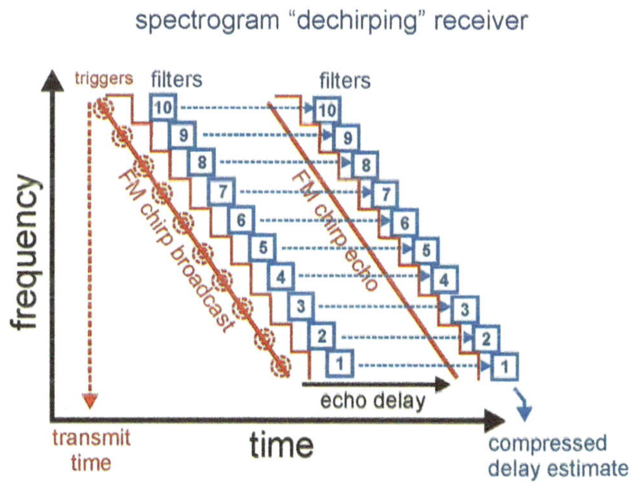

Fig. 8.8 The spectrogram correlation chirp receiver. Instead of using a fixed set of delay lines to dechirp the frequency-modulated (FM) sweep, the spectrogram correlation receiver segregates the frequencies in the FM sweep into different channels and then determines echo delay separately in each channel. The overall echo-delay estimate is obtained by summing all of the individual frequency delays into a single compressed estimate. Whereas the trigger for conventional pulse compression is the time the broadcast is sent out (Fig. 8.7), in spectrogram correlation, the trigger is the occurrence of each of the individual frequencies in the broadcast. This model accommodates changes in the FM broadcast sweeps without requiring the receiver to be "rewired" for each new signal. However, it requires retaining the delay estimates for higher frequencies until the estimates for lower frequencies have been made, so they can then be summed together

way. This method distributes the process of delay estimation across frequencies by retaining the delay of the echo as the family of times that elapse between the broadcast and the echo at all of the frequencies. The overall delay of the echo, and the compressed pulse, can be reconstructed by summing the individual delay estimates over the individual frequencies (Fig. 8.8), much like the summation that takes place in a conventional receiver (Fig. 8.7; Saillant et al. 1993; Wiegrebe 2008).

8.4.2 Spectrogram Transformation Theory

Spectrogram correlation as a means of pulse compression touches on one final issue that was taken up by early theories of echolocation—why do bats use FM sounds. Glaser (1974) explicitly proposed that bats must use some form of pulse compression; otherwise, the complexity of generating the FM sweeps would go unexploited. McCue (1966, 1968) examined a range of pros and cons of pulse compression theory and concluded that bats very likely did use some form of pulse compression because their behavior suggested that they could determine echo delay accurately enough that other kinds of theories would fall short of explaining how it was done. However, van Bergeijk (1964) offered an alternative explanation to the idea that a pulse compression receiver would be necessary to exploit the FM sweeps and their bandwidth. The spectrum of echoes from targets is more complicated than the spectrum of the original broadcast because the object's shape imposes frequency dependence on its acoustic reflectivity that is added to whatever complexity might originate in the larynx and vocal tract. Acquiring the frequency profile of a target's echoes would make it possible for bats to distinguish one object from another if their shapes were sufficiently different. Experiments done by Griffin and his colleagues (Griffin et al. 1965) revealed that bats could in fact distinguish between a mealworm and such shapes as spheres and discs, so some means had to be found to determine their shapes. Both Griffin (1958) and van Bergeijk (1964) advanced the idea that the FM sweeps in the bat's broadcasts might be used to scan across the target's reflectivity spectrum to identify the echo spectrum so the target's shape could be perceived.

This aspect of echolocation theory was brought back to life by Richard Altes (1984), who considered how spectrograms, already implicated in pulse compression models, could address the ability of bats to distinguish between targets that differ in shape, especially in circumstances where the echo spectrum was probably the information the bats used. A composite theory of echolocation—the spectrogram correlation and transformation theory (SCAT)—was developed to fold together the advantages of pulse compression for delay determination and target discrimination from the spectrum (Saillant et al. 1993; Ming et al. 2021).

Figure 8.8 shows the spectrogram version of a pulse compression or cross-correlation receiver for producing sharp registration of echo delay. The temporal order of frequencies is explicitly extracted and represented along the auditory system's internal frequency, or tonotopic, scale, which originates in the cochlea and is projected upwards along the auditory nervous pathway to dominate the spatial dimension of acoustic information. Just as the timing of the individual frequencies on the FM sweep is registered, so is the amplitude of the echo at each frequency. Several experiments make it clear that FM-emitting bats perceive targets as having different shapes from variations in the echo spectrum across frequencies (Yovel et al. 2011). The spectrum of echoes from targets used in early studies of shape perception during aerial captures (Griffin et al. 1965) contains amplitude minima and maxima—ripples in strength across frequencies—that serve as acoustic cues for shape (Griffin 1967; Simmons and Chen 1989). However, these ripples are a large-scale multidimensional representation of time, frequency, and amplitude. The target itself contains several discrete reflecting parts whose reflected broadcasts add together to create the spectral ripple. Its representation would be greatly simplified if it were transformed from the spectral domain or overlapping reflections and ripples to the low dimensionality of the distance to each of the small number of reflecting parts. The SCAT model implements this assumption by explicitly converting the pattern of spectral ripples into estimates of the time separation of the reflections and then attaching them to the overall estimate of echo delay (Saillant et al. 1993; Ming et al. 2021). This computational approach is supported by the perceptions that bats actually achieve from spectrally rippled echoes (Simmons 2012). The implication is that the bat's echolocation is itself a model—of the acoustics involved in forming echoes.

8.4.3 Pulse-Echo Ambiguity and Frequency Hopping

Echolocation was discovered in experiments with North American and European FM bats, such as *Myotis* and *Eptesicus*, and then slightly later with the long-CF/FM bat, *Rhinolophus*, suggesting that there must be more than one type of echolocation to be found in bats (Griffin 1958; Neuweiler 2000). Following these initial discoveries, Griffin's student Alvin Novick set out to examine the possibility of echolocation being used by different species of bats in both New World and Old World tropical environments (Novick 1969). The motivation was that many more families and species of bats are found in the tropics than in the north temperate zone, so it seemed desirable to learn whether they used echolocation, too. Even with the limited and very ponderous electronic capabilities of that era, Novick found an astonishing ubiquity of echolocation across many groups of bats, including different combinations of long- or short-CF components combined with FM components of various frequencies.

This newly revealed variety of bat echolocation figures into yet another link between the history of research on echolocation and contemporaneous work in signal technology. Recall that an important consideration in early theorizing on echolocation had been the possibility for echoes to be received at short delays while the broadcast was still being emitted, so that simultaneous reception of echoes and broadcasts might pose a source of interference for bats. This particular concern was largely settled when recordings of interceptions in acoustically uncluttered environments revealed that FM bats progressively shortened their broadcasts to avoid pulse-echo overlap. However, the larger problem of overlap between entire streams of echoes from successive broadcasts, or when the bat was echolocating in a cluttered environment containing both target and nontarget objects, was not considered. This is a well-known problem in radar (Skolnik 1980). If a bat is operating in complex surroundings with echoes from nearby objects such as insects or obstacles at short delays while echoes are also arriving from objects located farther away, the bat would have to wait until all of the echoes of one broadcast arrive before emitting the next broadcast. Otherwise, short-delay echoes of the second broadcast might be misinterpreted as long-delay echoes of the first broadcast—a problem known as pulse-echo ambiguity in sonar and radar (Skolnik 1980). The problem arises when the immediate requirements of aerial interception or obstacle avoidance make it desirable to emit sounds more rapidly than waiting until all echoes have been received before emitting the next broadcast.

The most prominent technological solution to this problem is to change the frequencies of successive broadcasts so their corresponding echo streams are distinguished by their frequencies. That way, one pulse-echo stream can overlap with another without becoming ambiguous. This is one version of a solution called frequency hopping, a method widely used in modern spread-spectrum technology and traced to a seminal patent by the American film actress Hedy Lamarr (Hedy Kiesler Markey) and the composer George Antheil (Markey and Antheil 1942). Figure 8.9a illustrates frequency hopping by the neotropical bat, *Molossus*, which searches for insects by emitting a series of very shallow-sweeping FM sounds at different frequencies from one sound to the next. The gap between the frequencies of successive sounds is wide enough that neither the broadcasts nor the echoes they produce can overlap in frequency. The only condition is that the sequences of echo-delay estimates made by the bat be kept separate according to frequency.

In Fig. 8.9a, the streams of corresponding pulses and echoes are kept separate and can overlap in time while not crossing each other in frequency. This is an example in a single species of the surprising diversity of echolocation sounds that Novick found in his world-spanning travels in the late 1950s (Novick 1969). Frequency-hopping strategies have now been documented in several bat species (Barclay 1983; Simmons et al. 2019).

The example of frequency hopping in Fig. 8.9a seems clear enough. The pulse-echo streams produced by the slightly offset broadcast frequencies retain their identities in spite of being overlapped in time for echo delays shorter than the interpulse intervals of roughly 50–60 ms. (These frequencies are low enough that echoes from large objects, such as vegetation or the ground, at distances greater than 8–10 m will return as audible to the bat, so pulse-echo ambiguities will indeed occur unless the frequency separations can be used to resolve them.) Resolving pulse-echo ambiguity by frequency hopping is more difficult to achieve, however, if the echolocation sounds are entirely wideband FM sweeps that cover mainly the same frequencies

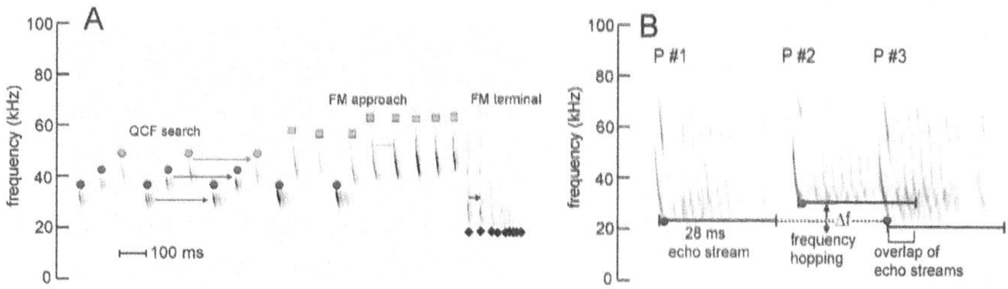

Fig. 8.9 Frequency hopping and pulse-echo ambiguity. (**a**) Spectrograms for a series of biosonar sounds emitted by the neotropical bat *Molossus* during an aerial capture of an insect (recording courtesy of Laura N. Kloepper). The bat is searching for flying prey with short, very shallow downward frequency-modulated (FM) sweeps, effectively short-constant frequency (CF) sounds (quasi-CF or QCF), surrounded by vegetation in the vicinity. These sounds are emitted at a series of three different, nonoverlapping frequencies (36 kHz red circles, 39 kHz blue circles, and 42 kHz green circles). After several such pulse sequences, it appears that when the insect is detected, the bat starts emitting wider FM sweeps, interspersed with the lowest-frequency QCF sounds, then it uses entirely higher-frequency wideband FM sweeps to approach (light blue squares) and lower-frequency FM sweeps in the terminal buzz stage (black diamonds). The frequency hopping in QCF sounds preserves frequency segregation for pulse-echo sequences that overlap in time (Simmons et al. 2019). (**b**) Spectrograms for wideband FM broadcasts emitted by *Eptesicus fuscus* flying in dense, extended cluttered surroundings where nearby short-range objects pose imminent collision hazards while longer-range objects return strong echoes that overlap for the short interpulse intervals between successive broadcasts necessary for reacting to the nearby objects. When the next broadcast is emitted before the 28-ms echo stream is completely received, the bat raises the terminal FM frequency of the first sound (blue circle) several kilohertz above the lower terminal FM frequency of the second sound (second red circle), which is near the terminal FM frequency of the broadcast that yields a nonoverlapping stream of echoes (first red circle). Segregation of the terminal FM frequencies appears sufficient to render the two largely overlapping pulse-echo sequences sufficiently apart for the pulse-echo ambiguity to be resolved, even though most of the FM frequencies do overlap (Ming et al. 2020)

in all of the broadcasts. Figure 8.9b shows an example of frequency hopping by *E. fuscus*, which uses just such problematic wideband sounds. Here, the bat shifts upward the low-frequency end of its FM sweeps in the first of two sounds that elicit echoes overlapping with the echoes from the second sound. This small shift does little to segregate the overall frequency bands of the overlapping echo streams, but experiments have shown that the bat starts its processing of echoes for their delay at the low-frequency end of the sweep (Ming et al. 2020). When echoes that lack the lowest frequencies are received, the bat's auditory delay-processing mechanisms are prevented from registering the echoes at all. For the critical case of longer-delay echoes from the first broadcast being misinterpreted as shorter-delay echoes of the second broadcast, the absence of the lowest frequencies in these echoes would be enough to avoid pulse-echo ambiguity. This part of the history of research on echolocation becomes entangled with the history of signal-processing leading up to modern spread-spectrum technology.

8.5 Summary

The discovery of echolocation in bats around 1940 prompted the development of several theories about how bats processed the echoes of their FM broadcasts to determine target range from echo delay. The motivation for one strain of theory was the implausibility that the bat's auditory system might operate directly on the ultrasonic frequencies contained in their sounds. They offered ways for the bat to demodulate the ultrasonic frequencies down into the human hearing range where familiar mechanisms related to auditory perception could be invoked to explain how echolocation "worked." The other type of theory was inspired by contemporaneous developments in radar, particularly pulse compression, or "chirp," receiver designs. The underlying structure of pulse compression is a method of converting the time–frequency pattern of frequencies in FM sounds into their most compressed form, the cross-correlation function of echoes with broadcasts. The way bats change the FM sweeps in their broadcasts implicates one particular form of pulse compression—spectrogram correlation—as the most viable theory to consider. In the course of the development of echolocation theories, the question arose as to why bats used FM signals. In one argument, because bats use FM sounds, they must also use pulse compression; otherwise, there is no reason to generate such complicated sounds. In the other, bats use FM sounds because they sweep across a wide band of frequencies to scan the spectrum of echoes for information about the target's shape. The dual requirement for target ranging from echo delay and target shape perception from the echo spectrum led to a combined spectrogram correlation and spectrogram transformation model that fits the current understanding of auditory mechanisms in bats. The complexity of echo-delay perception in bats leads to an even closer examination of signal theory and its relation to contemporary spread-spectrum methods in electronic communications.

Acknowledgments This review was carried out as part of work supported by ONR MURI Grant # N00014-17-1-2736 to the authors.

Conflict of Interest James A. Simmons declares no conflict of interest.
Andrea Megela Simmons declares no conflict of interest.

References

Altes RA (1984) Texture discrimination with spectrograms. IEEE Trans Sonics Ultrason 31:407–417. https://doi.org/10.1109/T-SU.1984.31521

Au WWL, Suthers RA (2014) Production of biosonar signals: structure and form. In: Surlykke A, Nachtigall PE, Fay RR, Popper AN (eds) Biosonar. Springer, New York, pp 61–105

Barclay RMR (1983) Echolocation calls of emballonurid bats from Panama. J Comp Physiol 151:515–520. https://doi.org/10.1007/BF00605468

Blauert J (1996) Spatial hearing. MIT Press, Cambridge

Buderi R (1996) The invention that changed the world. Simon and Schuster, New York

Busnel R-G (1966) Animal sonar systems: biology and bionics. NATO Advanced Study Institute. Laboratoire de Physiologie Acoustique, INRA-CNRZ, Jouy-en-Josas

Cahlander DA, McCue JJG, Webster FA (1964) The determination of distance by echolocating bats. Nature 201:544–546. https://doi.org/10.1038/201544a0

Charnock H (1981) Echo-ranging: L. F. Richardson's contribution. Weather 36(11):316–322

Dallos P (1972) Cochlear potentials: a status report. Audiology 11:29–41. https://doi.org/10.3109/00206097209072579

Fenton MB, Jensen FH, Kalko EKV, Tyack PL (2014) Sonar signals of bats and toothed whales. In: Surlykke A, Nachtigall PE, Fay RR, Popper AN (eds) Biosonar. Springer, New York, pp 11–59

Fenton MB, Grinnell AD, Popper AN, Fay RR (eds) (2016) Bat bioacoustics. Springer, New York

Galambos R (1942) Cochlear potentials elicited from bats by supersonic sounds. J Acoust Soc Am 14:41–49. https://doi.org/10.1121/1.1916201

Galambos R, Griffin DR (1942) Obstacle avoidance by flying bats: the cries of bats. J Exp Zool 80:475–490. https://doi.org/10.1002/jez.1400890308

Glaser W (1974) Zur hypothese des optimalenlpfangs bei der Fledermausortung. J Comp Physiol 94:227–248

Griffin DR (1958) Listening in the dark. Yale University Press, New Haven

Griffin DR (1967) Discriminative echolocation by bats. In: Busnel R-G (ed) Animal sonar systems: biology and bionics. Laboratoire de Physiologie Acoustique, Jouy-en-Josas, pp 273–300

Griffin DR, Galambos R (1941) The sensory basis of obstacle avoidance by flying bats. J Exp Zool 86:481–506. https://doi.org/10.1002/jez.1400860310

Griffin DR, McCue JJG, Grinnell AD (1963) The resistance of bats to jamming. J Exp Zool 152:229–250. https://doi.org/10.1002/JEZ.1401520303

Griffin DR, Friend JH, Webster FA (1965) Target discrimination by the echolocation of bats. J Exp Zool 158:155–168. https://doi.org/10.1002/jez.1401580204

Grinnell AD (1963a) The neurophysiology of audition in bats: intensity and frequency parameters. J Physiol 167(1):38–66. https://doi.org/10.1113/jphysiol.1963.sp007132

Grinnell AD (1963b) The neurophysiology of audition in bats: temporal parameters. J Physiol 167(1):67–96. https://doi.org/10.1113/jphysiol.1963.sp007133

Grinnell AD (1995) Hearing in bats: an overview. In: Popper AN, Fay RR (eds) Hearing by bats. Springer, New York, pp 1–36

Grinnell AD (2018) Early milestones in the understanding of echolocation in bats. J Comp Physiol A 204:519–536. https://doi.org/10.1007/s00359-018-1263-3

Grinnell AD, Gould E, Fenton MB (2016) A history of the study of echolocation in bats. In: Fenton MB, Grinnell AD, Popper AN, Fay RR (eds) Bat bioacoustics. Springer, New York, pp 1–24

Hartridge H (1920) The avoidance of objects by bats in their flight. J Physiol 54:54–57

Hartridge H (1945) Acoustical control in the flight of bats. Nature 156:490–494. https://doi.org/10.1038/156490a0

Henson OW (1966) The perception and analysis of biosonar signals by bats. In: Busnel R-G (ed) Animal sonar systems: biology and bionics. Laboratoire de Physiologie Acoustique, INRA-CNRZ, Jouy-en-Josas, pp 949–1003

Kay L (1961) Perception of distance in animal echo-location. Nature 190(4773):361–362. https://doi.org/10.1038/190361a0

Kay L (1962) A plausible explanation of the bat's echo-location acuity. Anim Behav 10(1–2):34–41. https://doi.org/10.1016/0003-3472(62)90128-8

Kay L (1966) Enhanced environmental sensing by ultrasonic waves. In: Busnel R-G (ed) Animal sonar systems: biology and bionics. Laboratoire de Physiologie Acoustique, INRA-CNRZ, Jouy-en-Josas, pp 757–782

Kick SA, Simmons JA (1984) Automatic gain control in the bat's sonar receiver and the neuroethology of echolocation. J Neurosci 4:2705–2737. https://doi.org/10.1523/jneurosci.04-11-02725.1984

Klauder JR (1960) The design of radar signals having both high range resolution and high velocity resolution. Bell Syst Tech J 39(4):809–820. https://doi.org/10.1002/j.1538-7305.1960.tb03943.x

Klauder JR, Price AC, Darlington S, Albersheim WJ (1960) The theory and design of chirp radars. Bell Syst Tech J 39(4):745–808. https://doi.org/10.1002/j.1538-7305.1960.tb03942.x

Luo J, Simmons AM, Beck QM, Macías S, Moss CM, Simmons JA (2019) Frequency-modulated up-chirps produce larger evoked responses than down-chirps in the big brown bat auditory brainstem. J Acoust Soc Am 146:1672–1684. https://doi.org/10.1121/1.5126022

MacArthur CWP (2000) Louis Jurine's papers on the hearing of bats and on their parasites: a retrospective with new information. Le Rhinolophe 14:1–35

Markey HK, Antheil G (1942) Secret communication system. U.S Patent No. 2292387. U.S. Patent Office, Washington, DC

McCue JJG (1966) Aural pulse compression by bats and humans. J Acoust Soc Am 40:545–548. https://doi.org/10.1121/1.1910116

McCue JJG (1968) Signal processing by the bat, *Myotis lucifugus*. MIT Lincoln Laboratory Report, Cambridge

Metzner W, Müller R (2016) Ultrasound production, emission, and reception. In: Fenton MB, Grinnell AD, Popper AN, Fay RR (eds) Bat bioacoustics. Springer, New York, pp 55–91

Ming C, Bates ME, Simmons JA (2020) How frequency hopping suppresses pulse-echo ambiguity in bat biosonar. Proc Natl Acad Sci U S A 117:17288–17295. https://doi.org/10.1073/pnas.2001105117

Ming C, Haro S, Simmons AM, Simmons JA (2021) A comprehensive computational model of animal biosonar signal processing. PLoS Comput Biol 17(2):e1008677. https://doi.org/10.1371/journal.pcbi.1008677

Möhres FP (1953) Über die ultraschallorientierung der Hufeisennasen (Chiroptera Rhinolophidae). Z Vergl Physiol 34:547–588

Neuweiler G (2000) Biology of bats. Oxford University Press, Oxford

Nordmark J (1960) Perception of distance in animal echo-location. Nature 183(4755):1009–1010. https://doi.org/10.1038/1881009a0

Nordmark J (1961) Perception of distance in animal echo-location. Nature 190(4773):363–364

Novick A (1969) The world of bats. Holt Rinehart Winston, New York

Pierce GW, Griffin DR (1938) Experimental determination of supersonic notes emitted by bats. J Mammal 19:454–455. https://doi.org/10.2307/1374231

Pollak GD, Casseday JH (1989) The neural basis of echolocation in bats. Springer, New York

Pye JD (1960) A theory of echolocation by bats. J Laryngol Otol 74:718–729. https://doi.org/10.1017/s0022215100057170

Pye JD (1961) Perception of distance in animal echo-location. Nature 190(4773):362–363

Pye JD (1966) Discussion of theories of sonar systems in relation to biological organisms. In: Busnel R-G (ed) Animal sonar systems: biology and bionics. Laboratoire de Physiologie Acoustique, INRA-CNRZ, Jouy-en-Josas, pp 1121–1136

Saillant PA, Simmons JA, Dear SP, McMullen TA (1993) A computational model of echo processing and acoustic imaging in frequency-modulated echolocating bats: the spectrogram correlation and transformation receiver. J Acoust Soc Am 94:2691–2712. https://doi.org/10.1121/1.407353

Sanderson MI, Neretti N, Intrator N, Simmons JA (2003) Evaluation of an auditory model for echo delay accuracy in wideband biosonar. J Acoust Soc Am 114:1648–1659. https://doi.org/10.1121/1.1598195

Schnitzler H-U (1973) Control of Doppler shift compensation in the greater horseshoe bat *Rhinolophus ferrumequinum*. J Comp Physiol 82:79–92. https://doi.org/10.1007/BF00714171

Schnitzler H-U, Henson OW Jr (1980) Performance of airborne animal sonar systems: I. Microchiroptera. In: Busnel RG, Fish JF (eds) Animal sonar systems. Plenum, New York, pp 109–181

Schuller G, Beuter K, Schnitzler H-U (1974) Response to frequency shifted artificial echoes in the bat *Rhinolophus ferrumequinum*. J Comp Physiol 89:275–286. https://doi.org/10.1007/BF00696191

Scientific American (1912) The sixth sense of bats. Sir Hiram Maxim's contention. The possible prevention of sea collisions. Sci Am Suppl 7:148–150. https://archive.org/details/newsystemforprev00maxirich/page/13/mode/1up
Simmons JA (1973) Resolution of target range by echolocating bats. J Acoust Soc Am 54:157–173. https://doi.org/10.1121/1.1913559
Simmons JA (1974) Response of the Doppler echolocation system in the bat *Rhinolophus ferrumequinum*. J Acoust Soc Am 56:672–682. https://doi.org/10.1121/1.1903307
Simmons JA (1993) Evidence for perception of fine echo delay and phase by the FM bat, *Eptesicus fuscus*. J Comp Physiol A 172:533–547. https://doi.org/10.1007/bf00213677
Simmons JA (2012) Bats use a neuronally implemented computational acoustic model to form sonar images. Curr Opin Neurobiol 22:311–319. https://doi.org/10.1016/j.conb.2012.02.007
Simmons JA (2014) Temporal binding of neural responses for focused attention in biosonar. J Exp Biol 217:2834–2843. https://doi.org/10.1242/jeb.104380
Simmons JA (2017) Theories about target ranging in bat sonar. Acoust Today 13:43–51
Simmons JA, Chen L (1989) Acoustic basis for target discrimination by FM echolocating bats. J Acoust Soc Am 86:1333–1350. https://doi.org/10.1121/1.398694
Simmons JA, Grinnell AD (1988) The performance of echolocation: acoustic images perceived by echolocating bats. In: Nachtigall P, Moore PWB (eds) Animal sonar systems: processing and performance. Plenum Press, New York, pp 353–385
Simmons JA, Simmons AM (2011) Bats and frogs and animals in between: evidence for a common central timing mechanism to extract periodicity pitch. J Comp Physiol A 197(5):585–594. https://doi.org/10.1007/s00359-010-0607-4
Simmons JA, Gaudette JA, Warnecke M (2017) Biosonar inspired signal processing and acoustic imaging from echolocating bats. In: Ballieri A (ed) Biologically inspired radar and sonar: lessons from nature. IET Press, London, pp 5–36
Simmons JA, Hiryu S, Shriram U (2019) Biosonar interpulse intervals and pulse-echo ambiguity in four species of echolocating bats. J Exp Biol 222. https://doi.org/10.1242/jeb.195446
Skolnik MI (1980) Introduction to radar systems, 2nd edn. McGraw-Hill, New York
Sterbing-D'Angelo S, Chadha M, Chiu C, Falk B, Xian W, Barcelo J, Zook JM, Moss CF (2011) Bat wing sensors support flight control. Proc Natl Acad Sci U S A 108:11291–11296. https://doi.org/10.1073/pnas.1018740108
Stewart JL (1979) The bionic ear. Covox Co., Santa Maria
Stewart JL, Kasson JM (1968) Simulating mechanisms in animal echoranging. AMRL-TR-68-192, Aerospace Medical Research Laboratories, Wright-Patterson AFB, Dayton
Strother GK (1961) Note on the possible use of ultrasonic pulse compression by bats. J Acoust Sci Am 33:696–697. https://doi.org/10.1121/1.1908771
Strother GK (1966) Comments on "aural pulse compression in bats and humans". J Acoust Soc Am 40:545–548
Teeling EC, Jones G, Rossiter S (2016) Phylogeny, genes, and hearing: implications for the evolution of echolocation in bats. In: Fenton MB, Grinnell AD, Popper AN, Fay RR (eds) Bat bioacoustics. Springer, New York, pp 25–54
Thurlow WR, Small AM (1955) Pitch perception for certain periodic stimuli. J Acoust Soc Am 27:132–137. https://doi.org/10.1121/1.1907473
Van Bergeijk WA (1964) Sonic-pulse compression in bats and people: a comment. J Acoust Soc Am 36:594–597. https://doi.org/10.1121/1.1919012
Wiegrebe L (2008) An autocorrelation model of bat sonar. Biol Cybern 98:587–595. https://doi.org/10.1007/s00422-008-0216-2
Woodward PM (1953) Probability and information theory with applications to radar, 2nd edn. Pergamon Press, New York
Yovel Y, Matthias OF, Stilz P, Schnitzler H-U (2011) Complex echo classification by echolocating bats: a review. J Comp Physiol A 197:475–490. https://doi.org/10.1007/s00359-010-0584-7

Chapter 9
Central Auditory Processing in the Mammalian System

Yi Zhou and H. Steven Colburn

9.1 Introduction

The central auditory system (CAS) has been studied and speculated about for millennia, and although our understanding of auditory processing has a long way to go, especially when the complexities of central processing are considered, much has been learned over time. This chapter is organized roughly historically, starting with observations and speculations from the Greeks and Romans and progressing to the early 1970s. The focus of this review is to revisit the work that led to the rise of neuroscience in the early twentieth century. Within different time periods, we attempt to describe what was accepted as knowledge at that time and how this knowledge has influenced the trajectory of auditory and hearing research.

The major resources for our study came from historical reviews, recollections, and biographical memoirs including many valuable volumes from those published by the National Academy of Sciences and the Society of Neuroscience (e.g., Squire 1998). These resources have allowed us to look through the lens of history and to gain an understanding of the intellectual environments that fostered brain-mind inquiry across a range of classical disciplines, including mathematics, chemistry, philosophy, physics, psychology, engineering, and biology. This cross-border inquiry that preluded the dawn of neuroscience has shaped the approaches taken to understand the brain. For example, the great neuroanatomist Ramón y Cajal was often also known as a great chemist and the great mathematician Norbert Wiener was also a well-known faculty member in engineering. To be fair to history, a

Y. Zhou (✉)
College of Health Solutions, Arizona State University, Tempe, AZ, USA
e-mail: yizhou@asu.edu

H. S. Colburn
Department of Biomedical Engineering, Boston University, Boston, MA, USA
e-mail: colburn@bu.edu

D. R. Ketten et al. (eds.), *A History of Discoveries on Hearing*, Springer Handbook of Auditory Research 77,
https://doi.org/10.1007/978-3-031-41320-9_9

thorough survey of experimental approaches and published results should be conducted; however, we adopt a more realistic goal. This chapter highlights the empirical observations and theoretical models that we consider to have played major roles in developing the auditory research field. We emphasize the scientists who laid the foundations in anatomy, physiology, behavior, and modeling to advance knowledge. Their influences are often most evident in the people they have trained and in the departments and research centers they helped develop.

The chapter starts the history of central auditory processing with ideas related to human processing of sounds from early Greek and Roman times and progresses through the mid-twentieth century (Sect. 9.2). The chapter then discusses several lines of work that manifested the rapid developments made in neurological and theoretical inquiries of the auditory system in the mid-twentieth century (Sects. 9.3 and 9.4). The chapter concludes with remarks about our current state of understanding of central auditory processing in the context of topics discussed in Sect. 9.4.

The chapter puts a special emphasis on the development of theories of hearing because hearing studies have historically revealed the *computational* needs of the auditory system. From the start, the processing of auditory information is different from the processing of other sensory information, such as visual or tactile inputs, because the initial analysis at the cochlea does not directly reveal the source of stimulation. In contrast to vision, where photoreceptors are spatially distributed on the retina to encode the spatial source of stimulation, the dominant tuning in peripheral auditory neurons is frequency dependent as opposed to spatially dependent. It follows that the mapping of neural activity to space must be extracted and established by central processing.

While this chapter is organized historically, special attention has been paid to the studies of the mechanisms of binaural stimulus processing and studies of cortical functions in the period between 1940 and 1970, as described in Sects. 9.3 and 9.4. This selection "bias" is largely driven by what we consider the desired state of a scientific environment, where the empirical and theoretical investigations reinforce each other in understanding the underlying mechanisms. In our opinion, the selected work demonstrates both rigor and openness for the clashes of ideas across disciplines as shown in the written records of dialogs made between experimentalists and theorists during several symposia (e.g., Symposium on Principles of Sensory Communication 1961, edited by Rosenblith (1961)). To recapture critical developments of CAS research, for studies mentioned in Sects. 9.3 and 9.4, we try to extract the historical information by asking these inter-related questions:

- How were the empirical observations of neural activity described based on the knowledge of the physical stimulus at the time?
- What mechanisms were proposed, and how did empirical investigations benefit from model hypotheses?
- What was the proposed model for the neural basis of behavior?

9.2 Historical Summaries

9.2.1 Early Ideas from the Greeks and Romans

Some of the earliest discussions of the operations of the central nervous system were captured in the writings of Greeks, Romans, and Egyptians. We summarize these discussions briefly here, including some early ideas of Pythagoras (around 550 BC) on musical sounds; the accessible work about mind and brain by Aristotle (about 350 BC) in Greece; the contributions of Herophilus (335–280 BC) and Erasistratus (310–250 BC) who worked in Alexandria, Egypt; and the contributions of Galen (around 200 AD) who studied in Alexandria and practiced in Rome.

In the sixth century BC, the ancient Greek philosopher Pythagoras wanted to know why some combinations of musical sounds seemed more beautiful than others, and he found answers in terms of numerical ratios representing the harmonic overtone series on a string. He is reported to have observed that when the lengths of vibrating strings are expressed as ratios of integers (e.g., 2/3 or 3/4), the tones produced sound harmonious, and that the smaller the integers, the more harmonious the sounds.

Around 340 BC, Aristotle (384–322 BC) described the coding of sensory information in his work "On the Soul" (*De Anima*) (Aristotle 1973). He discussed sound as vibrations that are transmitted through the air. He noted that when sounds are perceived, they generate motion in the ear, with the sensitivity of the organism seen in the responses to the received information. He described the response of people to sound, called hearing, explicitly as being mediated through the mechanical response in the ear, generating a sensation in the organism, and interpreted by the "soul" [cf., *De Anima*, Book II, Chapter 8 ("distinctions about sound and hearing")] (Aristotle 1973). As conceptualized by Aristotle, "the soul" represents conscious awareness. In *De Anima*, the sense of hearing is compared with vision, taste, smell, and tactile inputs, and the differences between imagined sounds and perceived sounds are explicitly discussed as fundamentally different in their basis in the stimulation of receptors or not. The activity of the CAS does not come up explicitly, although the idea of internally generated sensations is discussed and voice is considered a sound expressed by the soul.

In Aristotle's discussions of sensory processing and parts of the body, speculations about the roles of the brain and the heart arise, presumably because of the prominence of these structures in the gross anatomy of human beings and the obvious connectivity of arteries throughout the body to the centrally located heart. Aristotle speculated that the brain was primarily a useful structure for heat dissipation. This cardiocentric view was similar to those of ancient Egyptians who believed that emotions and intelligence arise from the heart rather than the brain. However, this notion was later refuted by Herophilus, who attributed the fourth ventricle of the brain as the center of the nervous system. Galen, who lived approximately 600 years later, furthered this idea by arguing that the brain was responsible for mediating sensation and action.

Contrasts in the thinking of Aristotle, Herophilus, and Galen regarding the organ of the mind have intrigued numerous historians interested in the history of neuroanatomy in Western medicine (see reviews in Dobson 1925, 1927; Wills 1999; Acar et al. 2005; Elhadi et al. 2012). The aforementioned historical accounts reveal several significant findings. Notably, Herophilus, Erasistratus, and Galen were all renowned physicians who adhered to the Hippocratic rationalist method. They also received their education from the medical school of Alexandria, one of the oldest empirical institutions in the history of medicine.

Herophilus and Erasistratus obtained their knowledge of the cranium through scientific investigations of the human body. Herophilus, in particular, made a significant contribution by distinguishing between blood vessels and nerves and proposed that the physiology of nerves was responsible for voluntary movement. Similarly, Erasistratus separated the functions of sensory and motor nerves and compared the human brain with those of other animals, concluding that human intelligence was linked to the greater number of foldings in the human brain. Despite the fact that these observations date back to the third century BC, they remain relevant to modern neuroscience inquiries. Unfortunately, the written works of Herophilus and Erasistratus were lost entirely after the Royal Library of Alexandria was burned down around 48 BC. As a result, what we know about their work is mainly based on references given in Galen's writings.

Galen, a Greek physician and philosopher, was credited with tracing most of the cranial nerves and recognizing that injuries or compression of the brain could cause loss of sensation and movement. Galen's comprehensive anatomical knowledge enabled him to reject Aristotle's cardiocentric view and instead argue that the brain served as the central organ for controlling sensory and motor functions. Unlike those of Herophilus and Erasistratus, Galen's writings have survived, and his anatomical views dominated Western medical science for over a millennium until Andreas Vesalius published illustrations of human dissections in 1543.

9.2.2 Anatomy and Auditory Processing Ideas During the Second Millennium

9.2.2.1 Study of Anatomy During the Renaissance

The Renaissance was a period in European history that lasted from the fourteenth to the seventeenth centuries. During the Renaissance, there was a renewed interest in classical culture, art, and humanism. Not only the arts flourished but also the pursuit of knowledge in seeking truth. Through the work of the Renaissance scientists, more and more detailed anatomical information became available.

Leonardo da Vinci (1452–1519) is widely regarded as a key figure in the development of the Renaissance culture in arts, sciences, and engineering. Da Vinci's drawings of the human body are both beautiful and amazing, and they are insightful sources of details of anatomy, including neural pathways and the brain. These drawings are part of a wide collection of notes, drawings, and observations that were entered into his notebooks while he was also painting his world-renowned masterpieces like the Mona Lisa.

Andreas Vesalius (1514–1567) was considered the founder of modern anatomy. Vesalius was a Professor of Surgery at the University of Padua, Italy. His teachings emphasized human anatomy as an empirical science. Like Herophilus and Erasistratus, Vesalius believed that anatomic knowledge could be gained through dissection of the human corpse and not through speculations. His collection of work "Seven Books on the Structure of the Human Body" (*De Humani Corporis Fabrica Libri Septem*), which was published originally in 1543 (Vesalius 1985), contains detailed descriptions and illustrations of the anatomy of the human body including the brain, the cranial nerves, and the organs of sense.

In 1664, the London physician Thomas Willis (1621–1675) published an anatomical study of the brain and led the world to think about the brain and its function in deeper ways (Willis 1664). Willis was particularly interested in the role of anatomy in shaping man's ability to think, and he spent great effort on details of the anatomy of the human brain, its relation to the brains of animals, and the role of the brain in thinking. The work of Willis, *Cerebri Anatome* (1664), which was originally published in Latin, described anatomical details from postmortem brain dissections and speculated about the relation of the brain to human abilities to think and analyze information.

These aforementioned anatomical observations were incorporated into the thinking about neural processing and the associated mechanisms at the time. A major stimulus to understanding neurophysiology came from the work of the Italian Luigi Galvani (1737–1798) who was the first to recognize the electrical nature of neural activity. In 1791, he described the effects of electrical stimulation on nerves and muscles in a series of experiments. His work, published late in his life, stimulated controversy, and he was challenged in his time (Piccolino 1997). Galvani's realization of the connection of neural and muscle activity to electrical currents was profound and correct.

9.2.2.2 Fourier and Helmholtz's Ideas About Sounds

In the late eighteenth and early nineteenth centuries, the development of tools for understanding sound and vibration continued with the important development of mathematical tools that allow the quantitative analysis of sound and its processing. For example, the work of Joseph Fourier (1768–1830) is the basis of modern signal characterization and analysis and is based on the representation of signals in terms of their frequency content, now called Fourier analysis. This is fundamental to our thinking about and modeling the physiological processing through the auditory periphery and brainstem and shapes almost all modern work on the CAS.

The work of Hermann von Helmholtz (1821–1894) in Germany provided ways of thinking and insights about hearing that are still important today. In his 1863 book *On the Sensation of Tone as a Physiological Basis for the Theory of Music*,

Helmholtz traced the progression of the response to acoustic stimulation from stimuli to perception and integrated physical and mathematical descriptions with perceptions of stimuli including music. This book, originally published in German, is available in English (Von Helmholtz 1912) and is a useful way to get insight into thinking in the late 1800s. It describes cochlear vibrations and the impulses on sensory nerves in an attempt to understand sensation. This book remains a useful reference for understanding the basis of the analysis of sound, sound perception, and abilities to use sound for reliable distinctions. In Sect. 9.3.2 of this chapter, we discuss the extension of this work to the analysis of sound processing, including decision-making strategies that were developed in the mid-twentieth century. This work is described and discussed in the context of current approaches.

Around 1850, Helmholtz made measurements of neural responses to stimulation and was the first to measure the velocity of neural propagation. He found that the nerve impulse was measurable and had a velocity of approximately 27 meters per second. This provided a direct functional description of neural activity and directly contradicted claims from "vitalists" like Johannes Müller, who believed that there are life energies transmitted through the nervous system (animal spirits) and that the transmission of sensory information was instantaneous.

9.2.2.3 Neuroanatomy at the Turn of the Twentieth Century

In the late nineteenth and early twentieth centuries, major developments in understanding neural processing were stimulated by the anatomical work of Camillo Golgi (1843–1926) and Santiago Ramón y Cajal (1852–1934). They were both physicians and pathologists, and both were neuroanatomists who made important contributions to our understanding of the nervous system. Both Cajal and Golgi were involved in detailed drawings of neural structures based on Golgi silver straining techniques and provided figures that are still key references in our illustrations for neural structures.

In 1906, Cajal and Golgi were awarded the Nobel Prize in Physiology or Medicine in recognition of their work on the structure of the nervous system; however, they disagreed on a number of points. One of their points of disagreement was the discovery by Cajal, ultimately verified and supported by the field, that neurons are not continuously connected (the reticular theory championed by Golgi) but rather that the neurons are separate entities and neural activity progresses via non-continuous connections called synapses (the neuron doctrine pioneered by Cajal). Cajal explained the distinction between the reticular theory and the neural doctrine in his memoir—"*Recollections of my life*" (Cajal 1989). Figure 9.1 contrasts continuity vs. contiguity connections as shown in I and II, respectively. This original illustration by Cajal is part of the Santiago Ramón y Cajal Exhibit at the National Institute of Health (http://bit.ly/3JI818M). Other major contributions of Cajal to science include studies of the possibility of forming new neural connections/synapses in adult brains and the implied possibilities for learning (Cajal 1959).

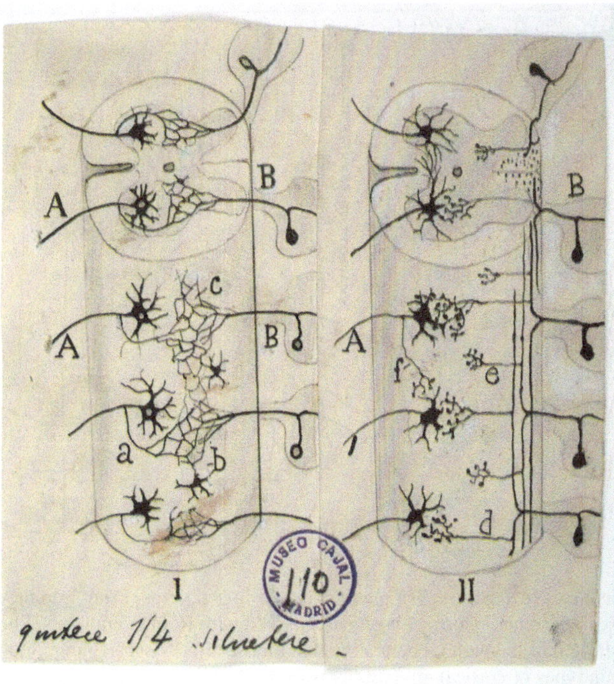

Fig. 9.1 Diagrams of comparison of Golgi's conception of the sensory-motor connections in the spinal cord (I) and Cajal's observation (II). *A*, anterior roots; *B*, posterior roots; *a*, collections of a motor root; *b*, cell with a short axon which Golgi believed to take part in the formation of the net; *c*, diffuse interstitial network; *d*, one of long collaterals in contact with the motor cells; *e*, short collateral. The annotations of labeling were based on the original legend for Fig. 33 from "Recollections of my life" by Cajal (1989). The photograph was retrieved online with permission from the Santiago Ramón y Cajal Exhibit at the National Institute of Health. (Courtesy of the Cajal Institute, Spanish National Research Council or CSIC©)

Cajal's last, and likely most distinguished, student was Rafael Lorente de Nó (1902–1990). Lorente de Nó was an assistant at the Cajal Institute from 1921 to 1929. In the early 1930s, he immigrated to the USA, where he held the position as the head of the Neuro-Anatomical Laboratory at the Central Institute for the Deaf (CID) in St. Louis, then a member of the Rockefeller Institute in 1941 and a Professor of Rockefeller University in 1953. Considered one of the most prominent neurophysiologists of the twentieth century, Lorente de Nó carried out studies of neurons in general and the central nervous system in particular. He authored one of the pioneering studies of the anatomy of the eighth nerve, which detailed the innervation patterns of the inner ear (Lorente de Nó 1933a, b).

Lorente de Nó made many contributions to our understanding of the functional units of the central nervous system and their structures. Figure 9.2 shows the drawing by Lorente de Nó (1938) illustrating the arrangements of pyramidal cells in the neocortex, which has been widely used as a foundational reference for

Fig. 9.2 Lorente De Nó's drawings of pyramidal cells in neocortex. From Lorente De Nó (1938), Fig. 71, "Cortical afferents." Cortical layers are marked with Roman numerals. Shown on the left side is an illustration of the parietal cortex of adult mouse after Nissl staining. Shown on the right end are the main types of cortical afferent fibers (*a, b*, specific thalamic afferents; *c. d*, unspecific or pluriareal afferents; *e, f*, association fibers). The labeled neurons on the left illustrate the bodies and dendrites of representative types of cells with descending axons (1, pyramids of layer II; 2, 3, pyramids of layer III; 4, large star pyramids; 5, star cells; 6, small star pyramids; 7, 8, 9, long deep pyramids; 10, short pyramids; 11, medium pyramids; 12, 13, short pyramids of layer VIa; 14, long spindles; 15, medium spindles; 16, short spindles; 17, deep star cells)

understanding cortical neuroanatomy. He presented the modular (i.e., columnar) organization of the cerebral cortex, using it to explain synaptic delay and nerve volume conduction in general. He also emphasized both synaptic summation and neural feedback possibilities.

These ideas are still active today and are incorporated into much of our current analysis, as implied by the discussions in the rest of this chapter. In this context, we note that the structure of the cortex is still not understood with respect to the detailed, sequential processing of information. When we consider the interpretation of sounds, other inputs, and expectations, we confront how little we understand about the CAS, but our thinking about these problems as described in the rest of this chapter is built on the work described here.

9.2.2.4 First Brainwaves by Hallowell Davis

Hallowell Davis (1896–1992) was a pioneer in the development of understanding of the CAS. Davis studied and worked at Harvard, with S.S. Stevens and others whose work is described below, until he left in 1947 to become the Director of Research at the Central Institute for the Deaf. In the words of Robert Galambos in his memoir written for the American Academy of Science in 1998 about Hallowell Davis, "The first brainwaves seen on the American continent came out of his own head in 1933, recorded by his graduate students using equipment he had designed. At about the same time, he was among the first anywhere to record animal cochlear potentials, human evoked brain potentials, and the activity of single nerve cells at work inside an animal brain." (Galambos 1998b, page 117). Davis' landmark work was reported in a pair of papers published in 1939 led by his first wife Pauline Davis and himself. These two papers are the first systematic reports of human electroencephalogram recorded on the waking and sleeping brain (Davis 1939; Davis et al. 1939).

9.3 Sensory Processing and Modeling of the CAS

In this section, a collection of theoretical work done in the early twentieth century that had a big impact on our thinking about and modeling of the CAS is discussed. We start with specific, classic work related to binaural processing in sound localization and then discuss more general work linked to cybernetics and information theory, leading to signal detection theory. The work that grew out of cybernetics and information theory had an impact on a broad spectrum of auditory modeling and experiments. The broad impact of this work at several laboratories is also summarized since that work had extended impacts on empirical and theoretical studies, and these laboratories and their leaders/directors had impacts on the field that continue today. We note again that the general emphasis in this chapter is on physiological and anatomical data and on the theoretical analyses for understanding these data; we are minimizing our coverage of psychoacoustic data, even though these data are clearly related to the physiology and anatomy, which are shaped by their ultimate function of providing and processing auditory information for behavior. For an efficient overview of psychoacoustic and behavioral data, see Yost (2015).

9.3.1 Sound Localization: Rayleigh's Duplex Theory, Jeffress's Model of Interaural Time Difference Processing, and the Experiments of Goldberg and Brown

One of the most important functions of hearing is to detect the location of an unexpected threat or event before seeing it. The brain's ability to realize the location of a sound source based on nearly identical signals at the two ears has been a fascinating question. One of the earliest formulations of mechanisms, known as the "Duplex Theory," came from Lord Rayleigh. *Lord Rayleigh* was the title borne by John William Strutt (1842–1919). Rayleigh won the Nobel Prize in Physics in 1904 and he made enormous contributions to the field of acoustics. In his monumental work

The Theory of Sound (Strutt 1877), Rayleigh reviewed the previous knowledge of the field and presented his own copious contributions. Rayleigh's work emphasized the experimentalist's perspective and had a long-lasting influence on acoustics research and education in the twentieth century (Garrett 2021). In his later paper on sound localization, Rayleigh (Strutt 1907) postulated that the localization of pure tones depends on two different interaural cues: time differences and level differences; this is called the duplex theory of sound localization.

According to Rayleigh's original duplex theory, low and high frequencies have different mechanisms at work in localization. For low frequencies, the directional information is provided by interaural time difference (ITD) at the two ears. ITD is an acoustic marker because it manifests the travel-time differences between sound waves to the left and right ears. For high frequencies, directional judgment is based on the interaural level difference (ILD) between the two ears. ILDs arise from the head shadow effect, which is also directional and frequency dependent. As a mass, the head can block and attenuate the sound waves coming to the ear, if the wavelength of a sound is sufficiently smaller than the size of the head. For these two sets of disparity cues, midline locations render zero ILDs and ITDs, whereas off-midline locations render nonzero ILDs and ITDs. The exact magnitude of each cue not only depends on the laterality of the sound source but also depends on the frequency of sound stimuli and the size of the head, which varies among species.

Rayleigh speculated that the frequency cutoff for useful low-frequency timing was around 500 Hz, but we now know that human listeners can use interaural time delay to localize tonal stimuli up to at least 1500 Hz, and because of synchrony to envelopes, temporal cues can also be used for higher frequencies. ILDs also provide useful, usually dominant, cues at high frequencies. Unlike vision, the auditory periphery does not directly encode spatial information of a sound, and auditory spatial perception arises from central mechanisms that interpret the pattern of ILDs and ITDs across frequency. Binaural disparities, as a biophysical mechanism, provide important insights into the functions of neural circuits that support binaural hearing.

While the psychophysical basis of horizontal localization has been well established in the literature (e.g., Strutt 1907), it was not until 1948 that Lloyd Jeffress proposed a neural mechanism for ITD detection in the landmark paper, "A Place Theory of Sound Localization" (Jeffress 1948), which elegantly integrated the structure and function of a hypothetical ITD-detection circuit.

Jeffress (1900–1986) was a Professor of Experimental Psychology at University of Texas, Austin. He was an active figure in research discussion at all levels, especially in topics related to psychological and physiological acoustics, and his extensive work in experimental measurements of hearing capabilities is widely recognized. The Jeffress model has three operational components—conductance delays of nerve fibers to offset ITDs, coincidence detection as a mechanism to register the simultaneous arrival of the two action potentials, and a spatial arrangement in the network of coincidence detectors to create a ladder-line structure for ITD detection based on place of excitation. The classic model of Jeffress is shown in Fig. 9.3 (from Jeffress 1948). In simple terms, Jeffress envisioned that a given

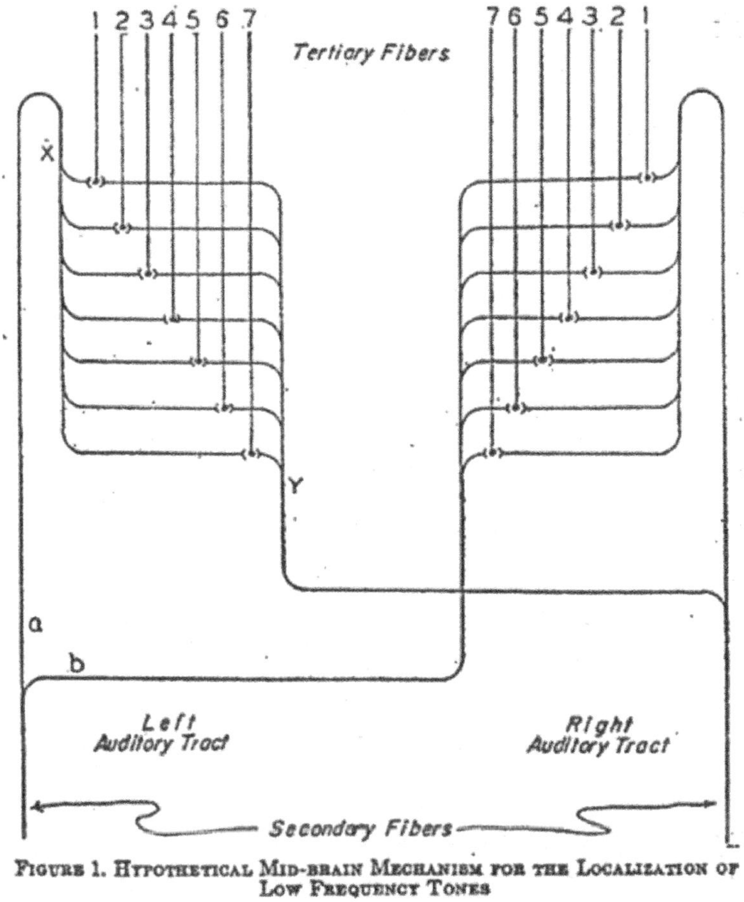

Fig. 9.3 Jeffress model of ITD detection of low-frequency sounds. From Jeffress 1948, Fig. 1, "Hypothetical Mid-brain Mechanism for the Localization of Low Frequency Tones." The model shows seven coincidence detectors on each side of the brain. A coincidence detector receives inputs from both left and right auditory tracts, where *a* marks the uncrossed and *b* marks the crossed tracks. The ipsilateral (X) and contralateral (Y) inputs are arranged based on the delay lines and the maximal response of a coincidence detector is triggered by the coincident arrival of inputs from the two sides. (The resources were published by the American Psychological Association and now in the public domain)

external travel-time difference between the two ears (i.e., ITD) could be counterbalanced by an internal conductance delay difference. The place for the coincident arrival of the ipsilateral and contralateral inputs in the neural network shows the maximal summation of nervous impulses and manifests the given ITD.

The Jeffress model is important because it deals directly with the question of how anatomy supports physiology in understanding neural function (i.e., from delay line to coincidence detection to ITD processing). To provide a context for his work,

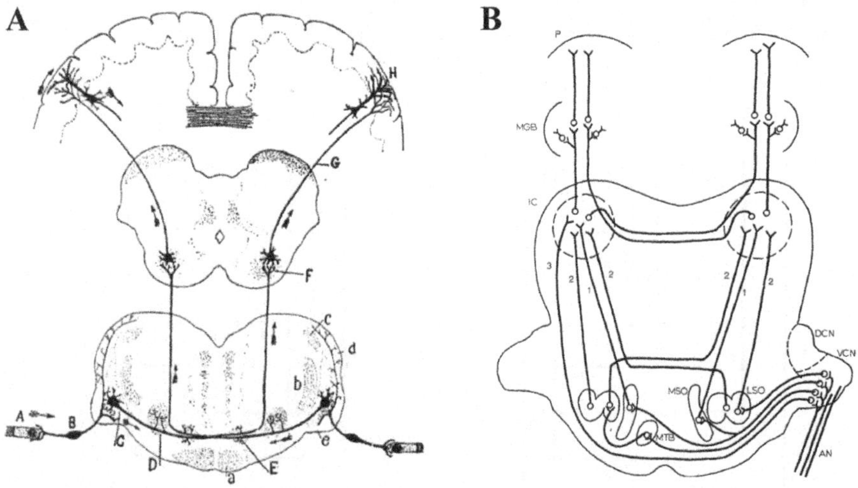

Fig. 9.4 Diagrams of the ascending auditory signal pathway published in the early and mid-twentieth century. (**a**) Flow of impulses in the auditory system (Fig. 26, Cajal 1911). *A*, hair cells of the organ of Corti; *B*, bipolar cochlear neurons of the spiral ganglion; *C*, ventral cochlear nucleus, which receives cochlear nerve endings; *D*, lateral superior olive; *E*, nucleus of the trapezoid body; *F*, lateral lemniscus, ending in the inferior colliculus; *G*, central auditory pathway, ending in the temporal cortex; *H*, terminal arborization of the auditory pathway, contacting pyramidal cells in the cortex; *a*, pyramidal tract; *b*, section through the descending root of the trigeminal nerve; *c*, inferior cerebellar peduncle; *d*, dorsal cochlear nucleus; *e*, ventral cochlear nucleus. (**b**) Schematic drawing of the principal ascending connections of the auditory system with respect to the acoustic nerve of the right side (Fig. 25, Harrison and Howe 1974). MSO medial superior olive, LSO lateral superior olive, MTB medial nucleus of the trapezoid body, DCN dorsal cochlear nucleus, VCN ventral cochlear nucleus, MGB medial geniculate body, P primary auditory cortex. Axons type *1* and *2* project from bilaterally innervated nuclei and type *3* from unilateral projections. (Reprinted from Cajal 1911 and Harrison and Howe 1974, with permission)

in the 1940s, although the basic arrangement of the auditory pathway has been speculated from early histological methods, detailed neural mechanisms for ITD-based sound localization were completely unknown at the time. Figure 9.4a shows Cajal's "circuit diagram" of the auditory system, in which sound information was presented as impulse signals sent from hair cells to the auditory cortex through four to five stages of synaptic transmission (Cajal 1911). This representation is far from being complete with regard to the bilateral innervation of structures at and above the superior olivary complex (SOC). Figure 9.4b shows the diagrammatic drawing of the mammalian auditory system based on what was known in the 1970s (Harrison and Howe 1974).

The realization of the complex connecting routes in the auditory pathways is attributed largely to lesion and electrophysiology studies starting to be developed in the 1940s. Notably, Galambos and Davis (1943) made the first recordings of single-unit activity in the auditory system from the cochlear nucleus of cats. Barnes et al. (1943) found that some fiber tracts leaving the cochlear nucleus ascend ipsilaterally and some ascend contralaterally and both terminate in the SOC, suggesting that the SOC is the first region in the auditory system where inputs from both ears converge. In this context, Jeffress made the following speculations on the possible location of the coincidence detection circuit in his 1948 paper:

> Our need is for a location where connections from both sides are to be found and where the delays due to prior synapses have been equal, so that the impulses that occur will be in phase where the sound at the two ears are in phase. The most obvious place would be the superior olivary nucleus. (Jeffress 1948, page 38).

Jeffress concluded his paper by making the following suggestion on the possible experimental procedures to verify his hypothesis:

> The existence of such a mechanism would be strongly suggested if one could find a region …, where impulses could be obtained with binaural stimulation under one phase condition and would drop out where the phase was shifted radically. (Jeffress 1948, page 39).

As history unfolds, we now know that not only was Jeffress right about the SOC as a key location of the ITD mechanism, but he also correctly predicted the way in which SOC neurons respond to the phase change between signals at the two ears. Jeffress restated his postulate in 1958 in a letter to the editor of *The Journal of the Acoustical Society of America* (Jeffress 1958). However, the definitive confirmation of Jeffress' coincidence detectors had to wait another 20 years for validation. In their review, Yin et al. (2019) provide a thorough and insightful recount of the history of studying this perhaps most famous auditory region to this day.

To fully characterize the coincidence mechanism, precise control of the acoustic stimuli and quantitative analysis of temporal patterns of peripheral neuronal responses are two needed techniques, both of which were introduced and established in the 1960s in auditory nerve fiber recordings at the laboratory of Nelson Kiang (1929–2023) at the Eaton-Peabody Laboratory (EPL) at the Massachusetts Eye and Ear Infirmary (MEEI) in Boston as described later in this chapter (e.g., Gerstein and Kiang 1960; Kiang 1965) and at the laboratory of Jerzy Rose at the University of Wisconsin, Madison (e.g., Hind et al. 1967; Rose et al. 1967). [See also the review by Edward F. Evans (1975).] While these technical developments and auditory-nerve recordings have contributed to an enhanced understanding of the peripheral auditory system under a wide variety of stimulus conditions, direct recordings from the anatomically confirmed SOC were rather limited. Of particular interest is the medial superior olivary nucleus (MSO) because it receives bilateral projections from the left and right cochlear nuclei, anatomically suited for the ITD-detection mechanism. Several attempts were made to record the MSO in cats, and only small numbers of cells were collected (see early review by Guinan et al. 1972a, b). Research shows that the SOC is a deep brainstem structure with thin sheets of cells and large background local potentials, making it notoriously difficult to record and isolate single-unit activities.

A breakthrough was made in the laboratory of Jay Goldberg (1935–2019). Goldberg was a student of Dewy Neff at the University of Chicago and later with Jerzy Rose at the University of Wisconsin. Goldberg joined the University of

Chicago as a faculty member in 1963 and remained there for the rest of his career. In two sequential papers in the 1960s (Goldberg and Brown 1968, 1969), Goldberg and his student Paul Brown published the anatomical and physiological results that were successfully taken from the SOC in dogs. The MSO in dogs is an unusually thick structure folded into a U-shaped ribbon when compared with the thin ribbon of MSO seen in cats (Goldberg and Brown 1968). Goldberg and Brown carefully manipulated the phase between signals at the two ears and compared the monaural and binaural phase sensitivities of the low-frequency cells (Goldberg and Brown 1969). Six of 15 phase-sensitivity cells are located in the MSO. The exemplar MSO cell (67-82-5) showed phase sensitivity as the Jeffress model predicted. "The discharge rate to binaural stimuli is maximum when the excitations emanating from the two ears arrive at the binaural neuron in phase and is minimum when the excitations are 180° out of phase. Hence, the observations are in agreement with the central feature of the theory, namely, that the binaural neurons are functioning as coincidence detectors." (Goldberg and Brown 1969, page 632).

While coincidence detection as evidenced by the binaural phase sensitivity made Goldberg and Brown (1969) one of the most cited papers in auditory physiology, their study also tested several important features of MSO that are relevant to the perceptual aspects of binaural hearing, including the tonotopic organization in the SOC, the role of average sound intensity, the interactions between ITD and ILD, and the presence of inhibition in the MSO. The prevalent view of binaural hearing is that the low-frequency cells in the MSO are responsible for ITD encoding, whereas the high-frequency cells in the lateral superior olive (LSO) are responsible for ILD encoding. The classic view is that the cells in the MSO have excitatory inputs from ascending fibers from both the left and right ears, making them "EE cells," and LSO cells receive excitatory inputs from the ipsilateral ear and inhibitory inputs from the contralateral ear, making them "IE" cells" (See diagram illustrations in Fig. 9.4b).

But such a dichotomy is not entirely consistent with the observations that ITD sensitivity remains for high-frequency sounds through phase sensitivity to sound envelopes (e.g., Henning 1974) and low-frequency ILDs can alter the laterality of an auditory image as observed in human perception (e.g., Domnitz and Colburn 1977). Goldberg and Brown (1969) revealed that both MSO and LSO are tonotopically organized and both nuclei contain high- and low-frequency cells. Importantly, there are high-frequency EE cells in the MSO, whose responses were predicted by average intensity, not ILD, in response to best-frequency tones. They did not reveal, however, whether high-frequency EE cells could encode ITDs of sound envelopes (including onsets) in narrowband stimuli at high frequencies.

Goldberg and Brown also tested the influences of ILDs on the rate-ITD response and binaural phase sensitivity. They concluded that many MSO cells received mixed excitation and inhibition and that the role of inhibition is most evident in the response to out-of-phase inputs. As they pointed out, "phase sensitivity depends not only on an in-phase facilitation, as Jeffress supposed, but also on an out-of-phase inhibition. The synaptic mechanisms of the inhibition are obscure." (Goldberg and Brown 1969, page 632). Well, the role of synaptic inhibition in the MSO as an alternative mechanism to the Jeffress theory was discussed in early 2000 (Brand et al. 2002) and the debate is still ongoing. (See review by Yin et al. 2019).

Perhaps the most important contribution of the Goldberg and Brown studies is that they offered a thorough description of methodology they used and a detailed explanation of observations made in their experimental work. The level of experimental details and nearly complete survey of response properties of a single cell is a gem in archived auditory research. Their study of MSO offers time-tested values for future work, especially for model simulations.

9.3.2 An Interactive Empirical and Theoretical Environment for Developing Models of Auditory Signal Processing

As suggested above by the work of Helmholtz (Sect. 9.2.2.2) and Jeffress (Sect. 9.3.1), an integrated understanding of hearing, including peripheral and central auditory processing and functional hearing abilities as measured in psychoacoustic experiments, is promoted by mathematical models and theoretical computations. Important developments in mathematical modeling, including cybernetics and decision theory, took place in the 1940s and have led to modern approaches toward an integrated understanding of hearing and the CAS.

The sensory modeling research that flourished at MIT is related to the presence and work of Norbert Wiener in "*Cybernetics: Or Control and Communication in the Animal and the Machine*" (Wiener 1948) and to the independent but related work of Claude Shannon in "*Information Theory*" (Shannon 1948). The nature of information, as formulated by Wiener and by Shannon, can be represented abstractly in terms of discrete or continuous signals and formulations of sets of alternatives. These formulations can be used to describe the operations involved in making choices and decisions, as well as combining sources of information. Their work and the work of the people who were attracted to follow their examples led to an environment of great discussion, exploration, and interactions between empirical and theoretical science.

In 1951, Jerome Wiesner, who later became the president of MIT, was the Director of the Research Laboratory of Electronics (RLE) at MIT, which was formed in 1946 and continues today. Wiesner, building on the impact of the work of Wiener and Shannon, recruited several important neuroscientists to come to MIT in the early 1950s, including Jerome (Jerry) Lettvin and Walter Rosenblith in 1951.

We first consider the role of Walter Rosenblith in the formation of a significant subdivision of RLE, called the Communications Biophysics Group or CBG. During the early years of CBG, Walter Rosenblith (1913–2002) built on the research (and the reputation) of Norbert Wiener in the area of sophisticated computations and formed a committee of researchers in auditory physiology and perception (including Larry Frischkopf, Bob Brown, and Nelson Kiang) to approach the administration of the Massachusetts Eye and Ear Infirmary (MEEI) and discuss the establishment of a research laboratory for auditory physiology. This led to the establishment of the Eaton-Peabody Laboratory for Auditory Physiology, called the EPL. The EPL, under the directorship of Nelson Kiang, became one of the world-renowned laboratories in auditory physiology and led to deep interactions between the researchers at CBG and those at EPL. This important lab continues today at the MEEI.

In general, the development of CBG created an interactive environment that included mathematical modeling and auditory research in neurophysiology and perception, areas of deep interest to Walter Rosenblith. Before coming to MIT, Rosenblith did research at the Harvard Psychology Department, working in the Psycho-Acoustical Laboratory (PAL) with S.S. Stevens (the founder of PAL) and with J.C.R. Licklider and future Nobel Prize recipient Georg von Békésy. Rosenblith worked on the impact of noise on hearing and then on physiology and psychoacoustics. His work concerned "how sounds affect the brain," and he famously said, "I got sucked into the brain through the ear" (Gray 2010). By developing the CBG at RLE and bringing an interesting collection of researchers to MIT, Rosenblith stimulated interactions that provided a blending of insights from many areas of focus. Rosenblith was the mentor of several young researchers at MIT who became important contributors to MIT and EPL; these included William Peake, who mentored Moise Goldstein before he left MIT for Johns Hopkins University (JHU), and Murray Sachs, who also went on to become a faculty member at JHU after his MS with Goldstein and then PhD with Kiang. Sachs recounted the research and interactions at EPL and CBG in his interview with the IEEE History Center (http://bit.ly/3KwsOwz).

Jerome Lettvin (1920–2011) came to MIT from Chicago, along with Lettvin's Chicago colleagues Warren Sturgis McCulloch and Walter Pitts in 1952. These researchers were attracted to MIT in part by Wiener and his work in cybernetics, although they did not maintain ongoing collaborations with Wiener. Their work was an interesting combination of empirical measurements and modeling. The 1943 paper by McCulloch and Pitts (1943)—"A Logical Calculus of the Ideas Immanent in Nervous Activity"—was an early example of cybernetics analysis and an excellent example of neural network modeling that considers the brain as a logical machine.

Lettvin pursued neural measurements and associated analysis in the optic nerve of frog within RLE. In contrast to the logical mathematics of McCulloch and Pitts, Lettvin suspected that logical machines might not provide useful models for perception. In an interview given in 1994 ("Talking Nets: An Oral History of Neural Networks," edited by Anderson and Rosenfeld 2000), he described his then intuition, "Everything I had found so far and had read about suggested that the world was reported by natural language rather than logical language, that is, in terms of things and their relations rather than sense data to be processed into the concepts of things" (page 12).

Lettvin found evidence to support his intuition through thoughtful analyses of neural responses in studies of the optic nerve and of more central nuclei, optic tectum, of the frog, as described in his classic paper with several colleagues, "What the

Frog's Eye Tells the Frog's Brain" (Lettvin et al. 1959). They found four types of optic fibers and each processes a different type of visual patterns such as moving edges that can be found in the frog's visual world. These four sets of encoded feature maps by optic nerves project in parallel onto the tectum at different depths; thus, all four maps are in registration by the brain. Their important finding suggests that the processing in the frog's retina gives the content of perception to a frog, and the frog's retina is an analog, not logical, system.

Inspired by this work, Frishkopf, Capranica, and Goldstein at MIT studied frog's ears, "Neural coding in the bullfrog's auditory system, a teleological approach" (Frishkopf et al. 1968). They found that the responses of auditory nerve fibers in frogs encode the spectral and temporal information of mating calls of frogs. Together, these two frog studies demonstrate that sensory experience affects the function of the nervous system, starting with the retina and auditory nerves. Their work was instrumental in the development of the frog model in the field of neuroethology (see Narins et al., Chap. 4).

Both quantitative and analytical areas of auditory research at MIT were stimulated by the formation of CBG within RLE, and researchers in CBG interacted with researchers in related areas. These include EPL and the Speech Communications Group in RLE. Specifically, Kenneth Stevens and Noam Chomsky were both members of RLE and were part of the overall interactions on topics of auditory processing in the speech domain. Stevens was head of the Speech Communications group in RLE for many years, and the techniques applied in CBG to sensory processing were also applied to speech studies by Stevens.

These developments at MIT were also summarized by Thomaso Poggio, who was in the Department of Brain and Cognitive Sciences at MIT. In a blog on *Times of Israel* on June 2, 2012 (http://bit.ly/40xJgmC), Poggio says, "Back in the 1950s and '60s, vigorous intellectual activity developed around the new engineering and physics of electronics and information processing. MIT was a major Institution contributing to this expanding whirlwind of ideas. Building 20 was at its center, the focus of such diverse fields as information theory, cybernetics, neural networks, linguistics, neuroscience, and computer science. Researchers roaming the building included Claude Shannon, Norbert Wiener, Warren McCulloch, Walter Pitts, Noam Chomsky, Jerry Lettvin, and Marvin Minsky. The intellectual ferment eventually ignited an ambitious attempt to understand intelligence and replicate it in machines."

As specific auditory examples of general mathematical formulations of decision making based on elements of information theory and decision theory from Wiener and Shannon, we consider the work of William Siebert (1925–2015) and Nathaniel Durlach (1927–2016), both of whom came to MIT's RLE in the early 1960s; they came from separate groups at MIT's Lincoln Laboratory, a laboratory organized independently of the academic structure of MIT.

Models based on cybernetics and information theory predict auditory performance abilities with mathematical operations on the input waveforms, which can be considered to be the patterns of activity of the auditory-nerve fibers. Following the assumptions of decision theory and information theory, the decision alternatives are typically a finite list, such as a choice of Stimulus A versus Stimulus B, and the models naturally predict performance abilities in pairwise discrimination tasks or in masked detection tasks. The computational processing of the waveforms leading to the behavioral judgments could be considered a model of internal auditory processing, including the processing of the CAS. Thus, these models can be considered models of central auditory processing and may suggest operations that can be tested physiologically and that can be used to predict the relationship of performances in different experiments.

Siebert's work in the 1960s was fundamentally an integrative study that quantitatively connected physiological data at the level of single primary neurons to the perceptual abilities of the fully functioning organism. His paper on optimal processing of stochastic neural patterns in the auditory nerve (Siebert 1965) compares predicted performance with observed frequency- and intensity-discrimination abilities. This work is an auditory example of the general principles discussed by Wiener and Shannon in the 1940s, and it provides an early example of the modeling of perceptual abilities based on the statistical properties of neural activity as measured by Kiang at EPL (Kiang 1965). This work was possible with neural data from peripheral auditory nerves because of the extensive amounts of detailed data gathered and made available by Nelson Kiang. Specifically, in Siebert's 1965 paper, the auditory nerve firing patterns were explicitly modeled as stochastic point processes, and decision processes based on the generated patterns were formulated with the ideas from cybernetics and information theory, particularly the analysis methods of statistical decision theory. The form of the processing depends on the task, of course, so that for some tasks the spike count is key, for others the temporal patterns of the spikes, and for others the distribution of activity levels across the neural population is critical. This approach is easily generalized, so for example, binaural tasks involve comparisons of right and left neural patterns, and can be formulated in similar ways. Examples of this basic approach applied to the processing of patterns for binaural decisions can be found in Colburn's papers (e.g., Colburn 1973, 1977). It is notable that the optimum processing algorithm, subject to some constraints, leads to a processor that can be implemented with the Jeffress coincidence mechanisms, as documented by Goldberg and Brown (1969).

The mathematical analysis of decision-making can also be applied to psychophysical experiments in hearing (psychoacoustics) with imperfect processing or internal noise included, which leads to limitations in performance that can be compared with human performance. This kind of modeling attempts to understand performance in a variety of psychophysical tasks and stimulus parameters based on a few assumptions about what processing is available internally. This approach is illustrated by the work of Durlach in this general area in the 1960s, when he moved to CBG from Lincoln Lab. Durlach's work included rigorous modeling and experimental studies of hearing, with special focus on binaural detection and intensity discrimination.

One of Durlach's most important and most quoted papers was written in 1963, about the time that he was moving from Lincoln Laboratories to RLE at MIT. This paper, "Equalization and cancellation theory of binaural masking-level differences" (Durlach 1963), introduced the "EC model," which is still the basis of much current modeling of source isolation and of intelligibility benefits of binaural listening. EC stands for equalization–cancelation; fundamentally, the model hypothesizes internal processing of the left and right input waveforms, corrupted by internal noise, in order to equalize and then cancel (by subtraction) interfering sound components. When the masking sound and the target sound come from different directions, they will have different ITDs and ILDs; this allows the cancelation of the masker components without the cancelation of the target components. This model has been successful in predicting a large amount of psychophysical data, for experiments predicting binaural benefits in detection, intelligibility, and localization. Durlach was an important advisor to many students, including Colburn and Louis Braida at MIT, and Durlach remained involved in this research and was a major contributor to auditory psychophysical research and modeling from the early 1960s until his death in 2016.

In roughly the same time interval, the 1960s and 1970s, David M. Green wrote his influential book with John Swets (Green and Swets 1966), which laid out the way that signal detection theory can be applied to describe psychophysical experiments, especially detection and discrimination (Yost et al. 2021). As noted above, Green was on the faculty at Harvard University from 1973 to 1985 (Chair of Psychology 1978–1981). In the words of William A. Yost (2015): "*Signal Detection Theory and Psychophysics* by Green and Swets (1974/1966) was essentially 'required reading' for any aspiring psychoacoustician in the 1960s and 1970s" (page 50).

The discussion in this section has emphasized the combined research activities that took place in the Boston area, including theoretical modeling, physiological measurements, and psychophysical measurements. This discussion is not intended to minimize the excellent work in other areas, including Europe, and the reader is encouraged to explore the work of Eberhard Zwicker, for example. *Psychoacoustics: Facts and Models* (Fastl and Zwicker 2007) (first edition published in 1990) summarized important works done in Zwicker's labs at the Institute of Telecommunications in Stuttgart and at the Institute of Electroacoustics in Munich. Also, the work of Jens Blauert, especially his book, *Spatial Hearing* (Blauert 1997) (published first in German as Räumliches Hören in 1974), provides an excellent review of the psychophysics of human sound localization.

9.4 Cortical Processing of Sounds

9.4.1 Functional Anatomy, Physiology, and Behavioral Roles of Auditory Cortex

It is notable that the role of central processing in the work described in the previous section was generally taken to be optimal processing, using all of the relevant information from the brainstem nuclei. Explicit modeling of the cortical neurons was not necessary to obtain quantitative predictions for performance. The simple model of ideal central processing was considered adequate at the time in part because the tasks that were being modeled, such as the precise location of a known stimulus in the horizontal plane of a known environment, do not require complex processing of the brainstem outputs. But it was expected that more sophisticated tasks would necessitate the involvement of cortical processing.

Between 1940 and 1970, new insights into the structure and function of the auditory cortex were acquired through cortical mapping and ablation techniques in various species, primarily in cats. The two most active research groups were led by Clinton Woolsey and Jerzy Rose at University of Wisconsin and by Dewy Neff at University of Chicago. Their studies concerned (1) the architectonic borders and tonotopy of the auditory cortex and (2) its role in auditory detection, discrimination, and localization tasks. The work of Rose, Woolsey, and Neff had a lasting impact on the growth of auditory neuroscience in the USA over the last half century. One of their shared contributions to science is the people they have trained; some of them became influential intellectuals in their own fields (Nelson Kiang, Jay Goldberg, Irving Diamond, Mike Merzenich, Jon Kaas, to name just a few).

Granted, the research community has gained more detailed knowledge of cortical anatomy and physiology through the rapid development and expansion of neuroscience technology since the 1960s. This section attempts to highlight the "thinking" part of their influences; their speculations on the structure–function of the auditory cortex remain relevant to current research.

Woolsey (1904–1993) was a pioneer in applying functional mapping techniques to understanding the structure of the neocortex. He worked with Phillip Bard at JHU, where Bard was the Director of the Department of Physiology, before accepting a professorship position at University of Wisconsin in Madison in 1948. Together with another colleague, Wade Marshall, they obtained the first cortical representations of body parts in the somatosensory cortex using the new evoked potential techniques (Marshall et al. 1937). Woolsey soon extended these techniques to mapping the auditory and visual cortices.

In the 1940s, it was known that different tone frequencies evoke responses in different parts of the cochlea, but the functional arrangements of the auditory cortex with respect to frequency representation remained unknown. Woolsey (1942) observed that highly localized evoked potentials in the cerebral cortex occurred when selectively stimulating small bundles of cochlear nerve fibers in cats. They defined two auditory regions below the suprasylvian sulcus. In the first region, the responses evoked by stimulating the basal (high-frequency) nerve fibers were found in the rostral areas of the ectosylvian cortex, and those by stimulating the apical (low-frequency) fibers appeared more caudally. In the second region, lying immediately ventral to the first, the order of cochlear representation was reversed. Their work was the first clear demonstration of multiple tonotopic representations in the auditory cortex.

The borders of these two auditory regions (named AI and AII) along with the surrounding areas were further studied by Rose based on Nissl cytoarchitectonic structure (Rose 1949), a technique that dominated the field at the time for parcellation of the cortex based on cell-type and cell-density distributions across different layers of the neocortex. Rose (1949) summarized his findings as follows:

> This region consists of a central field (first auditory area) possessing a characteristic structure of its own and a peripheral belt which represents a transition zone. The first auditory field is characterized by general smallness of cells, dense cell population in granular and supragranular layers, small variations in cell size, and marked blurring of lamination. The peripheral belt which possesses some characteristics of the first auditory field and some of the other adjoining fields is separated into three sectors: (1) the suprasylvian fringe sector, (2) the second auditory area, and (3) the posterior ectosylvian area (pp 426–427).

This cytoarchitectonic division of the auditory cortex was a milestone finding, as it helped guide many lesion studies at the time to reveal the thalamic projection patterns and to discern the role of the auditory cortex in behavioral tasks. Rose and Woolsey (1949) used the method of retrograde degeneration (which used retrograde thalamic atrophy as a means of studying functional anatomy) to reveal the thalamocortical connections of the auditory region of the cerebral cortex in the cat. As a major contribution to auditory research, their work introduced the concept of essential and sustaining projections. That is, the thalamocortical connections could either be restricted (essential projections) or distributed more widely to several functional and cytoarchitectonic areas (sustained projections).

Figure 9.5 shows a well-known layout of the Rose and Woolsey's auditory regions in the cat neocortex (Woolsey 1960). It is worth noting that in the 1950s, Woolsey and other researchers had already identified auditory evoked responses outside the auditory regions including the pre-motor and visual cortices. This observation did not go unnoticed and promoted heated discussions between Rose, Woolsey, Tunturi, and Galambos on the inclusion and exclusion criteria for defining an auditory field during the Symposium on Neural Mechanisms of the Auditory and Vestibular System in 1960 (Rasmussen and Windle 1960).

A bit more on Rose and his colleagues is relevant here. Rose (1909–1992) was an awe-inspiring name in many memoirs from this era. In the 1950s, like Woolsey, Rose was also a faculty member of the Department of Physiology at JHU, and in 1959, he moved to the University of Wisconsin and joined Woolsey. At JHU, Rose was considered the "resident genius" of the department (Kruger 1993). In those days, making microelectrodes with a tiny tip of a few microns was considered a work of art in the pursuit of stable recordings of single-unit activities. The techniques of electrode-making were honed through collaborations among a few labs. The lab of Galambos at the PAL at Harvard and at the Walter Reed National Military Center was one of those having this technique.

Galambos (1914–2010) was a great force in auditory research over a wide range of topics including bat echolocation, single-unit electrophysiology, and human auditory systems. His recounts of electrophysiology experiences reflect the unbiased investigation style of neurophysiologists from this great pioneer generation (Galambos 1998a).

> My plan was simple. The cats and I would converse, with me asking the questions by delivering clicks and tones to their eardrums, and they replying, one brain cell at a time, through a microelectrode. No theory, no preconceptions; just simple experimental facts (page 195).

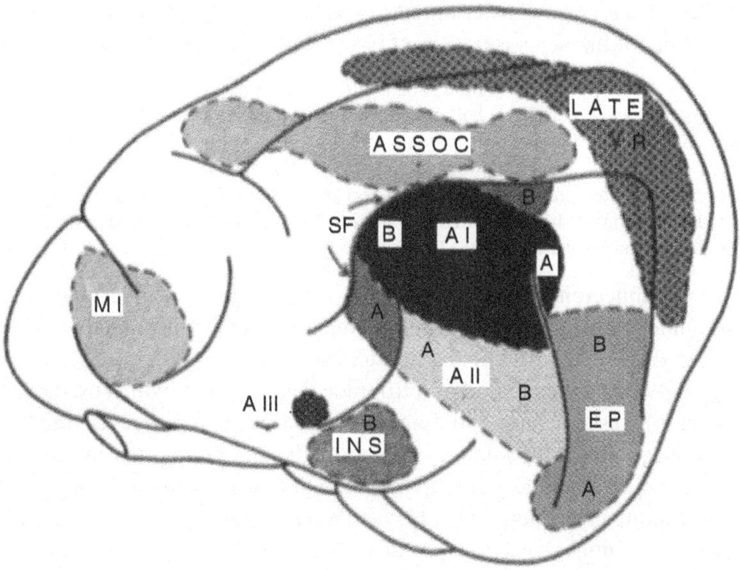

Fig. 9.5 Auditory cortical regions. Original figure legend with minor modifications for Fig. 128 in Woolsey (1960). Summary diagram showing four central areas with cochlea represented antero-posteriorly from apex (*A*) to base (*B*) in the suprasylvian fringe sector (*SF*), from base to apex in *AI*; from apex to base in *AII*. In posterior ectosylvian area (*Ep*), representation is base above, apex below. In insula (*INS*), evidence suggests base represented above, apex below. *AIII* is Tunturi's third auditory areas. "Association" cortex (*ASSOC*) and precentral motor cortex gave responses to click with 15-msec latencies under chloralose. Visual area II (*LATE*) gave responses with 100-ms latency, also under cholralose. (The resource is in the public domain)

Rose worked with Galambos to study the medial geniculate body (Galambos et al. 1952; Rose and Galambos 1952) and cochlear nucleus (Rose 1959) using microelectrodes. According to Galambos, Rose liked the electrodes they used and carried them back to Baltimore, scotch-taped inside the rear window of his car. Together with Robert Dowben, a Hopkins physical chemist, they devised the low-impedance electrodes by filling a micropipette with molten indium and coating it with gold and platinum (Dowben and Rose 1953). As it was said, "whereupon the famous Dowben-Rose metal version and the Johns Hopkins laboratory entered the single unit business" (Galambos 1998a, page 196).

Looping back to Rose-Woolsey's auditory cortical maps, in the early twentieth century, brain ablation was often used as a research tool for understanding how specific brain regions affect behavior. The correlational evidence found in behavioral deficits was used to infer the functions of the removed areas in the brain. The delineation of the core region of the auditory cortex and surrounding (belt) areas in the neocortex provides a guide for making brain lesions to study auditory cortical functions.

Neff (1912–2002) contributed to the research community a behavioral–anatomical–electrophysiological paradigm for testing the effects of brain ablation on animals' abilities to detect and discriminate stimulus features and to localize a sound source. His work mainly focused on the auditory cortex in cats, but expanded to several subcortical areas. Neff joined the faculty of the Department of Psychology at the University of Chicago in 1946. After a brief stay at Bolt Beranek and Newman Inc. (BBN), a prestigious company in architectural acoustics, Neff moved to the University of Indiana at Bloomington in 1963. Soon, he was made the director of the newly formed Center for Neural Sciences.

Neff, together with Irving Diamond and John H. Casseday, who studied with Neff at the University of Chicago and University of Indiana, respectively, wrote a comprehensive summary of the behavioral work in the auditory system (Neff et al. 1975). This classical read contains a historical review of the developments of prerequisites, such as improved aseptic surgical procedures in the 1940s, for conducting behavioral work in lesioned animals. More importantly, they provided a comparative analysis of the extent of behavioral deficits among cat, monkey, and human due to either cortical or subcortical lesions. They argued that the major goals that are important to behavioral studies of the auditory system are "to identify the special contribution of each part of the whole" and "to identify features of the stimulus that excited different neural units." As such, "at the organ level, anatomical, electrophysiological and behavioral studies are all aspect of a single structure-function inquiry." They were forthright about the limitations of the behavior-ablation method:

> The behavior-ablation method must not only be content with an incomplete picture of the auditory system; it is also limited by the fact that a deficit after ablation does not reflect simply a function of the ablated part. On the contrary, we see, in an animal deprived of a part, the functioning of the remaining system, and we recognize that each of the remaining parts is itself more or less disturbed by the ablation (Neff et al. 1975, page 311).

In this context, Neff and his colleagues addressed the importance of the intact auditory cortex in a wide range of detection, discrimination, and source-localization tasks based on undisturbed, relearned, or unlearned animal behaviors. They found that bilateral cortical ablations can lead to behavioral deficits in temporal pattern discrimination (Diamond and Neff 1957; Goldberg et al. 1957) and stimulus duration discrimination (Scharlock et al. 1965); however, the animal's ability to detect the onset of a sound or changes in frequency (Goldberg and Neff 1961) and intensity (Neff 1961) remained unaffected. Neff proposed a neural model to explain this difference (Neff 1961). The model is largely rooted in the place-code theory by comparing spatial and temporal patterns of neural responses across different neural units. The model posits that cortical deficits are associated with the dependence of recruiting "new units" for discriminating two stimuli.

Figure 9.6 shows the hypothesized model responses in four different behavioral tasks (Neff 1961). Neff speculated that, for each of the discriminations (onset, intensity, and frequency) that can be made after lesion, "the positive stimulus produces excitation in new neural units, that is to say, the units that were not excited by the negative stimulus. In the pattern discrimination, on the other hand, the positive stimulus and the negative stimulus excite the same neural units, and the difference is in

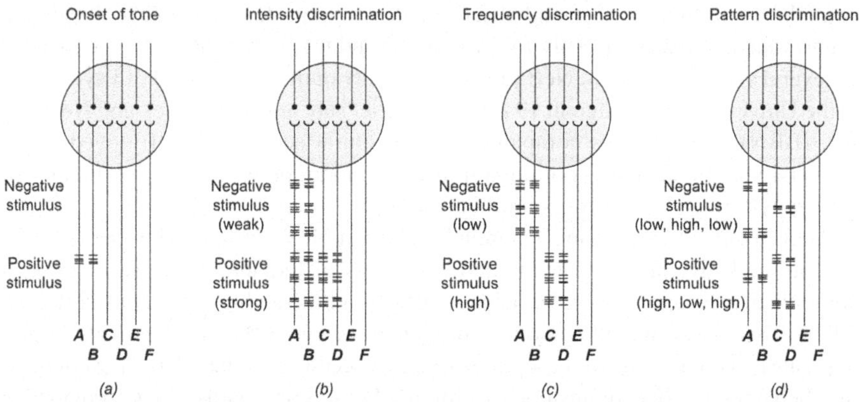

Fig. 9.6 Neff's neural model for auditory discrimination. Based on Neff (1961), Fig. 9, "Diagram of response to positive and negative stimuli." The model depicts the temporal events of responses of neural units in the afferent auditory system (labeled as *A*, *B*, *C*, *D*, *E*, *F*) in responses to negative (or control) and positive (or test) stimuli over time (labeled as vertical lines) in four different tasks (**a**) onset of tone detection, (**b**) intensity discrimination, (**c**) frequency discrimination, and (**d**) pattern discrimination. In tasks ***a***, ***b***, ***c***, positive stimuli evoke responses from "new" units that do not respond to negative stimuli, whereas in task ***d***, positive and negative stimuli evoke responses in the same units (*A*, *B*, *C*, *D*)

the order or sequence of excitation or in the total amount of excitation produced in certain units." (Neff 1961, page 273). He extended this analysis to explain the deficits in duration discrimination after ablation to support this model.

Neff and his students also developed a behavioral procedure for testing cats' abilities to localize a sound source by walking to the correct sound source location to obtain food rewards (Neff et al. 1956). This setup has been adopted into automated versions for animal behavioral work and is still widely used. In a series of studies (Neff 1961; Neff et al. 1975), they found that bilateral cortical ablation of core and belt areas resulted in a severe deficit in the ability of cats to walk to the source of a sound in a left-right, two-choice task. Several extensions of these experiments were designed to address whether the cortical deficits are sensory, motor, or associative in nature because sound source localization is not purely an auditory discrimination task.

Ravizza and Diamond (1974) tested the localization performances of two nocturnal mammalian species, hedgehogs (Erinaceinae) and bushbabies (Galagidae). They found that bilateral cortical ablation caused greater deficits for localizing short-duration stimuli than long-duration stimuli. They creatively tested whether this deficit was caused by deprived sensory information (with short stimuli) by modifying the procedure used for long-duration stimuli. Instead of walking right after stimulus onset, animals were held in a holding box listening to stimuli and allowed to approach the target location only after a stimulus was turned off. While animals performed normally if allowed to approach the sound source during sound presentation, severe deficits were found in the new condition. As they put it, many factors are involved in this seemingly simple listen-hold-go task. The requirements of the tasks contain three parts—identify the locus of the sound source, store this spatial information, and move toward the sound source. They suggest that "the intact mammal *knows* where its body is relative to the objects of the environment, *knows* the position of the objects relative to each other and the meaning of 'knows' implies that the organism can manipulate these objects including changes its own position relative to them. It may be just this high level of integration that is disrupted by removal of auditory cortex" (Ravizza and Diamond 1974, page 1919). As current auditory research continues to pursue the organizing principles of the auditory cortex, the views held by Neff, Diamond, and their colleagues are extremely valuable for understanding the role of working memory, sensorimotor interaction, and associative functions of the auditory cortex, especially for tasks requiring active sensing in a closed-loop nervous system.

9.4.2 Single-Unit Activity and Neural Encoding of Sounds in Auditory Cortex

Auditory neurophysiology carries a strong tradition of applying a quantitative approach to study the detailed response patterns of auditory neurons collected at various stages along the ascending auditory pathway. Empowered with fine stimulus control and single-unit recording techniques, quantitative neurophysiology allowed a deeper understanding of how individual neurons encode sound information, such as frequency and intensity over time in the auditory cortex. The work by Moise Goldstein and his group demonstrated a weak representation of columnar organization while raising the issue of heterogeneity in the auditory cortex.

One of the major achievements of sensory neuroscience in the twentieth century was the discovery of columnar organization as the essential building block of the neocortex. Lorente de Nó (1938) first proposed that there were functional columns in the sensory cortical areas based on the observation that cortical cells are organized in vertical (radial) chains. The electrophysiological confirmation of the columnar organization was found in the somatosensory cortex (Mountcastle 1957) and visual cortex (Hubel and Wiesel 1962). Abeles and Goldstein (1970) applied quantitative measures to seek the columnar organizations in the auditory cortex of cat. They compared data from radial and off-radial penetrations of the cortical surface, where radial penetrations were made at angles normal to the cortical surface. Their major result is an organization along the radial chains of neurons in which narrowly tuned units are located according to the best frequencies, but broadly tuned and multipeaked units do not reveal consistent clustering patterns across depth.

Goldstein and colleagues also provided strong evidence on the extent of variability of single-unit responses across the cortical depth. Early cortical studies (Erulkar et al. 1956) showed that anesthetics have large effects on sensitivity and patterns of responses of single units in the primary auditory cortex (AI), where a sizable fraction of units in anesthetized cat's AI were unresponsive to acoustic stimulation. With this knowledge, Goldstein and colleagues made recordings in AI of unanesthetized, muscle-relaxed cats. They showed that the portion of unresponsive units was below 5%, about 20% of units exhibited labile response properties, and the majority of cortical neurons ("the other 80%") showed repeatable responses to pure tone, clicks, or noises (Goldstein et al. 1968). It might appear a bit puzzling to see that these simple acoustic stimuli with limited spectral and temporal features could still evoke strong and reliable responses in the cortex as in the auditory periphery. However, unlike the homogenous firing patterns of auditory nerve fibers, cortical neurons do not respond at the same time as the stimulus is turned on and off. While some units responded at the onset or offset of a stimulus, others were observed to respond throughout the stimulation duration. Described as "on," "off," "through," or "on-off" patterns, Abeles (who was a postdoc at JHU and later the creator of the synfire chain theory—Abeles 1991) and Goldstein found that the temporal patterns of the same unit could change when varying stimulus parameters in frequency and intensity and when additional sound was added by using a two-tone interaction method (Abeles and Goldstein 1972). This suggests that *pattern detection* (i.e., detecting tone patterns shown in Neff's neural model; see Fig. 9.6) is potentially manifested by neural response patterns in the auditory cortex. Their assessment was made based on single-unit responses to a pair of tone bursts with varying frequency, intensity, and timing relationships.

Figure 9.7 shows the responses of one narrowly tuned and one broadly tuned unit to a pair of partially overlapping tone bursts (data taken from Abeles and Goldstein (1972)). For both units, the leading conditioning tone suppresses unit response to the best frequency (BF) tone over either a narrow frequency range (top panel, Fig. 9.7) or a broad frequency range (bottom panel, Fig. 9.7) centered on BFs. Nevertheless, for both units, different temporal patterns of responses are observed for different frequency combinations of the tone pair.

Abeles and Goldstein (1972) also identified a subpopulation of units with multi-peaked frequency tuning ranges. These "multirange units" showed summation responses when stimuli covered the different frequency ranges, a precursor for combination sensitivity. They stated, "It is in the detection of frequency transitions that we have our only evidence of a special selectivity at the cortical level. Suppose we have two tone bursts occurring sequentially with the time between bursts small. A multirange unit responding to the first burst will fail to respond to the second if the frequencies of both bursts are close (within the same range) but will respond vigorously if the frequencies of the bursts are in different ranges" (Abeles and Goldstein 1972, page 350).

The observation of diverse response patterns of single units made Goldstein believe that the auditory cortex might implement different computational principles in encoding sensory information than other sensory cortices. But he recognized that the one-unit-at-a-time approach has severe limitations in obtaining a systematic picture of the functional aspects of the cortical population. "The response patterns of cortical neurons are so diverse that it is inappropriate to consider them as a population in the sense that physicists and mathematicians treat a group of elements with

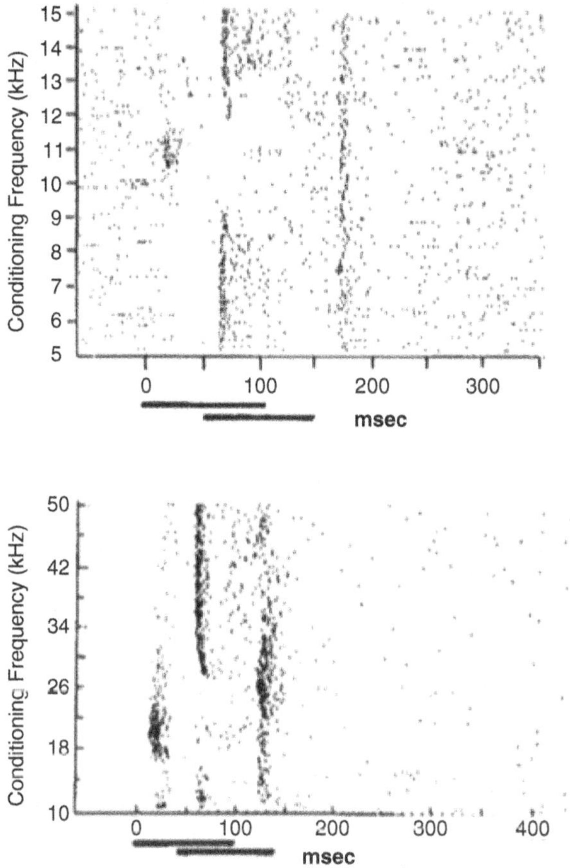

Fig. 9.7 Single-unit response to a pair of partially overlapping tone bursts. This figure is combined from Fig. 4B and Fig. 8C in Abeles and Goldstein (1972). Top panel shows a narrowly tuned unit; bottom panel shows a broadly tuned unit. For both units, the leading conditioning tone had varying frequencies (top, from 5 to 15 kHz; bottom, from 10 to 50 kHz); the lagging test tone had a fixed frequency at the unit of best frequency (top, 10.7 kHz; bottom, 22 kHz) and the onset of the lagging tone was delayed relative to that of the leading tone (top, 50 ms; bottom, 40 ms). The conditioning and test tones each had a duration of 100 ms. The results show that the responses to the test tone is suppressed by conditioning frequencies over a various range. (Reprinted from Abeles and Goldstein 1972, with permission)

homogenous properties. We are more in the position of the social scientist who is investigating a population of individuals" (Goldstein and Abeles 1975).

What is learned from these early investigations is that the auditory cortex seems to stand apart from the visual and somatosensory cortices and also from subcortical auditory structures. The auditory cortex contains less organized selectivity maps, and auditory cortical neurons show more diverse response patterns. To understand the computational goal of the auditory cortex, which remains a longstanding and unresolved question, the early investigations on the response characteristics of the auditory cortex deserve ongoing attention.

9.5 Concluding Remarks

Our abilities to navigate in loud environments, appreciate music, and communicate thoughts through speech, especially in multisource environments, are tightly coupled to the ways in which the brain processes sounds. For millennia, the sense of hearing has attracted a variety of scientists to study sensory processing in both humans and animals. This chapter reviews topics that were selected to reflect the work of individuals who have shaped our field in understanding central auditory mechanisms. We hope that this review can provide insights into how certain research topics have helped establish a trajectory of scientific inquiries into what we now regard as classical studies and traditional approaches. What we intend to emphasize is the "tradition" in scientific practice that holds time-tested values in experimentation and theory generation. After all, the battlefield has not changed much between the beasts of brain puzzle challenges and the conquerors with their weapons of scientific investigation. The most fundamental aspects of science—specifying facts and validating theories—are also not changed. What has changed since the 1960s, however, is technology. New knowledge in biology has propelled scientific inquiries to cellular, molecular, and genetic levels of mechanisms. New biomaterials, advanced imaging, and large-scale recording technology have empowered scientists with new tools to capture brain activities with improved stability and precision. These advances yield new types of data, which seem to make understanding the brain (including the auditory part), finally an attainable goal.

As work on central auditory processing progresses, we believe that it is important to develop experiments and theories that can deal with complex interactions between internal states of the brain and external acoustic environments including multiple sources. We must also explicitly include combined processing that incorporates the challenges of processing the sounds of an immediate impact, in which case the organism needs to detect, discriminate, and locate the signals, make decisions, plan a response, and execute it. In natural environments with multiple sound sources, source localization relies on correctly sorting out the order of events originating from each individual source. This task of pattern recognition can be complicated by internal states of the brain (intention, attention, expectation, memory, etc.), especially when they are intertwined with temporal dimensions of stimuli and when behaviors can reset the internal states through feedback signals.

Thus, the overall processing of auditory signals must include complex mechanisms to allow the central intelligence to focus on sources and content critical for local decision-making and interpretation. These mechanisms would naturally include not only feedback to the peripheral processing stages (such as the brainstem nuclei), but more importantly mechanisms for focusing the CAS processing on stimuli that are relevant for immediate interpretation or decisions based on acoustic inputs. These factors must all be combined in order to understand the most central aspects of auditory processing. With this in mind, we consider it important to continue the development of the mathematics of decision theory and information theory as described in Sect. 9.3.

As we consider these complex demands on auditory processing, it becomes apparent that our understanding of the complex processing of information in the real world is still at an early stage. This processing almost certainly affects the organization of the neural processing, including efferent mechanisms, but has not been studied in much detail and is not yet understood, even at a gross level. Nevertheless, this processing allows a focus on relevant signals and aspects of signals within complex acoustic environments and may help us to understand the structure of the auditory pathways, including feedback (efferent as well as afferent), and may be important for our understanding of the neural pathways and complex observations. Work in this area is still very primitive, but is proceeding in ongoing studies.

Acknowledgments The authors acknowledge the many useful inputs from the editors; we appreciate the constructive style with which their suggestions were provided. We also want to thank several colleagues who were willing to read earlier versions of the manuscript in order to provide large-scale suggestions about content and structure, as well as some detailed recommendations. We specifically thank Eric Young and Theo Goverts for their detailed readings and suggestions. The time spent on this work was supported by NIH NIDCD (Grant R01 DC000100 for H. Steven Colburn and R01 DC019278 for Yi Zhou).

Compliance with Ethics Statement: H. Steven Colburn declares that he has no competing interest. Yi Zhou declares that she has no competing interest.

References

Abeles M (1991) Corticonics: neural circuits of the cerebral cortex. Cambridge University Press, Cambridge

Abeles M, Goldstein MH (1970) Functional architecture in cat primary auditory cortex: columnar organization and organization according to depth. J Neurophysiol 33(1):172–187

Abeles M, Goldstein MH (1972) Responses of single units in the primary auditory cortex of the cat to tones and to tone pairs. Brain Res 42:337–352

Acar F, Naderi S, Guvencer M, Türe U, Arda MN (2005) Herophilus of Chalcedon: a pioneer in neuroscience. Neurosurgery 56(4):861–867

Anderson JA, Rosenfeld E (2000) Talking nets: an oral history of neural networks. MIT Press, Cambridge

Aristotle (1973) De Anima, Book II. In: McKeon RP (ed) Introduction to Aristotle. University of Chicago Press, Chicago

Barnes WT, Magoun HW, Ranson SW (1943) The ascending auditory pathway in the brain stem of the monkey. J Comp Neurol 79(1):129–152

Blauert J (1997) Spatial hearing: the psychophysics of human sound localization. MIT Press, Cambridge

Brand A, Behrend O, Marquardt T, McAlpine D, Grothe B (2002) Precise inhibition is essential for microsecond interaural time difference coding. Nature 417(6888):543–547

Cajal SR (1911) Histologie du système nerveux de l'homme et des vertébrés, vol 1. Maloine. English edition: Cajal SR (1995) Histology of the Nervous System of Man and Vertebrates (trans: Swanson N and Swanson, LW). Oxford University Press, Paris

Cajal SR (1959) Degeneration & regeneration of the nervous system, vol 1. (Trans and Ed: May RM). Hafner Publishing Company, New York

Cajal SR (1989) Recollections of my life, vol 8. (trans: Craigie EH and Cano J). MIT Press, Cambridge

Colburn HS (1973) Theory of binaural interaction based on auditory-nerve data. I. General strategy and preliminary results on interaural discrimination. J Acoust Soc Am 54(6):1458–1470

Colburn HS (1977) Theory of binaural interaction based on auditory-nerve data. II. Detection of tones in noise. J Acoust Soc Am 61(2):525–533

Davis PA (1939) Effects of acoustic stimuli on the waking human brain. J Neurophysiol 2(6):494–499

Davis H, Davis PA, Loomis AL, Harvey EN, Hobart G (1939) Electrical reactions of the human brain to auditory stimulation during sleep. J Neurophysiol 2(6):500–514

Diamond IT, Neff W (1957) Ablation of temporal cortex and discrimination of auditory patterns. J Neurophysiol 20(3):300–315

Dobson JF (1925) Herophilus of Alexandria. Proc R Soc Med (18):19–32

Dobson JF (1927) Erasistratus. Proc R Soc Med 20:825–832

Domnitz R, Colburn H (1977) Lateral position and interaural discrimination. J Acoust Soc Am 61(6):1586–1598

Dooling RJ, Prior NH (2017) Do we hear what birds hear in birdsong? Anim Behav 124: 283–289. doi:10.1016/j.anbehav.2016.10.012

Dowben RM, Rose JE (1953) A metal-filled microelectrode. Science 118(3053):22–24

Durlach NI (1963) Equalization and cancellation theory of binaural masking-level differences. J Acoust Soc Am 35(8):1206–1218

Elhadi AM, Kalb S, Perez-Orribo L, Little AS, Spetzler RF, Preul MC (2012) The journey of discovering skull base anatomy in ancient Egypt and the special influence of Alexandria. Neurosurg Focus 33(2):E2

Erulkar S, Rose J, Davies P (1956) Single unit activity in the auditory cortex of the cat. Bull Johns Hopkins Hosp 99(2):55–86

Evans E (1975) Cochlear nerve and cochlear nucleus. In: Keidel WD, Neff WD (eds) Handbook of sensory physiology, Auditory system: physiology (CNS) Behavioral studies, psychoacoustics, vol V/2. Springer, Heidelberg, pp 1–108

Fastl H, Zwicker E (2007) Psychoacoustics: facts and models. Springer, Berlin Heidelberg

Frishkopf LS, Capranica RR, Goldstein M (1968) Neural coding in the bullfrog's auditory system a teleological approach. Proc IEEE 56(6):969–980

Galambos RC (1998a) Robert Galambos. In: Squire LR (ed) The history of neuroscience in autobiography, vol 1. Oxford University Press, pp 178–220

Galambos RC (1998b) Hallowell Davis 1896–1992. Biogr Mem Natl Acad Sci 75:117–137

Galambos RC, Davis H (1943) The response of single auditory-nerve fibers to acoustic stimulation. J Neurophysiol 6(1):39–57

Galambos RC, Rose JE, Bromiley RB, Hughes JR (1952) Microelectrode studies on medial geniculate body of cat: ii. Response to clicks. J Neurophysiol 15(5):359–380

Garrett SL (2021) One-hundred years of English-language acoustics textbooks. Acoust Today 17(3):22–30

Gerstein GL, Kiang NY-S (1960) An approach to the quantitative analysis of electrophysiological data from single neurons. Biophys J 1(1):15–28

Goldberg JM, Brown PB (1968) Functional organization of the dog superior olivary complex: an anatomical and electrophysiological study. J Neurophysiol 31(4):639–656

Goldberg JM, Brown PB (1969) Response of binaural neurons of dog superior olivary complex to dichotic tonal stimuli: some physiological mechanisms of sound localization. J Neurophysiol 32(4):613–636

Goldberg JM, Neff W (1961) Frequency discrimination after bilateral ablation of cortical auditory areas. J Neurophysiol 24(2):119–128
Goldberg JM, Diamond I, Neff W (1957) Auditory discrimination after ablation of temporal and insular cortex in cat. Fed Proc 16:204–205
Goldstein MH, Abeles M (1975) Single unit activity of the auditory cortex. In: Handbook of sensory physiology: auditory system: physiology (CNS) Behavioral studies, psychoacoustics, vol V/2. Springer, Heidelberg, pp 199–218
Goldstein MH, Hall JL, Butterfield BO (1968) Single-unit activity in the primary auditory cortex of unanesthetized cats. J Acoust Soc Am 43(3):444–455
Gray PE (2010) Walter Alter Rosenblith 1913–2002. A Biographical Memoir by National Academy of Sciences
Green DM, Swets JA (1966) Signal detection theory and psychophysics, vol 1. Wiley, New York
Guinan JJ, Guinan SS, Norris BE (1972a) Single auditory units in the superior olivary complex: I: responses to sounds and classifications based on physiological properties. Int J Neurosci 4(3):101–120
Guinan JJ, Norris BE, Guinan SS (1972b) Single auditory units in the superior olivary complex: II: locations of unit categories and tonotopic organization. Int J Neurosci 4(4):147–166
Harrison J, Howe M (1974) Anatomy of the afferent auditory nervous system of mammals. In: Keidel WD, Neff WD (eds) Handbook of sensory physiology: auditory system: anatomy, phsyiology (ear), vol V/1. Springer, Heidelberg, pp 283–336
Henning GB (1974) Detectability of interaural delay in high-frequency complex waveforms. J Acoust Soc Am 55(1):84–90
Hind JE, Anderson DJ, Brugge JF, Rose JE (1967) Coding of information pertaining to paired low-frequency tones in single auditory nerve fibers of the squirrel monkey. J Neurophysiol 30(4):794–816
Hubel DH, Wiesel TN (1962) Receptive fields, binocular interaction and functional architecture in the cat's visual cortex. J Neurophysiol 160(1):106–154
Jeffress LA (1948) A place theory of sound localization. J Comp Physiol Psychol 41(1):35–39
Jeffress LA (1958) Medial geniculate body—a disavowal. J Acoust Soc Am 30(8):802–803
Kiang NY-S (1965) Discharge patterns of single fibers in the cat's auditory nerve. MIT Press, Cambridge
Kruger L (1993) Jerzy Edwin Rose, 1909–1992. J Comp Neurol 327:469–472
Lettvin JY, Maturana HR, McCulloch WS, Pitts WH (1959) What the frog's eye tells the frog's brain. Proc IEEE 47(11):1940–1951
Lorente de Nó R (1933a) The central projection of the nerve endings of the internal ear. Laryngoscope 43(1):1–38
Lorente de Nó R (1933b) Anatomy of the eighth nerve: III.—General plan of structure of the primary cochlear nuclei. Laryngoscope 43(4):327–350
Lorente de Nó R (1938) Cerebral cortex: architecture, intracortical connections, motor projections. In: Fulton JF (ed) Physiology of the nervous system. Oxford University Press, London, pp 291–325
Marshall WH, Woolsey CN, Bard P (1937) Cortical representation of tactile sensibility as indicated by cortical potentials. Science 85(2207):388–390
McCulloch WS, Pitts W (1943) A logical calculus of the ideas immanent in nervous activity. Bull Math Biophys 5(4):115–133
Mountcastle VB (1957) Modality and topographic properties of single neurons of cat's somatic sensory cortex. J Neurophysiol 20(4):408–434
Neff WD (1961) Neural mechanisms of auditory discrimination. In: Rosenblith W (ed) Sensory communication. MIT Press & Wiley, New York, pp 259–278
Neff WD, Fisher J, Diamond IT, Yela M (1956) Role of auditory cortex in discrimination requiring localization of sound in space. J Neurophysiol 19(6):500–512
Neff WD, Diamond IT, Casseday JH (1975) Behavioral studies of auditory discrimination: central nervous system. In: Keidel WD, Neff WD (eds) Handbook of sensory physiology: auditory system: physiology (CNS), Behavioral studies, psychoacoustics, vol V/2. Springer, Heidelberg, pp 307–400

Piccolino M (1997) Luigi Galvani and animal electricity: two centuries after the foundation of electrophysiology. Trends Neurosci 20(10):443–448
Rasmussen GL, Windle WF (eds) (1960) Neural mechanisms of the auditory and vestibular systems. CC Thomas, Springfield
Ravizza R, Diamond I (1974) Role of auditory cortex in sound localization: a comparative ablation study of hedgehog and bushbaby. Fed Proc vol 8:1917–1919
Rose JE (1949) The cellular structure of the auditory region of the cat. J Comp Neurol 91(3):409–439
Rose JE (1959) Microelectrode studies of the cochlear nuclei of the cat. Bull Johns Hopkins Hosp 104(5):211–251
Rose JE, Galambos R (1952) Microelectrode studies on medial geniculate body of cat: I. Thalamic region activated by click stimuli. J Neurophysiol 15(5):343–357
Rose JE, Woolsey CN (1949) The relations of thalamic connections, cellular structure and evocable electrical activity in the auditory region of the cat. J Comp Neurol 91(3):441–466
Rose JE, Brugge JF, Anderson DJ, Hind JE (1967) Phase-locked response to low-frequency tones in single auditory nerve fibers of the squirrel monkey. J Neurophysiol 30(4):769–793
Rosenblith W (ed) (1961) Sensory communication. MIT Press, Cambridge
Scharlock DP, Neff WD, Strominger NL (1965) Discrimination of tone duration after bilateral ablation of cortical auditory areas. J Neurophysiol 28(4):673–681
Shannon CE (1948) A mathematical theory of communication. Bell Syst Tech J 27(3):379–423
Siebert W (1965) Some implications of the stochastic behavior of primary auditory neurons. Kybernetik 2(5):206–215
Squire LR (1998) The history of neuroscience in autobiography, vol 1. Academic, San Diego
Strutt JW (1877) The theory of sound, vol 1. Macmillan and Company, New York
Strutt JW (1907) On our perception of sound direction. Philos Mag 13(74):214–232
Vesalius A (1985) De humani corporis fabrica libri septem. Hormon-Chemie, Belgium
Von Helmholtz H (1912) On the sensations of tone as a physiological basis for the theory of music. Longmans, Green
Wiener N (1948) Cybernetics or control and communication in the animal and the machine. MIT Press, Cambridge
Willis T (1664) Cerebri Anatome, London
Wills A (1999) Herophilus, Erasistratus, and the birth of neuroscience. Lancet 354(9191): 1719–1720
Woolsey CN (1942) Topical projection of nerve fibers from local regions of the cochlea to the cerebral cortex of the cat. Bull Johns Hopkins Hosp 71:315–344
Woolsey CN (1960) Organization of cortical auditory system: a review and a synthesis. In: Rasmussen GL, Windle WF (eds) Neural mechanisms of the auditory and vestibular systems. Charles C Thomas, Springfield, pp 165–180
Yin TC, Smith PH, Joris P (2019) Neural mechanisms of sound localization in the auditory brainstem. Compr Physiol 9:1503–1575
Yost WA (2015) Psychoacoustics: a brief historical overview. Acoust Today 11(3):46–53
Yost WA, Patterson RD, Feth LL (2021) David M. Green and psychoacoustics. Acoust Today 17(3):51–59

SPRINGER NATURE

GPSR Compliance

The European Union's (EU) General Product Safety Regulation (GPSR) is a set of rules that requires consumer products to be safe and our obligations to ensure this.

If you have any concerns about our products, you can contact us on ProductSafety@springernature.com

In case Publisher is established outside the EU, the EU authorized representative is:

Springer Nature Customer Service Center GmbH
Europaplatz 3
69115 Heidelberg, Germany

The manufacturer's authorised representative in the EU is Springer Nature Customer Service Centre GmbH, Europaplatz 3, 69115 Heidelberg, Germany. If you have any concerns regarding our products, please contact ProductSafety@springernature.com

Printed and bound by CPI Group (UK) Ltd, Croydon, CR0 4YY
30/09/2026
02225322-0008